# Wetlands Drainage, River Modification, and Sectoral Conflict in the Lower Illinois Valley, 1890–1930

John Thompson

# Wetlands Drainage, River Modification, and Sectoral Conflict in the Lower Illinois Valley, 1890–1930

Southern Illinois University Press
Carbondale and Edwardsville

To Jean and Joy


Printed in the United States of America

05 04 03 02 4 3 2 1

Publication partially funded by a subvention grant from the U.S. Geological Survey Upper Midwest Environmental Sciences Center (formerly Environmental Management Technical Center), U.S. Department of the Interior, La Crosse, Wisconsin.

*Frontispiece:* One of the steam dredges, with dippers of 2- to 3-cubic-yard capacity and reaches of 35 to 50 feet from the hull, that raised most of the original riverfront levees and excavated some drainage canals on the floodplain of the Illinois River. Photograph courtesy of Walter P. Hatton, Havana, Illinois.

Library of Congress Cataloging-in-Publication Data

Thompson, John, 1924–
Wetlands drainage, river modification, and sectoral conflict in the lower Illinois Valley, 1890–1930 / John Thompson.
p. cm.
Includes bibliographical references and index.
1. Drainage—Illinois—Illinois River Valley—History. 2. Illinois River Valley (Ill.)—Historical geography. I. Title.

TC977.I6 T46 2002
333.91'815'097735—dc21

2001018374

ISBN 0-8093-2398-2 (alk. paper)

The paper used in this publication meets the minimum requirements of American National Standard for Information Sciences—Permanence of Paper for Printed Library Materials, ANSI Z39.48-1992. ∞

# Contents

# Illustrations

## Figures



## Maps

# Tables

# Preface

FAMILIAR to many Illinoisans is the role of the Illinois River as an avenue for explorers and settlers and for the commerce that contributed much to the early growth of St. Louis, Peoria, and Chicago. The rise of towns along the navigable river and of contemporaneous settlement in the groves and prairies of the uplands beyond the valley's bluffs is generally understood. The subsequent contribution to regional development made by the spread of a dense railway network, by the opening of scatterings of shallow coal mines, and by the rise of manufacturing and commerce in the towns is also recognized.

Much less a matter of general record is the process by which the floodplain of the Illinois valley downstream of Peoria was transformed into an integral part of the settled landscape. This transformation of perennially to seasonally overflowed wetlands into exceptionally productive farmland occurred between the 1890s and 1930. It was accomplished by enclosing large segments of the bottoms with levees, behind which ditch and tile systems were built to convey unwanted water to pumps that discharged into the river. This development of the bottomland was the final phase of the land breaking process in the core agricultural area of the Prairie State. Thus, the bottoms along the first avenue of Anglo-American penetration and economic development in central Illinois were the last significant frontierlike area to become part of the intensive general farming and cash-grain scene.

While human modification of the lower valley began with indigenous people and was extended by pioneers bearing values and tools that evolved in colonial and newly independent America, the focus here is on the era when large-scale ventures in land drainage and other intensifying uses of the gifts of nature greatly transformed land, water, and habitat. The changes that occurred in the lower valley between 1890 and 1930 arose from the interplay of human activities and physical processes in the immediate area and in the watershed and because of events and circumstances arising at midcontinental and national levels. The larger contexts are sketched here to enrich the account of the heritage of the lower valley, while highlighting linkages that sharpen the larger perspectives. The antecedent settlement geography and history are not delineated because the focus is on the land drainage experience, when the landscape recognizable today was shaped.

The principal inducement to build levees and drain the wetlands was a build-up of water levels over the floodplain at a time when rising commodity prices placed a premium on the conversion of wetlands into permanent cropland. The elevation and spread of water beyond levels created by locks and dams built in the late nineteenth century were caused by several factors: the diversion into the Illinois River system of large volumes of water from Lake Michigan through the Chicago Sanitary and Ship Canal, which began to operate in 1900; the intensification of land drainage activities throughout the watershed; and the onset of a cycle of relatively wet years in 1902. Before long, the levee building and land draining activity itself contributed to elevated water levels by removing substantial areas of swamp and overflowed land from their natural functions and by constricting floodway channels.

The mounting inducements to reclaim the bottoms resulted in the formation of drainage and levee districts by associations of resident and absentee landowners. These public corporations were organized under the auspices of county courts within a legal framework dating from 1879. The elected officers of the districts were authorized to make assessments on the land, issue bonds, engage engineers and contracting firms to build the works, and exercise the rights of eminent domain. In making farmland that was relatively secure from overflow, seepage, and runoff, these local enterprises accelerated the modification of original relationships between water and land in the valley. While wetlands ecosystems were being destroyed inside the levee systems, they were also being disrupted and displaced by the spread of water outside the levees. Resident commercial fishermen and hunters and visiting sportsmen were upset to lose haunts where they were accustomed to making a living and finding recreation. Naturalists deplored the depletion of nature's recuperative powers. Meanwhile, numerous sportsmen's clubs acquired the best residual wetlands for the pleasure of their nonresident members, while excluding other folks. Resentment, poaching, and controversy were carried into state and federal courts, the General Assembly, and Congress. Most evident among the protagonists were the landowners, whose interests lay in land drainage for agricultural use; the Sanitary District of Chicago, which chose to treat a serious metropolitan sanitation problem by diverting water from Lake Michigan to flush waste into the Illinois River system; various governments, which found Chicago's taking of water to be detrimental to navigation and hydroelectric interests on the Great Lakes; and the valley's commercial fisheries sector, which tried to stop the destruction of spawning and feeding grounds. The state's naturalists, wardens, and attorneys worked to preserve the residual natural areas, as well. It took longer to jell effective opposition to pollution of the river.

Ultimately, the resolution of the conflicts arising from mutually incompatible uses of river and floodplain resulted from natural events and from judicial and political contests over which the valley's protagonists had little control. In the long run, the effort to transform wetlands into cropland and farmsteads prevailed, as did the maintenance of navigation interests with dams and locks. On the other hand, the Sanitary District of Chicago was required to curtail diversions of Lake Michigan water and to abate pollution. Commercial fishing and recreational hunting and fishing in the valley, like tourism, made do with the residual resource base. State and federal agencies became more effective in protecting residual natural areas as the drainage era ended and even extended them in succeeding decades.

The land drainage experience and outcomes on the floodplain of the Illinois River were affected by a number of regional and national events and trends. The relationship of the experience in the valley to the broader drainage effort on floodplains elsewhere in the interior of the nation is noteworthy. Also, it is well to recall that the venture in land drainage occurred at a time when the use of coal, steam power, and steel were ascendant; when Midwestern urban and industrial growth was vigorous; when large cities and corporations wielded power seemingly without check; and when relatively unrestrained exploitation of resources stirred countercurrents favoring interventionist governmental policies in fish and wildfowl protection, watershed planning, and municipal sewage treatment plant construction.

Contemporaneous with the land reclamation effort on the floodplains of the nation's interior were large-scale land drainage and irrigation efforts in subhumid and arid areas in the West. Although the drainers of wetlands received no direct federal help after the United States ceded the swamp and overflowed lands to the states in 1849 and 1850, irrigation development in the West received significant encouragement from Congress and the states. Nevertheless, private enterprise achieved a great deal in the large-scale draining of wetlands, especially after the entrepreneurs adopted power excavators for ditching and levee building in the 1880s. When such dredging was begun in the bottoms of the lower Illinois valley in the 1890s, it was a pioneer effort in floodplain drainage for the Midwest and the border states adjoining the Mississippi River.

The expansive era of land drainage, like that of irrigation project development in the West, ended with the crash in commodity prices in the early 1920s. Subsequent major flood events in the interior contributed to the end of the boom in land drainage and prompted the General Assembly and Congress to intervene in the restoration of the works of drainage and levee districts in Illinois. In similar fashion, the United States government and the state governments in the West aided financially distressed irrigation districts and drainage districts. To overcome the calamitous Great Depression, numerous federal programs were designed to assist agriculture, the unemployed, manufacturing, and commerce. Among these undertakings of the 1930s was the Illinois Waterway, a project of the State of Illinois with antecedence in the multipurpose Chicago Sanitary and Ship Canal. Federal completion of the Illinois Waterway came about because the State of Illinois lacked the resources to finish it. The circumstances were akin to the experience of states nationwide. In California, for instance, the Central Valley Project was federalized during the Depression.

The compartmentalization of large areas of the Illinois River floodplain within levee systems, and their dewatering, involved the application of new technologies in earth moving and water management. Excavators powered with steam and internal combustion engines made large-scale projects possible. Although men with teams, shovels, scrapers, and wagons contributed to the work, these older modes of earth moving lacked the capacity for timely completion of large-scale projects on swamp and overflowed land. Also, the prevailing dependence upon gravity flow to drain land was supplanted by a reliance on pumps powered first by steam and then by electric motors and internal combustion engines. At most leveed tracts, the unwanted water had to be lifted through or over levees into the elevated navigation pools of the lower Illinois River. Meanwhile, steam and internal combustion engines were adopted in farming, and electric power lines penetrated the lower valley. The technologies adopted between 1890 and 1930 benefited farmers and landowners, as they did the drainage engineers and contractors whose businesses were shaped by experience gained on the floodplain in west-central Illinois.

The primary focus of this historical geographer's exploration of the land drainage venture is reflected in the titles of most chapters. They variously describe the institutions that undertook land drainage, the drainage engineering and contracting sectors and technology, the engineered and natural challenges to effective reclamation, and the morphology and occupance patterns of the drained landscape. Inasmuch as contemporaneous reclamation and induced flooding of bottomlands greatly depleted the resource base that supported a sizable fishing and hunting economy, a comprehensive account of the harvesting of the gifts of nature is offered.

The account of the land drainage process on the floodplain between Peoria and the confluence of the Illinois River with the Mississippi River begins in the first chapter with a description of the valley's physical setting. The legal framework that facilitated the human modification of the valley's water bodies and swamp and overflowed lands is outlined subsequently, followed by a description of the general circumstances of commodity price and land value appreciations that drove the development process. These physical, legislated, and economic contexts guided the landscape changes that subsequent chapters delineate in detail.

Traced in the second chapter is the process by which drainage and levee districts were planned, developed, and maintained. The progress in the transformation of floodplain into farmland is reviewed in maps and text, and cost allocations are made. The degree to which the achievements of landowners, drainage practitioners, and engineers were realized through trial and error is suggested, and the encroachment upon the river is reviewed. In chapter 3, which deals with the integration of pumping facilities into the works of drainage, the learning experience is traced again. The machinery of land drainage and the nature of the contracting sector are described in chapter 4, "Shapers of the Drainage Landscape."

The challenges to the land drainage process resulting from the modification of the regime of the Illinois River by dams and locks for navigation and by the flow from Lake Michigan through the Sanitary and Ship Canal of Chicago are reviewed in chapter 5, "Challenges Engineered by Other Sectors." The circumstances that heightened the antipathy between people in the valley and in Chicago are delineated, as are Midwestern and national contexts for water resource development and control. The progression of floods from events that stimulated reclamation to calamities that brought the buoyant drainage era to an end in the 1920s is described in chapter 6, "Days of Reckoning for Reclamation." The chapter also recounts how the damages wrought by nature during the decade were compounded by the collapse of commodity prices nationwide in late 1921 and how the ultimate recovery of privately developed drainage and levee districts was assisted through the direct involvement of state and federal agencies—a new departure.

The nature of the agricultural system and landscape in the 1920s, together with a sense of the mix of occupations and the composition of households on the land and in small service centers of the study area, are described in chapter 7, "Agricultural Activity and Settlement in the 1920s." The intent is to reveal occupance patterns at the culmination of the era of large-scale land drainage, a time when the adoption of gasoline engines and electricity in the area signaled changes in the nature of farming, in the tasks and quality of rural life, and in the accustomed role of small service centers. For smallholders and tenants alike, the costs and risks to residing and farming on former wetlands had become daunting; moreover, commodity prices were low, and capital and credit were scarce at a time when the adoption of the tractor and motor vehicle required commensurately larger farm operations.

Another aspect to local economic activity during the era of large-scale reclamation is described in chapter 8, "Harvesting Nature's Endowment." The chapter describes commercial and recreational fishing and hunting, commercial mussel gathering, trapping, and timber extraction. The extractive activities engaged substantial numbers of the valley's gainfully employed residents for decades, and commercial and recreational hunting and fishing drew short-term visitors to the valley, as well. For those whose livelihood or recreation depended upon the bounty of aquatic and floodplain ecosystems, the spread of leveed cropland was a bane. Less overt were the destruction and displacement of wetlands by the surfeit of water that land drainage generally, and the Sanitary District of Chicago's water diversion from Lake Michigan particularly, conveyed into the Illinois River system. While nature's recuperative powers were severely taxed by the excesses of fishermen and hunters and by pollution, the land developers' reclamation of wetlands destroyed habitats where fish spawned and fed, nesting and migratory wildfowl were sustained, and timber flourished. The appreciable loss of wetlands caused commercial fishermen to challenge the land developers. The spread of water over unreclaimed areas of floodplain resulted in a scramble among outsiders to control residual wetlands for duck clubs, which alienated the unaffiliated, some of whom became poachers. The plaints about habitat loss, the conflicts between competing takers of the bounty of floodplain and water, and the excesses of fishermen and hunters sharpened sensitivities among elected and appointed officials and the public about the conservation and preservation of the state's natural bounty. Thus, the chapter relates how nonagricultural pursuits engaged the populace, modified the environment, and induced the State of Illinois and the United States government to assume larger roles in husbanding the natural endowment of the lower valley.

The final chapter reviews the experience of draining the floodplain in the lower valley to facilitate the expansion of general and cash-grain farming into the last reservoir of unbroken land remaining in the heart of Illinois. The chapter scans the spectrum of physical and human, local and external, circumstances that contributed to the transformation of both riverine wetlands and dry valley floor into the landscape evident today. The chapter recalls how the land drainage venture in west-central Illinois fits into the national experience in reclaiming floodplains.

In the broad spectrum of approaches to understanding the processes of drainage and settlement on floodplains in the United States during the late nineteenth and early twentieth centuries, this geographical monograph probably is most akin to the contributions of Robert W. Harrison (1961), Gary Lane McDowell (1965), and Philip V. Scarpino (1985), whose perspectives were honed in economics and history. The present study shares the focus of the first two upon institutions that effected large-scale land drainage and the broader areal and national historical contexts into which the reclamation ventures fit. While appreciative of the qualities of the earlier works, this study gives more attention to the roles of drainage engineering and contracting sectors, the technology applied, and the conflicts arising from competing uses of the river and wetlands. In those concerns, it is more akin to Scarpino's work, although the focal agencies and technologies of change are quite different. The description here of the texture of settlement and gainful occupations on the newly drained landscape departs substantially from the work of the other three. In that respect, it is more akin to numerous analyses by geographers of land use patterns on former regional wetlands—a genre that is well reflected in Hugh Prince's (1997) fine historical geography of drainage on the extensive Midwestern wet prairies associated with outwash plains of the Pleistocene. Among the historical geographies of floodplain settlement and development in the United States is my study of the delta of the Sacramento and San Joaquin rivers in California (1957), which was followed years later by studies of the institutions, technology, and machinery of land drainage that evolved in the Central Valley and in the greater San Francisco Bay area.

While studying the various aspects of large-scale land reclamation in the valleys of the Sacramento and San Joaquin rivers, I began to explore kindred matters in the Midwest. The expanded quest focused initially on early periodicals of agriculture, contracting, and civil engineering that described land drainage projects and technology. The quest focused, also, on accessing corporate archives of several one-time makers of dredges. The holdings related to land drainage in a number of Midwestern university and state historical libraries were studied, and the usefulness of drainage archives in select counties of Ohio, Indiana, and Illinois was assessed. Ultimately, major areas of former wetlands were considered as to their significance and potential instructiveness on the roles of the landowner groups, the drainage contracting and engineering sectors, the machinery of large-scale drainage, and the sectoral conflicts and environmental impacts that attended the regional land drainage venture. The attractiveness of the floodplain of the lower Illinois River for a case study was enhanced because the area had attracted little scholarly attention, yet had experienced an early and sustained venture with large-scale drainage efforts during the golden ages of steam power and of land reclamation in the United States.

Through the courtesy of the Water Resources Center, University of Illinois, my research resulted in *Case Studies in Drainage and Levee District Formation and Development on the Floodplain of the Lower Illinois River, 1890s to 1930s*, which was published by the center as Special Report 017 in May 1989. This report was the vehicle for presenting forty case studies with supportive documentation drawn largely from the archives of circuit courts and from newspapers of the period. In response to suggestions by reviewers of the larger manuscript to which the case studies had been appended, the research was extended in 1989 to more clearly reflect the relationship of the local experience to the national experience in land drainage and to provide a more comprehensive account of the use, abuse, transformation, and destruction of the wetlands. The present volume is the result of those efforts.

# Acknowledgments

THE research for this study was begun with support from the Water Resources Center and the Graduate College Research Board of the University of Illinois, Urbana-Champaign. The funds made available by the center were provided by the United States Department of the Interior as authorized under the Water Resources Act of 1985. This financial assistance enabled graduate students to sift through voluminous archives in a dozen county courthouses, to prepare illustrative maps, and to extend the review of public documents in the libraries of the university. The aid for fieldwork enabled the writer to access records in the possession of commissioners and attorneys for a handful of districts and to use the State Archives in Springfield and the archives of the Federal Land Bank of St. Louis. In essence, this archival research on forty drainage and levee districts provided the foundation for an analysis and synthesis of the land drainage experience. The data base was enriched by the writer's review of a number of weekly and daily newspapers that appeared in the county seats and riverside towns of the study area between 1898 and 1930. Helpful, as well, was access gained to raw census data of townships, precincts, and towns of the area through the university library's microfilmed copies of decennial censuses of the United States made in 1900, 1910, and 1920.

The gathering of data on the drainage and levee districts was facilitated through the courtesy of the clerks of the circuit courts in Carrollton, Havana, Jacksonville, Jerseyville, Lewistown, Mt. Sterling, Pekin, Peoria, Pittsfield, Rushville, Virginia, and Winchester. I was ably assisted in sorting through the archives by Ruaidhri McSharry, Paul A. Rollinson, and Tim Teddy, then of the Department of Geography, University of Illinois. Original maps were drawn by Lily Pan, then also in the department. The courteous help extended by commissioners and attorneys for various districts and by staff members of the State Archives and the Federal Land Bank of St. Louis led to the finding of invaluable material. The work was also advanced in many ways through the day-to-day support from the Department of Geography.

The published records concerning the drainage and levee districts; the Sanitary District of Chicago; the state and federal agencies concerned with public works, flood events, and wildlife and wetlands study and protection; and the private sector environmentalist interest in the Illinois valley were obtained with the assistance of the staffs in the libraries of the Natural History Survey and the Illinois Historical Survey and in the Geology, Law, Map and Geography, and Rare Book collections of the University of Illinois Library. Recurring assistance was afforded by the university's

Interlibrary Loan Service and the Illinois State Historical Library, Springfield, which provided the microfilm copy of newspapers published between 1898 and 1930 in the major towns of the study area. Helpful in various respects were the staffs of the public libraries in Beardstown, Havana, Lewistown, Meredosia, and Pittsfield. With pleasure, the writer acknowledges the courtesies received in the corporate archives departments of the Marion Power Shovel Division of Dresser Industries, Inc., in Marion, Ohio, and the Bucyrus-Erie Company, in South Milwaukee, Wisconsin.

It is a pleasure to acknowledge the forbearance and courtesy of several residents of the study area and of Champaign who provided insights and access to archival materials not ordinarily in the public domain: Shirley M. Gross, Norman Korsmeyer, and Milton McClure Jr. (Beardstown); Lew W. Cummings (Chambersburg); Jane Wilcox Villegas (Eldred); William A. Gettings Jr. and Richard Zipprich (Fieldon); Walter P. Hatton (Havana); Florence Hutchison (Jacksonville); Richard Bull (Liverpool); Ralph Guengerich (Manito); Ken McGann and John Van Ness (Peoria); William Klingner (Quincy); Louis Stone (Versailles); and R. Edward Frost (Winchester). Helpful, too, in facilitating access to records were Hugh A. Strickland (Carrollton), Thomas M. Atherton (Pekin), Brice Irving (Pittsfield), and Kathryn Mann Herring (Winchester), who serve as attorneys for various districts. Mary L. Pearlstein (Champaign) was helpful in identifying legal firms that represent the drainage and levee districts and in calling attention to investigations done in the 1930s by the Federal Land Bank of St. Louis.

The illustrations of historic scenes and activities herein are largely from the collections of John George Karl III and George and Virginia Karl of Havana, the Havana Public Library, and the collections of the Illinois Natural History Survey and the Map and Geography Library, Urbana. Their generosity in providing the material is gratefully acknowledged. Suggestions as to the richness of the collections were forthcoming from Stephen P. Havera and Frank C. Bellrose, Forbes Biological Station, Illinois Natural History Survey, in Havana, whose general encouragement and comments on the penultimate chapter were most helpful. Warm appreciation is due, too, for suggestions and encouragement by John M. Hoffmann, librarian of the Illinois Historical Survey, Urbana-Champaign. A source of recurring cheer for the writer was the skill and dispatch with which drafts of the manuscript were prepared by Barbara Bonnell of the Department of Geography.

I am most grateful to the editorial director and staff of Southern Illinois University Press for their roles in bringing this work to fruition. In this respect, too, I am deeply appreciative of the support that enabled publication provided by the U.S. Geological Survey Upper Midwest Environmental Sciences Center, U.S. Department of the Interior, La Crosse, Wisconsin.

# Wetlands Drainage, River Modification, and Sectoral Conflict in the Lower Illinois Valley, 1890–1930

1

# The Physical Setting and General Context of Land Drainage

THE transformation of the valley of the lower Illinois River from its perennial and seasonal wetlands condition was largely accomplished during the half century before 1930. The incorporation of relatively unchanged swamp and overflowed land into an intensively farmed segment of the Corn Belt was facilitated by a number of factors, chief among them a legislated institutional framework, favorable general and local economic circumstances, and application of new technologies to modify land and water relationships. A review of this early introduction of power machinery to levee building, ditching, and pumping on a floodplain occurs in subsequent chapters. Here, the first task is to describe the physical setting—valley and bluffs, water and cover—before it was modified by land reclaimers. Then, the framework of facilitative state laws and general economic circumstances that fostered large-scale ventures in draining, clearing, and breaking the remarkably productive bottoms are reviewed.

## The Physical Stage

### Valley and Bluffs

The valley of the lower Illinois River extends from Peoria to the confluence with the Mississippi River, about 160 miles to the south (map 1.1).[1] This lower valley has three distinct sectors: northern and southern ones that are each about 53 miles long, and a 56-mile-long intermediate sector.[2] The broad central sector of the valley spans as much as 5 to 8 miles but averages about 4.6 miles wide; the northern sector is about 1.25 to 3 miles across, and the southern sector is from 1.5 to 4 miles across. Both the northern and southern sectors receive drainage directly from about 7 or 8 percent of the entire watershed of the Illinois River, which is some 27,914 square miles. In these two sectors, relatively short tributaries drain small basins, except for the Mackinaw River (1,217 sq. mi.). The broader central sector of the lower Illinois Valley receives drainage directly from about 36 percent of the entire watershed, notably from the

Sangamon (5,670 sq. mi.), Spoon (1,870 sq. mi.), and La Moine (Crooked Creek—1,385 sq. mi.) river basins. Commonly, flood crests on these tributaries enter the lowland before the crests that are carried past Peoria by the Illinois River from the half of the watershed that arises in the distant Kankakee and Des Plaines rivers of northeastern Illinois, Indiana, and Wisconsin. The behavior of the lower Illinois River is affected, too, by flood crests of the Mississippi River, which can back water to Beardstown, in the middle sector of the lower valley. The extremely gentle gradient of the Illinois River averages 1.8 inches per mile between Peoria and Grafton; it is less than one inch per mile upstream of Peoria, from about La Salle. Throughout its course below La Salle, the river flows on aggraded material, not bedrock. A great flood, like that of June 1844, all but covers the entire valley floor.[3]

The principal features of the lower valley are the aggrading river and the floodplain, originally strewn with shallow lakes and ponds. Important, too, are the somewhat higher terraces of glaciofluvial, fluvial, and lacustrine origins that date from the late Wisconsinan and very early Holocene. At the valley margins, the floodplain and terraces are overlain with alluvial fans that head in the bluffs that rise on both flanks. By and large, the bluffs average about 150 feet above the valley floor, most of which lies between 420 and 460 feet elevation.

In its unmodified condition more than a century ago, the sluggish river increased in width from an average of about 600 feet to 1,000 feet between Peoria and Grafton. However, islands, bars, and narrow elongate lakes that represented former channels gave it a braided quality, especially in the central sector of the valley. Nevertheless, it was a relatively straight, rather than meandering, river that was deflected back and forth through the lowland by accumulating alluvium from tributaries such as Farm Creek (opposite Peoria) and the Mackinaw (below Pekin), Spoon (opposite Havana), and Sangamon (above Beardstown) rivers. The river's proximity to the western bluffs downstream of Valley City represents entrenchment of prehistoric origin and reflects larger watershed areas of left bank (east) tributaries than are found to the west, where the uplands are confined between the converging Mississippi and Illinois rivers. Deposition resulting from overbank flows by the tributaries and river formed natural levees that were about 4 to 10 feet above the floors of adjacent bottoms, where shallow ponds, elongate lakes, and sloughs were numerous. By and large, 60–70 percent of the ponds and lakes occupied less than 50 acres. The lakes were smaller and more infilled with sediment downstream of Beardstown than above. Some of the lakes represented meanders abandoned by the tributaries. Yet other linear and shallow lakes and sloughs occupied higher troughs, or paleochannels, associated with the terraced areas near and to the south of Beardstown. Some lakes were fed by springs.[4]

Local relief on the floodplain approached the imperceptible, except near the bluffs and watercourses. The river's banks averaged 7 to 12 feet above low water, increasing to

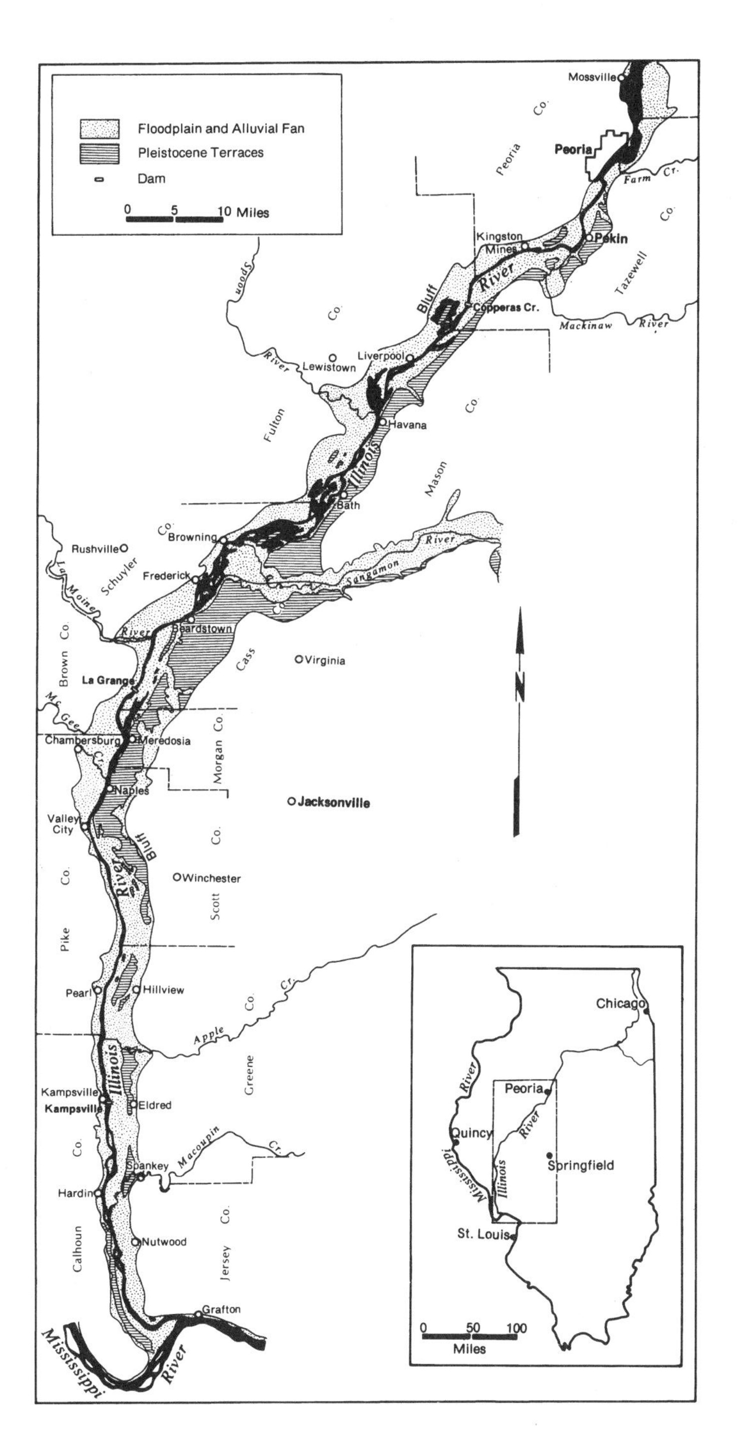


**Map 1.1.** Valley floor materials, lower Illinois valley

about 15 feet nearer the Mississippi River. Bank-full river stages, which occurred most frequently between March and June, usually were preceded by rises attributable to heavy precipitation and runoff from the bluffs and immediate tributaries. Sometimes contributing to elevated river stages was water backed up by flood stages on the Mississippi and Missouri rivers. By the time that bank-full stage was reached, water would have spread across the bottoms or flood basins through sloughs and tributary streams. Another 3 feet of rise covered the lower bottoms, and a rise of 7 feet inundated 80 percent of the floodplain. At such times, the lowest areas of Wisconsinan terraces were flooded; settlements such as Liverpool, Havana, Beardstown, and Meredosia were vulnerable. As the river subsided, the overflow retreated slowly and incompletely, many lakes and ponds persisting through the year without evident connection to the river,[5] except for innumerable seeps from the river banks.

A sense of the flood event experience on the pristine valley floor of 1858 is captured in published excerpts from a journal kept by a resident who farmed to the north of Apple Creek, in the present Keach Drainage and Levee District, in Greene County:

> The river began to grow uneasy in its banks from about the first of April, and soon came over Newport, mouth of Apple Creek and the lowest lands, and run up the sloughs, rising very slowly and at times falling back a few inches, until about the middle of May, when it reached the top of the sandridge and came over the Stubblefield place about the 20th of May. Gradually it came crawling over the cultivated bottom land and spreading out over the prairie. About the first of June it rose until it reached the highest stage on the 17th of June.
>
> It came over and carried one-half of my west string of fence, and two-thirds of my south string, most of my neighbors faring worse. It crossed the road at one place so deep that it ran into the wagon box of any who drove across, and the public made a road as high up as possible on the hillside to avoid the water. It was within six inches of the high-water mark of '44. It destroyed two or three acres of my corn and four or five acres of wheat.
>
> About the 18th of June the water fell about one and a quarter inches. The wind blew from the south-west for ten days, driving upon my field immense quantities of rails, cord wood, one hen-house full of chickens and one sled upon which were piled three plows.
>
> . . . The water was very cold and as clear as a crystal. . . .
>
> The water kept on falling slowly until on the first day of July, it disappeared beyond the sandridge, and by the middle of July it was within the river banks.[6]

There is a continuity of Pleistocene terraces to the east of the river for much of the distance between Farm Creek, opposite Peoria, and Macoupin Creek, opposite Hardin (map 1.1), except for breaks attributable to erosion by larger tributaries and places where the terraces are buried by the alluvial fans of lesser creeks. The surface elevations of these terraces vary from 5 to 10 feet, 25 to 35 feet, 40 to 55 feet, and more above the bottomland. The terraces are from one to four miles across. Those of intermediate and higher elevation undulate with sand ridges, hummocks, or dunes that rise another 10 to 80 feet. A much smaller incidence of terrace remnants is found along the northwestern margin of the valley floor, but they are relatively more evident in the valleys of tributaries that reach the Illinois River from the west. Terraces appear as well along the margins of tributary valleys that enter from the east. Although shallow, undrained depressions may occur, the terraces tend to be as well drained as the fans and colluvial slopes adjacent the bluffs.[7]

The steep bluff lines are scalloped deeply by numerous ravines and small valleys that head as much as two miles away; and the continuity of the bluffs is broken here and there for a half mile or more where larger tributaries enter the Illinois Valley. Beyond the valley's flanking bluffs are the gently rolling surfaces of the Galesburg Plain, where late Wisconsinan loess mantles older loess, till plains, and morainic ridges of the Illinoisan, which overlie bedrock. The larger accumulation of loess to the southeast of the valley results in somewhat higher bluffs and upland areas. Elevations generally are about 600 to 680 feet, upwards of 200 feet above the valley floor. Within the scalloped bluff margins, springs and seeps are common where Pennsylvanian shales underlie the glacial material. The bluffs are quarried here and there for the limestone road metal that is used in roadways across the lowlands and in the draws or valleys that head in the rolling loessial uplands.[8]

## Water and Cover

The behavior of water and nature of the floodplain along the Illinois River was summarized by Stansbury about 150 years ago: "The bottom lands extend from one to five miles on each side of the river, seldom rising more than a few feet above the level of the stream in its ordinary stages, and from the fact that they are constantly overflowed by every freshet, to a depth varying from one to fifteen feet, are now, and must ever remain, uninhabited. Hence, the river presents the appearance of flowing through an ancient, vast, and solitary forest, clothed with a foliage rich and luxuriant beyond description."[9]

During the remarkable flood of 1844, when the valley was in the "state of nature" seen by Stansbury, the waters extended nearly from bluff line to bluff line below La Salle. Over 80 percent of the entire inundated area comprised the floodplain between Peoria and Grafton. These 328,520 acres of bottoms in the lower valley included 40,380 acres (12 percent) where lakes and ponds reappeared on the plain as overflow waters subsided (table 1.1).[10]

**Table 1.1.** Water and land areas subject to overflow, by sector of the unmodified Illinois valley

| | River reach | | | Unmodified floodplain | | |
|---|---|---|---|---|---|---|
| | Length (mi.) | Surface (ac.) | Lakes & ponds surface (ac.) | Flooded average spring (ac.) | Flooded 1844 (ac.) | 1844 flood total area (ac.) |
| Above study area | 62.4 | 12,250 | 8,960 | 31,890 | 48,250 | 69,460 |
| La Salle–Peoria | | | | | | |
| Northern sector, study area | 53.4 | 4,120 | 14,550 | 52,030 | 72,510 | 91,180 |
| Peoria–Copperas Cr. Dam | 24.7 | 1,880 | 4,310 | 18,010 | 27,950 | 34,140 |
| Copperas Cr. Dam–Havana | 28.7 | 2,240 | 10,240 | 34,020 | 44,560 | 57,040 |
| Middle sector, study area | 56.2 | 5,540 | 18,300 | 55,960 | 119,080 | 142,920 |
| Havana–La Grange Dam | 30.1 | 2,900 | 12,820 | 34,480 | 65,680 | 81,400 |
| La Grange Dam–Valley City | 26.1 | 2,640 | 5,480 | 21,480 | 53,400 | 61,520 |
| Southern sector, study area | 52.4 | 6,580 | 7,530 | 28,990 | 80,310 | 94,420 |
| Valley City–Kampsville Dam | 20.9 | 2,860 | 4,820 | 15,520 | 39,540 | 47,220 |
| Kampsville Dam–Grafton | 31.5 | 3,720 | 2,710 | 13,470 | 40,770 | 47,200 |
| Total | 224.4 | 28,490 | 49,340 | 168,870 | 320,150 | 397,980 |

*Note:* River and lake/pond surfaces as measured at low water, 1901.

The shallow residual lakes and ponds with ephemeral surface ties to the river might shrink to marshy spots or mud flats by the fall, but those fed by springs or tributaries usually kept a link to the river. It was through such sloughs that water rose or fell with the river's stages. All lakes were akin to Thompson Lake, near Havana, in that their surface areas varied greatly with changes in river stage. Thompson Lake covered about 3,000 acres when the river was 6 feet above low water and about 4,300 acres when the river rose another 3 feet.[11]

Such bodies of water were particularly evident upon the floodplain in the 30 miles above and below Havana, the heart of the broad central sector of the lower valley. Lakes and ponds occupied close to 17 percent of the floodplain surface. They covered only 8 percent of the bottoms downstream of Meredosia and less than 13 percent of the bottoms in the 25 miles below Peoria.

The extent to which the floodplain was covered by ordinary flood stages was greatest in the same 60-mile stretch centering on Havana, which coincides with the valley sector that received tributaries that conveyed the runoff from over 35 percent of the river's entire watershed. To the north, in the first 25 miles of lowland downstream of Peoria, ordinary flood stages affected a smaller proportion of floodplain. Also, ordinary flood stages covered a relatively small part of the valley floor in the 50-plus miles above the junction with the Mississippi River. Implicit in the degree to which overflow, lakes, and ponds occurred in the three sectors was the relative attractiveness of the areas to pioneer farmers and to the later initiators of land drainage works.

As may be imagined from a setting that embraced a narrow lowland wherein river, natural levees, perennially and seasonally overflowed bottoms, Pleistocene terraces, and peripheral fans and colluvial slopes lay, there were fairly diverse patterns of natural vegetation. In general, corridors of forest occupied the active and abandoned natural levees of the river and tributaries. These forests were adjoined by prairies on higher ground or by marshes, ponds, and lakes on the lower ground. Prairies and open woodland tended to occupy the terraces. Except where frequent prairie fires reached up the gentler flanks of the valley, sustaining grassland cover, the bluffs and draws were forested. Wooded declivities marked the rolling prairie uplands.[12]

The cover has been modified greatly since Stansbury described a luxuriant mature forest. In Greene and Jersey counties, at least, the valley's grassy terraces were being broken near the bluffs in the 1820s, and the timber was being cut for cabins, fencing and fuel, and rafting to St. Louis. The process extended northward as settlers occupied the high ground along the foot of the bluffs and as firewood cutting was pursued to supply fuel for river boats and for the homes of St. Louis.[13] The cover was further modified when navigation pools were created by the locks and dams built on the river in the late nineteenth century. This permanent elevation of water surfaces and ground water levels was exacerbated in the two decades following 1900, when the Sanitary

District of Chicago diverted increasing volumes of water into the river system from Lake Michigan. Meanwhile, the spread of cultivated land toward the river in the 1890s was accelerated after 1904 by the success of large-scale drainage enterprises. All the while, commercial timber cutting occurred, as is traced later.

The prairie is gone, as is much of the timber and brush land, and the areas of marsh are greatly reduced. These wetlands abounded in various species of duck potato (*Sagittaria* spp.), smart weed (*Polygonum coccineum*), river bulrush (*Scirpus fluviatilis*), and other species. Remnants of timber occupy limited areas within and just outside the drainage and levee districts and on the islands and residual pockets of floodplain adjacent the river and along the bottoms of the tributaries, notably the Sangamon River. In these forests adjacent the watercourses, the most evident tree forms include black and white oaks (*Quercus velatina* and *Q. bicolor*), hickory (*Carya* spp.), green ash (*Fraxinus pennsylvanica*), silver maple (*Acer saccharinum* L.), and pecan (*Carya illinoensis*). Where artificial levees were allowed to develop a woody cover, maples and elms (*Ulmus americana* L.), and pin oaks (*Q. palustris*) tended to replace such early dominants as willows (*Salix* spp.) and cottonwood (*Populus deltoides* Marsh). On the rising land of the valley margins and flanks of the bluffs and tributary valleys and draws, the black oak and red oak (*Q. rubra*), sugar maple (*Acer saccharum*), and hickory species are most common in the woodland.[14]

Wetlands conditions aside, the cover that evolved in and near the Illinois Valley did so under an annual regime of about 35 to 38 inches of precipitation, over 60 percent of which fell during the growing season. The remainder occurred fairly evenly through the winter months. The seasons were marked by cold winter temperatures (26–32°F. January mean) and the warm (77–80°F. July mean) and humid summers characteristic of the continental interior. While snow cover generally is neither great nor prolonged over the study area, runoff from heavy spring rains over the watershed may be augmented to such a degree by melting snow and thawed groundwater that heavy flooding results along the valley.

## Legislated Institutional Framework for Drainage

The floodplain of the Illinois River was a part of the nearly 1.5 million acres of upland and bottomland swamp and overflowed areas ceded to the State of Illinois by the United States after 1850. Although Congress's first swampland cession act (1849) was designed to enable the State of Louisiana to levee and ditch wetlands by dedicating proceeds of their sale to that end, provision was made in the "Arkansas Act" (1850) for cession of federal swamp and overflowed lands to the states generally. Illinois, like most states that were granted swamp and overflowed land, transferred responsibility for sale and land drainage to the counties (1852).[15] The proviso that land ceded by the

United States to the states be improved with the proceeds from their sale to individuals became policy, too, when public lands in the subhumid and arid West were sold to settlers. The reclamation of Western lands through irrigation was facilitated by passage of the desert land act (1877), the Carey act (1894), the reclamation act (1902), and the irrigation district act (1916). To accomplish the intent of Congress that the ceded lands be reclaimed, the public lands states adopted legislation authorizing the formation of assessment districts that could incur debt (bond sales) to construct the works of drainage or irrigation.[16] Illinois adopted such legislation in 1879 to enable owners of wetlands to collaborate in drainage endeavors that had not been accomplished by individual landowners in the counties. In essence, the United States and the State of Illinois removed themselves from direct involvement in the land drainage effort. The task was left to the landowners.

The organization and operation of large-scale land drainage enterprises along the flood-prone lower Illinois River after 1890 was in accordance with guidelines prescribed by the General Assembly on May 29, 1879, and amended or clarified through judicial review. Of the two sets of kindred guidelines enacted in 1879, the Levee Act was designed to accommodate the needs of owners of bottomland, while the Farm Drainage Act (codified in 1885) served landowners in upland areas of wet soil.[17] Whereas land drainage was achieved in the uplands with tile lines and open ditches, reclamation of bottomlands required that levees be built to protect ditch systems and pumping plants from stream overflow. Pumping plants were essential to reclaimed tracts on the bottoms because the level of water in the Illinois River precluded gravity drainage from most drainage and levee districts.

To facilitate the installation of pumping plants at drainage and levee districts and to validate earlier installations, an act was approved by the General Assembly in 1905. It was an acknowledgment that land developers along the Illinois River were being injured by waters diverted from Lake Michigan into the watershed by the Sanitary District of Chicago, beginning in 1900. Less overt a factor in raising water levels above the planes of the pools created by locks and dams was the very process of enclosing areas of floodplain with levee systems. At the same time, improvement of land drainage on the wet prairie uplands and in the bottoms of tributaries like the Sangamon River added to the surfeit of water in the lower valley of the Illinois River. Additional legislation in the interest of pumping entities passed the General Assembly in 1907, 1911, and 1913, the last of these acts correcting features of antecedent legislation that were ruled unconstitutional in the courts.[18]

Thus, in the heyday of the development of drainage and levee districts on the floodplain, the entities on the bottoms were apt to be called "levee districts" or "pumping districts," as distinguished from "drainage districts," the entities organized in upland areas under the Farm Drainage Act to achieve drainage by gravity flow. Drainage and levee districts, or pumping districts, are the primary concern of this study. However, references will be made to drainage districts in the vicinity of the Illinois Valley for the sake of perspective. Also, it should be noted that the development of the Banner Special Drainage District, organized under the Farm Drainage Act, differs too little from the drainage and levee districts of the area to warrant separate treatment.

The drainage and levee districts and the drainage districts were alike in enabling a majority of landowners to coerce others in the prospective district to share the costs and benefits of works designed to transform seasonally or perennially wet land into artificially drained cropland. In addition to having the power to assess land to pay for construction and operating costs, these public, involuntary corporations could incur debt and own and manage property acquired through the exercise of the power of eminent domain. All such districts had to conform to the basic principles of drainage law in Illinois, which held that the natural flow of surface water could not be cast upon the land upslope and that lower landowners could not prevent the upper landowner from hastening the removal of water through ditches so long as the system did not tap a second natural drainage basin. Regardless of the enabling act, a district had jurisdiction over land to which it conferred the benefit of improved drainage. But the procedures adopted by landowners for a district to organize, construct, and operate drainage works had to conform to the prescriptions in either the Levee Act or the Farm Drainage Act.[19] Not until 1956 did a new drainage code meld the two sets of laws,[20] although there was strong sentiment among the districts and in the General Assembly to do so in 1911. The land drainage laws, it seems, "have apparently developed in the same disorderly progress as the drainage developments. They have been amended and added to from time to time to meet new phases and conditions of the drainage development, until they have become so voluminous that it is inconvenient and even difficult for engineers and commissioners to follow them without making mistakes. Not only this, but they are inadequate to meet the condition imposed by the broader and more comprehensive scale of developments now being proposed."[21] Nevertheless, many drainage and levee districts were organized to build levee and ditch systems and pumping stations on the floodplain of the Illinois River under the Levee Act of 1879, as modified by later acts and by decisions of the Supreme Court.

Most of the quasi-public institutions have remained viable entities for over 80 years. Only one, the Otter Creek Drainage and Levee District, was dissolved (1918) during the developmental phase, but its dissolution was preliminary to forming three successor districts. The Langellier Drainage and Levee District was separated (1905) from the Lacey Drainage and Levee District, and both survive. Three other districts have foundered since 1926; two failed because the cost of maintaining them for agriculture could not be sustained by income from the land, and the third, the Banner Special Drainage District, became a surface coal mining operation.[22]

## Organization of Drainage and Levee Districts

The drainage and levee districts in the lower valley were formed under the jurisdiction of the circuit court for the county in which lay all or most of the wetlands to be reclaimed. In practice, petitioning landowners represented the majority of landowners and the majority of acreage in a proposed district, although the Levee Act allowed districts to be formed where the majority of adult landowners owned but one-third of the land. In some instances, as at the Coal Creek and Hillview drainage and levee districts, opponents to organization were bought out by the principals. Other protesting parties might have their land excluded from a proposed district at the outset or following a hearing and inspection by disinterested parties, if they could demonstrate that their land would not be benefited.[23] For the most part, however, landowners who were resident in or near the proposed districts and who did not sign the petition to organize took no active role before the court.

Petitions to form districts were required to set forth the proposed name and boundaries of the entity, to describe the nature and source of the overflow problem, to define each proprietor's holdings, and to bear the signatures of petitioners whose aggregate acreage and whose numbers exceeded the required minima. Assuming that the court was satisfied that appropriate notice of the petition and of related public hearings were posted, mailed, and published, that the properties and ownerships were represented properly, and that the generally described works of reclamation would be effective, it appointed a commission of three disinterested persons to evaluate the proposition. This commission submitted a written report of its examination of the land, the proposed drainage plans and anticipated construction costs, the anticipated annual maintenance and operation costs, and the benefits and damages that would accrue to each land parcel involved. In the event that the preliminary study revealed that the benefits probably would exceed costs, the temporary commission was authorized to engage an engineer to make surveys, plans and maps, and a cost estimate. In practice, the engineer already had been advising the proponents of reclamation and might propose enlargement of the district to accommodate newly recognized areas of benefit. Also, owners of land located outside but adjacent the perimeter of the proposed district could petition to have their property included. There was no difficulty incorporating land into the proposed district so long as the property lay within the natural watershed of the area to be reclaimed.[24]

Ultimately, the temporary commissioner's feasibility report and the engineer's report were aired at a public hearing before the court. Objections tended to be resolved informally before the court concluded that the requirements of the law were satisfied, the district should be formed, and the engineer's plans adopted. Thereafter, a commission drawn from among the landowners reported to the court each July or

August, reviewing the district's achievements and problems, income and expenditures. As a rule, the itemized chronological accounts of monies received and expended were prepared by a salaried district treasurer. Much more cryptic were the minutes of commissioners' meetings, at least one of which was scheduled quarterly. The minutes books were kept by the commissioner who served as secretary.[25]

Once the circuit court ordered the district formed, the commission comprised of prime movers among the landowners began to acquire rights of way for the levee and ditch systems and to have the assessment roll prepared so as to underwrite construction. Such rolls were listings of 40-acre parcels, each being assigned anticipated damages and a proportional share of anticipated monetary benefits of drainage, as well as the owner's name. The benefits for each 40-acre parcel were determined on the basis of the parcel's average elevation, which was equated with susceptibility to inundation in the absence of levees. The completed assessment roll was reviewed by a court-appointed jury, which repaired to the tract in question to evaluate the appropriateness of all assessments of benefits and damages. The jury's report was given a public hearing; ultimate decisions were confirmed by the court. Although landowners were entitled to appeal to the Supreme Court of Illinois following action by the lower court, no appeals are known to have arisen concerning districts in the study area.[26]

Landowners might pay their construction assessments in full at the outset or in annual installments that included 8 percent interest from the time of district confirmation. However, most districts obtained the bulk of capital for construction through the sale of bonds that earned 6 percent per annum.[27] Under court auspices, the bond issues were prepared; amortization was scheduled for a period of 20 or 30 years. Landowners expected to retire their share of the debt with earnings from tenants or from their own farming enterprise.

Since the construction of drainage works usually was financed on the anticipated income from the reclaimed lands, it is understandable that nearly three-fourths of the studied drainage and levee districts issued bonds to cover upwards of 69 percent of projected construction costs. However, the Coal Creek, Mauvaise Terre, and Otter Creek drainage and levee districts issued no bonds to underwrite their initial projects, and landowners of the Meredosia Lake and Nutwood districts paid 55 to 60 percent of the costs of their first reclamation plans. In the Scott County district, cash underwrote 42 percent of initial costs. Nevertheless, virtually every district had issued bonds for construction purposes by 1929. For most districts, the bond issues underwrote at least half of all construction; the range was 37 to 97 percent. Particularly dependent upon the sale of bonds were the developers of the Keach, South Beardstown, Spring Lake, Crane Creek, and Spankey drainage and levee districts.[28]

Although the general practice in Illinois was for districts to give bonds to contractors for their services,[29] archival documents and assessor's maps indicate that it was not a common practice in the study area. Most bonds were placed with local and Chicago banks, with dealers in bonds, and among some landowners of means. The bonds were regarded as "high-class securities" because the law required that the proceeds be used only for construction and because investors were sure that the income from reclaimed land would readily bear the cost of borrowed money and of drainage works.[30]

As it turned out, the districts and bondholders were in very difficult financial straits by the mid-1920s due to calamitous floods and falling commodity prices, as will be reviewed later. Whatever was salvaged from the ensuing Depression of the 1930s was attributable to the Federal Land Bank of St. Louis, the Reconstruction Finance Corporation, state and federal assistance in levee rehabilitation, and the improved commodity prices of the late 1930s and 1940s.

The law forbade commissioners to incur construction costs beyond the total authorized in a construction assessment. Yet, the realities of nature and inflation and the foibles of commissioners, contractors, and engineers almost invariably elevated construction costs above first estimates. Consequently, second and later construction plans and assessments, and fresh assessment rolls, were required to complete the works of reclamation. The same procedural steps that carried first proposals to fruition were followed, the district's engineer and attorney providing guidance to the commissioners.[31]

## Operation of Drainage and Levee Districts

A district's operating and maintenance costs were paid through assessments made annually upon the land and were due September 1. These annual assessments averaged 30 to 50 cents per acre at first but increased to $1 or more by World War I. The rates were raised through petition and public hearing proceedings. For the most part, such assessments defrayed the costs of pumping, equipment maintenance and levee repair, ditch maintenance and weed and rodent control, sandbag and lumber purchases, engineering and legal services, wages, incidental expenses, and the allowances paid to commissioners for services to the district. Sometimes operating funds were used for new construction or to repay debts incurred through anticipation warrants, which were issued to underwrite a flood fight or other exceptional circumstance. By the same token, operating deficits were covered sometimes with money raised by assessments for new construction, for which a bond issue might be made.

Management of the routine affairs of the districts was in the hands of the three landowners who were commissioners, but the key person was the elected chairman. Duties might be shared among the commissioners, but it appears that the chairman assumed the principal role of hearing out the constituency, inspecting facilities, and ordering repairs or maintenance that the salaried stationary engineer supervised. The chairman was responsible for stockpiling emergency equipment, conducting flood fights, and managing the general business of the district. By and large, the commissioners served for many years, although appointments by the court were made for staggered three-year terms. Commissioners were replaced or succeeded in response to a petition from a majority of the landowners in the district.

Actually, the prescribed mode of operating a district was apt to be tempered by long associations and human foibles, as reflected in an anonymous insert to the minutes of an "average" meeting at Big Swan Levee and Drainage District:

> After a long delay and much misunderstanding about when and where it would be held, and evening arrived and one by one the Commissioners arrived all full of farming interests and community gossop [*sic*], maybe one or more had heard some good questionable jokes.
>
> So after every subject had been exhausted and no hint or allusion to what purpose might have been in mind relative to the gathering one fellow began to reach for his hat, and another without any hat having his hand on the door knob, the Chairman suggested that we might hear the minutes of the previous meeting if any, along with such and said bills paid or otherwise disposed of and so I begins to rumage [*sic*] through these notations, noting that one Commissioner had left and as I glanced up from the bill reading, I heard the lower door slam as the last commissioner had vacated the premises, and thereby assumed that the meeting was adjourned.
>
> ______________________ Chairman ______________________ Secretary[32]
>
> Sign here if you ever come back.

## The Economic Context

During the decade or so preceding and following 1900, speculative developments were widespread in the wetlands areas of the eastern half of the nation and in the irrigable lands of the subhumid and arid West.[33] Commenting in 1909 on the relative attractions of the West and the flood-prone bottoms of Illinois, the editor of the *Ottawa Fair Dealer* observed:

> [The] people of Illinois are going to the far west, southwest and northwest and are investing in irrigated lands or lands that can be irrigated, at fancy prices for both land and water, neglecting an opportunity that is now opening at their very doors, to serve the richest corn growing land in the country.
>
> Along the Illinois and Mississippi rivers are large areas of bottom lands, the richest it is possible to make, that has been subject to such overflow from these rivers, as to render them heretofore useless. But now a new system of heavy dikes and inside ditches has been devised that will make these lands garden spots, lands

> that will be perfectly drained during the wettest seasons and moist enough the dryest years.
>
> These dike[d and] drained bottom lands have some marked advantages over the high prairies and timber lands. In the first place they are richer, then no matter how wet the season, the surplus water can be moved off by the pumps, and no matter how dry it may get, there will be enough moisture from below to insure a crop, when the upland may be burning up.[34]

The editor might have added that commodity prices were soaring, as is suggested in table 1.2.[35] Moreover, markets were close by. The bulk of the corn went to livestock feeding operations near the bluffs and in the upland areas adjacent, while much of the wheat moved by rail and barge to elevators in Peoria and Pekin. Finished hogs and cattle were shipped by rail and water, too.[36]

There was pride among the drainers of land that theirs was "the work of individuals, unaided by state and federal government,"[37] in contrast to the case in parts of the West, where public funds were used to develop water for irrigation. "Land that was waste and worse than nothing, . . . we turned . . . into the most fertile land in the world."[38] Moreover, it was held, this was a homegrown enterprise. "Drainage lands in Illinois do not have to be populated by foreign immigration and people from other states."[39]

**Table 1.2.** Average quotations of selected commodities in downstate Illinois, five-year periods, 1890–1929

| | Average prices | | | Price if 1900–1914 = 1.00[a] | |
|---|---|---|---|---|---|
| Period | Corn ($/bu.) | Wheat ($/bu.) | Hogs ($/cwt.) | Corn ($/bu.) | Wheat ($/bu.) |
| 1890–94 | .42 | .69 | 4.47 | . . . | . . . |
| 1895–99 | .27 | .70 | 3.51 | .40 | .99 |
| 1900–04 | .43 | .74 | 5.32 | .51 | .87 |
| 1905–09 | .52 | .91 | 5.86 | .55 | .96 |
| 1910–14 | .58 | .92 | 7.44 | .58 | .91 |
| 1915–20 | 1.16 | 1.82 | 12.79 | .68 | 1.04 |
| 1921–25 | .71 | 1.21 | 8.56 | .46 | .78 |
| 1926–29 | .74 | 1.31 | 10.35 | .51 | .85 |

[a]Average corrected for changes in price level.

## The Promise of Bottomland

Land subject to overflow in the lower valley cost $.50 to $10 per acre around the turn of the century. Some of the bottoms were valued only for timber and hunting grounds, but where they were elevated enough to produce a crop every year or so, the land fetched $10 to $25 per acre or more. For example, well-improved 80- and 100-acre properties near the bluffs in Greene County sold for $65 and $80 per acre in 1902. Larger pieces were bought on speculation for $20 to $30 per acre, to judge from transactions involving 1,396 acres in the Hillview district (which sold for $21.49 per acre in 1905), and 600 acres of what became the Kelly Lake district (which fetched $28.33 per acre in 1912). Even as drainage work proceeded at the Hillview district, land sold there for $35 to $40 per acre. By 1906, reclaimed land was said to be priced at $50 to $75 generally, but very fine farms were sold from the Fairbanks Ranch in 1909 for $125 per acre (480 ac.) and for $140 per acre (240 ac.). In 1912, properties of 390 and 563 acres at the new Crane Creek district sold for about $85 per acre, while a 179-acre farm sold for $117 per acre. Such prices became more general by World War I, to judge from purchases at $109 to $155 per acre reported for farms of similar size in districts near and below Beardstown. A little later, land apparently sold for as much as $300 per acre in the South Beardstown and Spring Lake districts.[40]

The substantial increments in land value that resulted from drainage were a response to the handsome returns from crops and rents and, at least at first, the relatively low cost of land reclamation. There was money to be made in draining land costing less than $30 per acre. A bonanza was foreseen as early as 1896 for landowners in the Lacey district, whose investment of $60,000 in drainage works was expected to produce land aggregating $300,000 to $600,000 in value. More modest was the expectation of the landowners in the Banner Special Drainage District in 1910, whose benefits were projected to be double the investment of $182,597 in works of reclamation. Stockholders in the district at Spring Lake expected that an investment of $25 per acre in reclamation would result in land worth $100 to $150 per acre. The increment in "producing capital" for all of the reclaimed lands in Greene County was expected to be a tripling of the $1 million invested. Thus, very roughly, by investing as much in reclamation as one paid for acreage, one could expect to double or triple one's property value. Already in late summer 1909, the owners of land in the Nutwood district foresaw that a 30-bushel wheat crop and a prospective corn crop nearing 100 bushels would aggregate in value what it cost to reclaim the large tract. The expectation at the McGee Creek district was that the value of the 1910 crops alone would be double the cost of existing works of drainage. A bonanza reported for 1910 by a farmer and landlord in the Hillview district was an $18 per acre share of the first crop raised on a lake bed that had cost $52.50 per acre to purchase, reclaim, and tile. Although the tenant's net return was $45, the general average was more apt to fall in the $25 to $33 per acre range in 1910 and 1911.[41]

Viewed from the county seat, the transformation of bottomland into drainage and levee districts was a most salutary development. The editor of the *Carrollton Patriot,* in referring in 1906 to three projects on 29,000 acres of western Greene County, commented:

> For productivity purposes, this county will have practically annexed territory considerably larger than a congressional township when these levees are completed. . . . Suppose that the 29,000 acres advanced in value from $10 to an average of $60 an acre. That means an addition of $1,450,000 to the permanent wealth of Greene County. . . . When that land is all under cultivation and producing fifty bushels of corn per acre, at 40 cents a bushel, the crop will add $580,000 a year to the prosperity of Greene County.
>
> The one regrettable thing about the reclaiming of this bottom land . . . is that so much of it is owned in large tracts and by non-residents. That which most contributes to the prosperity of a county or community is homes—homes owned by their occupants, at frequent intervals along the highways and surrounded by small farms well tilled. This condition will not obtain in the levee districts. But, on the other hand, had the bottom lands remained in small holdings, what is now being accomplished would probably have been delayed many years longer. At all events the operation of these big ranches will give employment to many hands, and the increased value of the lands will add largely to property valuation in the county and consequently lower the general rate of taxation.[42]

The price of reclaimed land in the bottoms was within the range of $100 to $175 per acre associated with good land in the prairie uplands around 1905–7. The "almost fabulous" price of $200 per acre for prime upland properties to the northeast of Pekin seems not to have been reached for a decade. Nevertheless, the land of the bottoms was said to produce about a third more per acre than did land in the uplands.[43]

The tendency to underestimate and understate the cost of effectively reclaiming bottomland was recognized at least as early as 1904 at the Lacey district, for which a summary of development and operating costs is made in table 1.3. Even then, it was calculated, rental land costing the investor $5.24 per annum would yield rent of $7.50, for a net profit of $2.26 per acre.[44]

Reclamation assessments at the Lacey district were within the range of reclamation costs expected in the early 1900s, $15 to $25 per acre. Such costs were in the $20 to $40 range by 1913. An average reclamation cost of about $33.50 per acre may be calculated

**Table 1.3.** Estimated developmental and annual operating costs, Lacey Drainage and Levee District, 1904

| Development | $/ac. | Annual operation | $/ac. |
|---|---|---|---|
| Land cost | 5.00 | Interest on $54 | 3.24 |
| Reclamation assessment | 24.00 | Taxes, insurance, maintenance | 1.00 |
| Original | 14.00 | | |
| Additional | 10.00 | | |
| Clearing | 15.00 | Pumping assessment | 1.00 |
| Buildings, etc. | 10.00 | | |
| Total | 54.00 | Total | 5.24 |

for all districts functioning in 1914, but it conceals the relatively low costs incurred in the districts near and below Beardstown and the high costs incurred at districts from Havana upstream. A little over $25 per acre in reclamation costs were incurred in the districts near and below Beardstown, while costs as high as $54 per acre were incurred at the Lacey and the Pekin and La Marsh districts. Over $60 per acre were expended at the East Peoria district, where industrial and residential land uses vied with truck farming, and where a constricted river and unruly Farm Creek created special problems. Regardless of the tract, the cost of reclaiming the bottoms adjacent the lower Illinois River was much greater than the cost of land drainage in the prairie uplands. It was estimated in 1908, for instance, that upland farms could be provided with a system of open ditches for $3 to $5 per acre and that another $5 to $10 per acre covered the cost of developing a district's outlet ditch system. Thus, in the prairie uplands, ditch systems were obtained for $8 to $15 per acre; pumps were not needed. In both the bottoms and uplands, where field tiling was paid for by the individual landowner, it cost $6 to $12 per acre.[45]

Although land in the floodplain was far more costly to drain than was land in the prairie uplands, and the security of the bottoms from overflow was impaired by the progress of drainage in the uplands and by the diversion of water into the Illinois River system from Lake Michigan after 1900, the bottoms attracted investors for a third of a century beginning in the 1890s. At least at first, farmers and other investors found that the relatively inexpensive overflowed land would appreciate substantially as drainage works and other improvements were made. As fortune would have it, however, drainage costs seemed ever to mount, which matter is described in subsequent chapters. Meanwhile, commercial agriculture flourished.

## Contextual Summary and Foretaste

To understand the transformation of the seasonally to perennially wet bottoms of the lower Illinois valley into an artificially drained and leveed segment of the agricultural landscape of the Prairie State required introducing three background elements: the physical nature of the area, the legislated institutional framework for reclamation, and a sense of the prevailing economic conditions. The physical setting that was altered consisted of a floodplain of very low relief and numerous bodies of water; a river of low gradient but occasional unruly behavior; slightly elevated terraces; pronounced flanking bluffs; and marking each type of terrain, distinctive mixes of wet and dry woodland and prairie. The transformation of the land and water features of the valley floor into a landscape of leveed and wooded stream banks, multiple systems of drainage canals and ditches, extensive fields and scattered farmsteads, and a few villages and towns was facilitated (even accelerated) by the General Assembly's responsiveness to the aspirations of prospective reclaimers of the wetlands and to resource developers of a more viable and livable Chicago. The intensification of land use on the bottoms reflected, also, a time when development of agricultural land in the nation was drawing public and private funds into large-scale irrigation projects in the West and into large-scale drainage projects in the more humid East and Midwest. Commodity prices were such that investment in new cropland was attractive. It was a buoyant era. Contributing to the rapidity with which wetlands were transformed into fields was the adoption of power excavators and mechanically driven pumps, to say nothing of the growth of cadres of drainage engineers and contractors.

Given the national experience with resource development at the time that drainage and levee districts were created on the floodplain of the Illinois River; given the tendency of state and county authorities to facilitate the work of the entrepreneurial landowners in private drainage and levee districts; given the formative nature of drainage engineering and technology; and given the state of knowledge about runoff and erosion rates in the watersheds tributary to the Illinois River, it is understandable that the perception of the Illinois River watershed and the Illinois Valley as integral natural systems was not well honed. This larger perspective came to be recognized in the wake of conflicts that arose from competing uses of the Illinois River and floodplain. Simply put, the fisheries and wildlife interests were damaged by land reclaiming, and the interests of both of these sectors were affected adversely by the flow of sewage and water through Chicago, whose commercial interests sought to further intrude in the valley with the development of a deep waterway to the Gulf of Mexico. At the same time, a spectacular boom in petroleum production was occurring in the state. Clearly, a number of matters had to be addressed by the legislature.

The advisability of taking stock of the mineral and water resources of the state resulted in the formation of the Illinois Geological Survey Commission and the Illinois Internal Improvements Commission in 1905. The former agency was to survey petroleum potential and general economic geology especially; but it was expected to collaborate with the Internal Improvements Commission and federal agencies to make and publish topographic surveys and to place stream gauges in such rivers as the Kaskaskia, the Big Muddy, the Little Wabash, and the Spoon. Both agencies were to facilitate rational resource development. Insofar as river bottoms were concerned, the specific topographic studies served land drainage interests. The model was the 1902–4 survey of the Illinois and Des Plaines rivers by the U.S. Corps of Engineers.[46]

Actually, the primary object in creating the Internal Improvements Commission was to "investigate the various problems associated with a projected deep water way from Lake Michigan to the Gulf of Mexico."[47] Not the least of the problems was the failure of the United States government to rise to the opportunity for collaboration offered by the General Assembly in 1889 as part of enabling legislation for the Chicago Sanitary and Ship Canal. In 1908, at a White House conference of governors, Governor Charles S. Deneen yet urged that the waterway project be a cooperative one between the states and the federal government, "and that States and localities should contribute to the general program in some proportion to their exceptional advantages."[48] The federal government was not alone in being unenthusiastic about collaboration on the projected deep waterway. It seemed to landowners along the Illinois River that Springfield and Chicago conceded little to their interests.[49] This concern was a factor behind the organization of the Association of Drainage and Levee Districts of Illinois in 1911, one of whose spokesmen urged that state and federal agencies and the Sanitary District of Chicago "get together and put into execution a proper plan for the regulation of the Illinois River that will result in reclamation of the lands, prevent excessive duration of floods and aid navigation, all of which will be of great value to the State."[50] By that time (1911), over half of the drainage and levee districts were in place along the river, and more were being planned. The officers of the districts had come to understand the merits of the larger perspective on watershed management, but Springfield and Chicago were at loggerheads with them on priorities. Not until the 1920s did state, regional, and national economic and political circumstances begin to shape public action on the civil works of the lower Illinois Valley in accordance with the larger frame of reference. By 1921, when the Division of Waterways of the Department of Public Works was authorized to regulate the building of levees,[51] the works of all but five drainage and levee districts were in place.

# 2

# The Construction and Spread of Drainage Districts

For the most part, drainage and levee districts occupied segments of floodplain that extended for 5 to 10 miles along the river and for 2 to 4 miles between the river and rising land near the bluffs (map 2.1). The higher margin of the districts usually approximated the high-water mark reached in the flood of 1844. Although some districts were formed to protect land from the overflow of creeks in their midst, such as those adjacent Coon Run and Willow Creek (near Meredosia) and at Otter Creek (now Kerton Valley and Seahorn districts, to the southwest of Havana), the majority were designed with levee systems that kept the channels of larger creeks outside. Because the watersheds of larger creeks could exceed tens of square miles to more than 100 square miles, the streams were a recurring flood threat, as was the river. Therefore, the districts built levees against the river and the flanking creeks, commonly straightening, widening, and deepening the tributary channels in the process. The object was to have flood flows reach the Illinois River as quickly as possible. The disposal of runoff, or "hill water," that rushed into the districts from adjacent bluff areas was managed through ditch systems and pumps when it could not be done with external diversion ditches that discharged into the flanking creeks.[1]

The pioneer farmer's experience with encroaching water upon higher bottoms, recounted in the preceding chapter, was indicative of an experience that became more common as farmers ventured onto lower land in the 1890s. The overflows could be localized as well as general, and they could arise in any season. In this chapter, a sense of the situation some 80 to 100 years ago and the nature of the responses by landowners and farmers are described. The responses evolved from modest earth-mounding and ditching efforts by individuals to collaborative, large-scale projects of levee and ditch system construction and the installation of a sizable pumping capacity to dewater the land. The nature of the extensive earthworks and the chronology of their spread over the floodplain is recounted in detail. The role of pumps in freeing the bottoms of water is reviewed in the chapter following; and the subsequent chapter

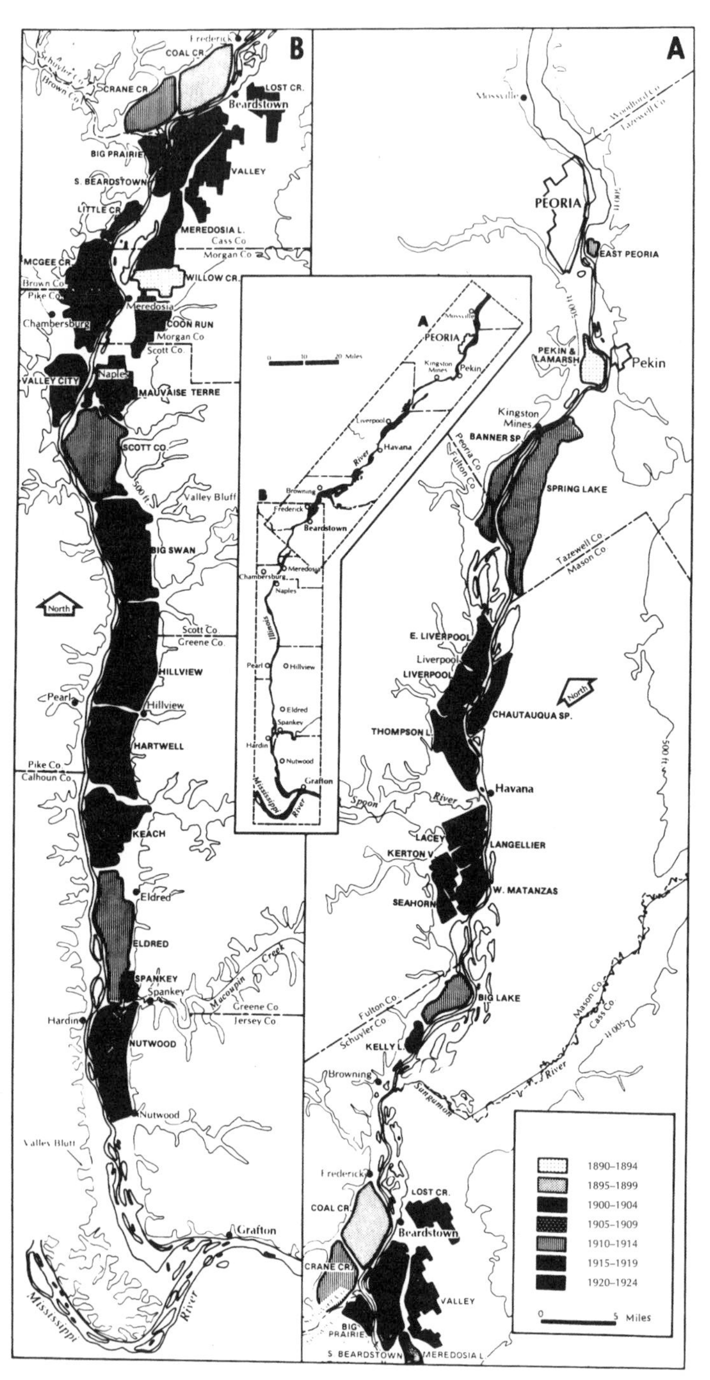


**Map 2.1.** Drainage and levee districts, lower Illinois valley

describes the drainage contracting and drainage engineering sectors and the operations that transformed wetlands into cropland and farmsteads.

## Floodplain Conditions at the Turn of the Century

Land clearing and cultivation encroached upon the bottoms from the foot of the bluffs by the latter 1860s.[2] By 1890, the progress of such land clearing varied, the bottoms from Peoria to about 10 miles below Pekin and to the south of Meredosia being as much as 50 percent cleared, while no more than 5 to 10 percent of intervening areas was cultivated.[3] By 1904, about 48 percent of the floodplain between Peoria and Beardstown was cleared, and downstream of Beardstown about 70 percent of the bottoms were cleared.[4] The areal differences in cleared land reflected relative security from overflow. Still, within these different areas, land clearing did not progress uniformly across the valley. For instance, the land just upstream of Havana later associated with the Chautauqua Lake district had only 1 percent of the natural cover removed; about 10 percent of the floodplain associated with the adjacent Thompson Lake and more northerly Spring Lake was cleared by 1904. Whereas lakes, ponds, and sloughs covered about 26 percent of the tract that became the Spring Lake district, about 51 percent of the tract that became the Thompson Lake district was covered with water. Relatively wetter was the area that became the South Beardstown Drainage and Levee District; 54 percent of it was under water in 1904. On the other hand, 80 percent of the land later identified with the Scott County district, below Naples, was clear of timber and brush in 1904 (map 2.2).[5] Most cropland lay on the high ground associated with Pleistocene terraces and on the more elevated, "second," bottoms. The more flood-prone vicinities of the myriad lakes, ponds, and sloughs largely remained in timber, brush, and wet prairie or marsh. These overflow-prone vicinities, such as Spring and Thompson lakes, were private hunting grounds, of which a number were distributed through the valley. Indeed, the lowest areas adjacent the navigation pools formed by the dams and locks at Copperas Creek (1877), La Grange (1889), and Kampsville (1893) were wetter than when they were in the pristine state, as will be discussed in chapters 5 and 8.

In 1902–4, bands of cleared land extended from the bluffs irregularly for a half mile to one or two miles across the valley floor. The continuity of cultivated land was interrupted fairly often by stands of timber and brush, which commonly adjoined undrained sloughs and shallow lakes and ponds.[6] As might be expected of a floodplain having modest local relief, 80 percent was awash in a 16-foot stage of the Illinois River, while a 12-foot stage filled the sloughs, lakes, ponds, and adjacent low bottoms.[7] Then, too, freshets arising in the bluff areas and beyond flowed across the fans and Pleistocene terraces into the bottoms. The localized events might presage the arrival of

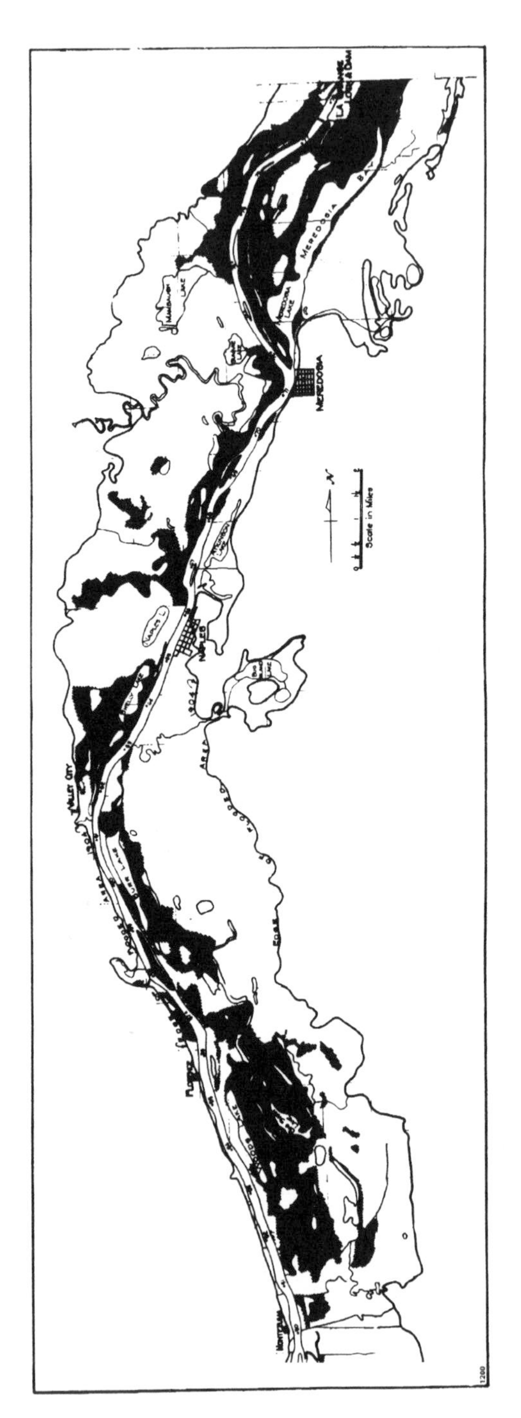

**Map 2.2.** Floodplain sector between Montezuma and La Grange, 1904. Within the irregular heavy lines appear shaded areas of timber and brush, lakes, and cleared land. Most of the latter is in drainage and levee districts. From Alvord and Burdick, *Illinois River,* fig. 29; courtesy of Map and Geography Library, University of Illinois.

the river's general flood stages that forced the evacuation of families, livestock, and worldly goods to high ground adjacent the bluffs, where camp was made. High waters arose in any season, to judge from events occurring between 1892 and 1907 at Browning and Frederick, which lie opposite the mouth of the Sangamon River. Fields on the bottoms were awash in January 1897, May 1898, May 1899, April and May 1900, July and August 1902, and April 1904. The events of 1902 put up to 3 feet of water over a corn crop that was laid by; and week-long rains in January 1897 that culminated in a 4-plus inch downpour so raised water over the fields that only the tasseled tops of shocked corn stood above the water. During intervening years, the floods of spring delayed planting and replanting. In 1899, for instance, farmers wondered if the corn would be planted before June 20. Even the earliest levee systems did not assure freedom from flooding, to judge from events at the Pekin and La Marsh Drainage and Levee District after the levee was completed in 1891. Aside from levee failures in 1892, 1902, 1904, and 1907, the structure's porosity, the runoff from the bluff, and a deficient pumping plant kept the tract ponded much of the time.[8]

The floods interrupted road traffic across the bottoms for a day or so to as much as two or three weeks at a time, sometimes longer. Especially sensitive to such circumstances were elevator operators, merchants, and elected officers of trading centers such as Beardstown, Havana, and Pekin, who rushed to restore westerly approaches to their toll bridges over the Illinois River. The assumption of responsibility for road maintenance in the adjacent counties followed upon the purchase of the right-of-way across the bottoms. Beardstown, for instance, constructed an elevated road through the bottoms of Schuyler County to Frederick in 1894 to eliminate travelers' dependence for one to six months each year on a lengthy ferry ride between the communities.[9]

At worst, cropland was overflowed in three of five growing seasons on the bottoms near Beardstown in the 1890s and early 1900s. More precisely, it was said of the bottoms in 1904 that "the higher parts yield crops about two years out of three, but the lower parts are flooded so frequently that their cultivation is unprofitable."[10] Such lower lands as those in Schuyler County were "sold time and again for taxes, and more for timber than anything else."[11] Similar conditions prevailed throughout the lower valley.

Understandably, within areas of cleared land, a large number of ditches and embankments were in place by 1902–4. The discontinuous earthworks were so common and so widely distributed that they probably were shaped by individual landowners over a period of years. Such work is thought to have accelerated during and immediately following the unusually dry 1890–1901 period, when summer rainfall was below the mean (except in 1896). At the time, planting on the bottoms flourished in response to the demand for corn and forage crops on the farms of the adjacent uplands, where livestock rearing and finishing was the principal source of income. Thus, the demand for feed, the generally good growing conditions, and the relatively high yields favored more cropping on the bottoms. Land prices advanced.[12] Having invested in improvement of very productive land, the owners were induced to raise levees, cut ditches, and install pumps because of the diversion of water from Lake Michigan in 1900, a cycle of rainy years that began in mid-1902, and the improvement of land drainage on the upland wet prairies of the watershed. As Harman put it:

> Excessive rains came in June and continued throughout the remainder of the year. A great deal of the lowland was flooded and the crops destroyed; moreover, there was so much rain during the harvest that many fields of wheat and oats were destroyed in the shock. 1903 was another wet year and there have been no dry years since. The return of the wet years has increased the demand for the completion of the drainage of all the wet lands in the state, and there has been great activity during the past few years. There has probably been as much expended for drainage in Illinois in the last five years as any other equal period. A great deal of attention is now being paid to the reclamation of the bottom land.[13]

Elliott, describing the value of overflowed lands in the Midwest generally, noted that had they "been successfully protected during the floods of 1902, 1903 and 1904," their value "would have been greatly increased, as uplands which are not as productive are now valued at $125 per acre."[14]

The drainage ditches were of two types, either short (quarter mile or so), narrow, and shallow conduits for runoff from the bluffs and its springs, or ditches extending several miles that were broad enough to have been cut by "drag boats" or dipper dredges, although no corroborating documentation has been found. These larger ditches followed the axes of troughs in the valley floor, and they cut across the grain of the floodplain. Notwithstanding the likelihood that the larger ditches were cut by machine, most ditches, embankments, and levees probably were shaped by men and teams pulling plows, drag scrapers, and wagons. Embankments and levees rarely exceed 6 feet above adjacent land and for much of their length were a good deal lower. Most of them were located for the first mile or so below where creeks issued from the bluffs, but there were miles of discontinuous embankments in the lower reaches of larger streams such as the Spoon and Sangamon rivers. Embankments appeared, too, across the axes of swales or troughs, as if designed to prevent runoff or the rising river from backing over cultivated areas.[15] In general, the alignments, length, and discontinuities of the embarkments suggest that they were raised to protect individual properties, if not pastures and fields. Some embankments subsequently were incorporated into levee systems and some became roadbeds. Yet others were destroyed as field and farm consolidation advanced.

To establish the primacy of development of an integrated system of embankments and ditches with one property or area is not possible, but a handful of ranchers and farmers in Greene County had carried the idea furthest by 1902–4, by which time pumps were in place at the lower end of three ditch systems. The first of these independently developed works began in 1888 at the northeast corner of the present Hartwell district, where Cyrus Hartwell is understood to have raised an embankment along Hurricane Creek to counter the effect on stream behavior of the embankment opposite, which carried the Chicago and Alton track past Hillview toward the bridge at Pearl. Another landowner, Louis Lowenstein, enclosed and ditched 125 acres adjacent and to the northwest of Hillview in 1888 or 1889 to protect his property from water that reached through the trestled Chicago and Alton embankment. This work may have been required because Hartwell had confined the creek on the south side. In any case, by 1902–4, Lowenstein had about 525 acres leveed, ditched, and furnished with a pump to the northwest of Hillview. Meanwhile, westerly and downslope of Hartwell's levee and ditch system stood a 5-mile loop of rudimentary levee, ditches, and pump. The adjacent systems comprised the first independently built defenses that spanned the bottoms from the bluff to the Illinois River.[16] The extent to which tile lines facilitated field drainage is unknown. In addition to the earthworks and ditches that spanned the bottoms adjacent Hurricane Creek in 1902, similar achievements evidently occurred in Peoria and Schuyler counties. They were the work of landowners who first organized the Pekin and La Marsh (Peoria County) and Coal Creek districts in 1891 and 1896, respectively.[17]

The wondrous transformation of bottoms resulting from the construction of integrated levee and ditch systems was described in the *Rushville Times* for the vicinity of Coal Creek in late May 1903. From atop the western levee of the Coal Creek district (map 2.3), a good 3.5 miles to the west of Beardstown, a reporter observed that the wheat crop outside the rampart was almost entirely lost and that farm work was at a standstill because of high water. Within the redoubt, the wheat "looked fine and thrifty" and "in every field farmers were at work and the ground was in perfect condition." These farmers expected to plant corn as early as many farmers on the upland.[18] Similar evidence of the merits of land reclamation must have been occasioned wherever embankments or levees held the overflow at bay.

## Levee Building

Artificial levees were raised on the natural levees of the river and on the relatively high ground built up in the form of fans and natural levees by creeks that were tributary to the Illinois River. The river's natural levees extended for about 200 to 300 feet from the low-water bank to troughs where backswamps lay. The natural levees of the creeks were narrower. In a few instances, railroad embankments were incorporated into the levee systems. The district paid to have trestled sectors filled in with dirt, and

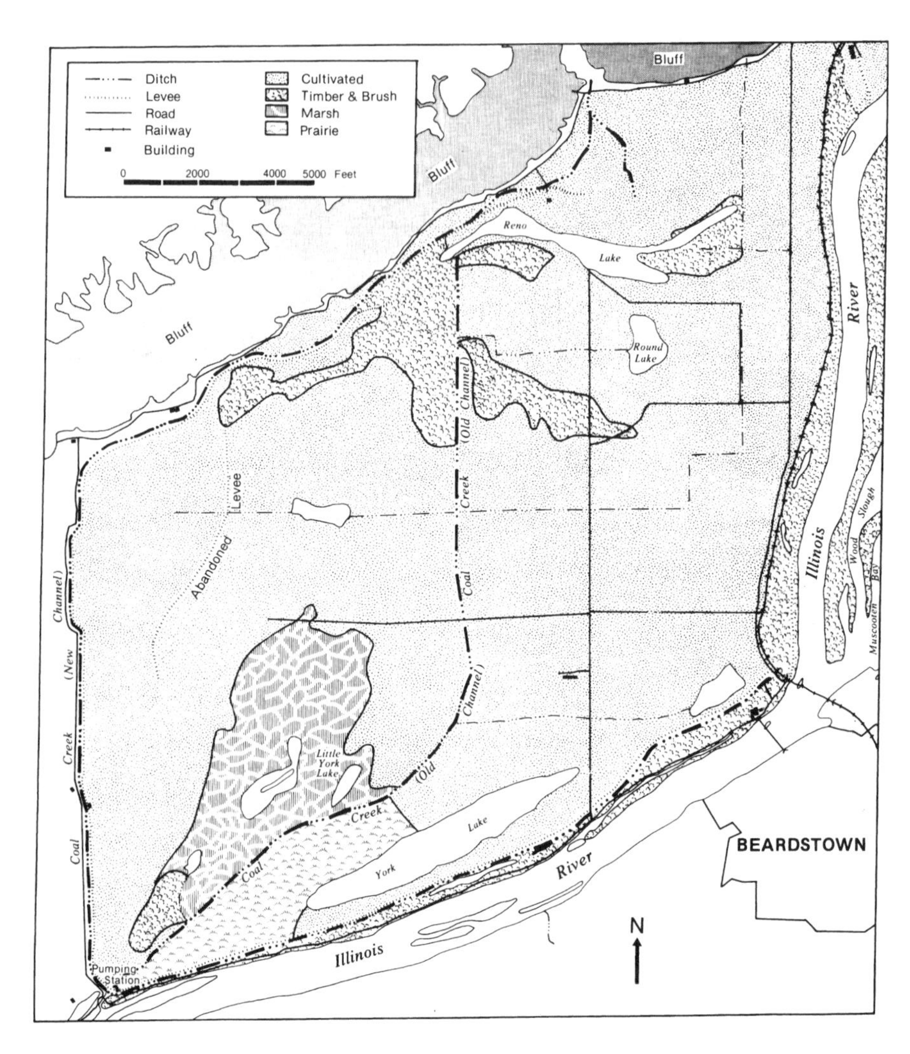


Map 2.3. Coal Creek Drainage and Levee District land use, 1902–1904. After Woermann, "Map of the Illinois and Des Plaines Rivers."

the railroads agreed to maintain the structure as a levee in exchange for being left off the assessment rolls. Embankments 0.9 to 4.0 miles long were used for levees at the Pekin and La Marsh, Coal Creek, and Hillview districts; the Lacey district took advantage of an abandoned railway embankment (maps 2.3, 2.4, and 2.5). In a couple of areas, the districts were advantaged by having adjoining levee systems, as is true to the west of Havana, where the Lacey, Langellier, West Matanzas, and Kerton Valley dis-

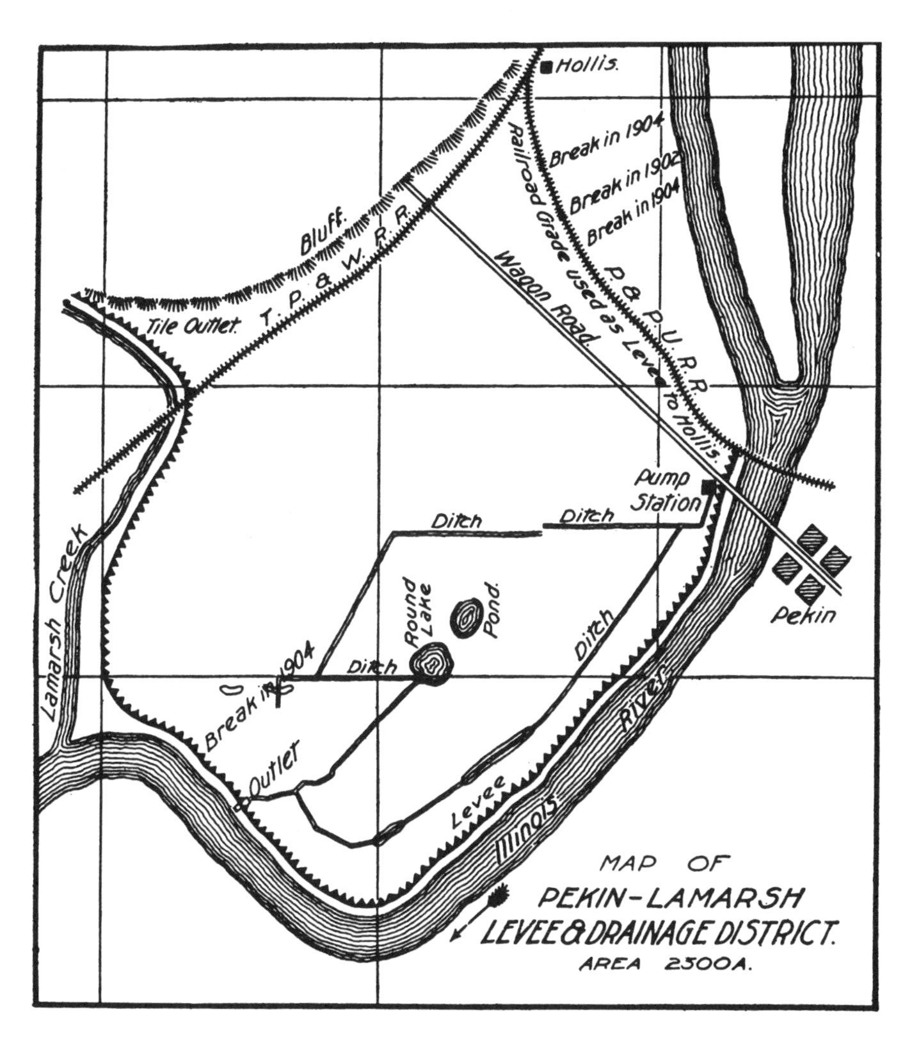


**Map 2.4.** Pekin and La Marsh Drainage and Levee District, 1904. From Elliott, "Drainage Investigations," 669.

tricts are located, and near Meredosia, where the Meredosia Lake, Coon Run, and Lost Creek districts adjoin. While the last two districts were formed to protect land from overflow by creeks in their midst, most districts are separated by floodways and opposing flank levees, some of which were set back cooperatively, such as those between the Crane Creek and Big Prairie districts following overtopping and breaching in June 1917.[19]

The earliest levee systems were built with team-drawn wheeled and drag scrapers.

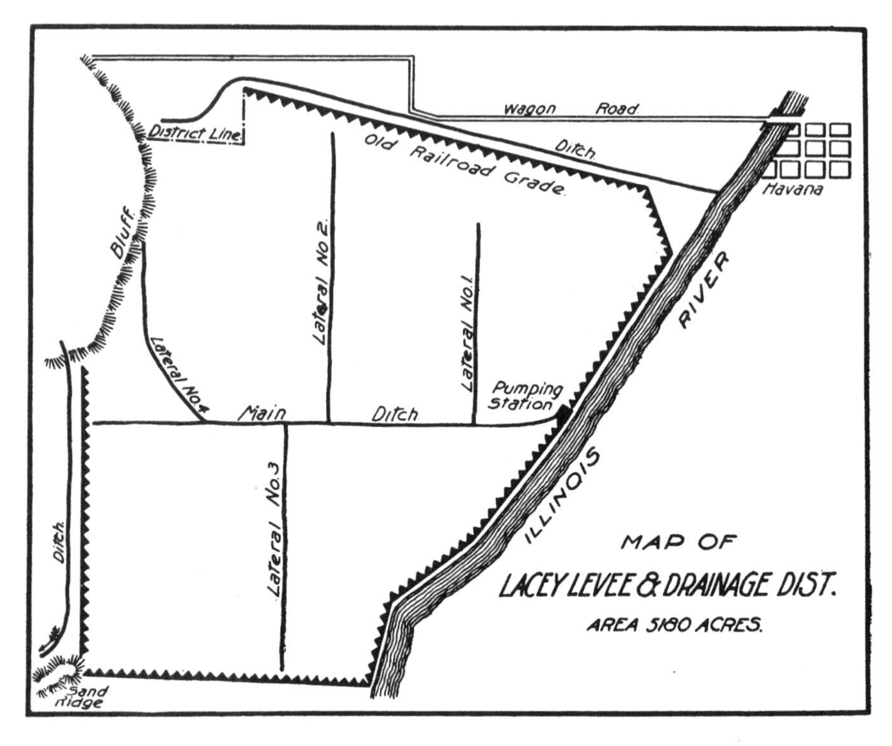


**Map 2.5.** Lacey Drainage and Levee District, 1904. From Elliott, "Drainage Investigations," 671.

These levees were raised close to the low-water banks of the river and creeks with earth excavated from trenches located on the land side of the structures, to judge from the Hartwell, Roberts, and Lowenstein ranches in Greene County and from the Pekin and La Marsh district. Such borrow pits or trenches fostered seepage from the river and threatened levee stability where the water stood against the interior levee toe. The problem was countered at the Pekin and La Marsh district between 1908 and 1914 by raising a new levee by dredge about 150 feet inside the first one. Eventually, all the internal borrow trenches were filled. At the Hartwell district, the filling was done with material excavated from the river bottom by a cutterhead hydraulic dredge.[20]

As a rule, construction of a district's levee system began before excavation of the ditch system. Ditching occurred when a sufficient levee system was in place to withstand nominal rises of the river and creeks or after the levee system was done. The practice reduced the chance that overflows would interrupt work and damage the ditches, requiring renewed effort by the contractor. As it was, work on a number of levee systems was interrupted and the structures damaged by high-water stages.

The prescribed order of tasks in levee building began as soon as the assessment roll for the district was confirmed in court. The district's engineer proceeded with transit and level crews and axmen to set the stakes that guided the contractor's crews and machines along the right-of-way for the structure. Within two or three weeks, the contractor's crew began to remove timber, stumps, and roots from a swath that extended about 50 feet beyond the prospective levee toes, filled and tamped the holes, and then plowed along the axis of the site to unearth roots. The timber was burned or was placed on the riverbank to be barged to a sawmill; also, cordwood was apt to be stacked for use in the dredges. Where sectors of the proposed levee were expected to have water adjoining for much of the year, a muck ditch was excavated in the natural levee along the central axis of the proposed artificial levee. These ditches were dug with teams and scrapers, by machine or hand, or were blown with dynamite. They were excavated to depths of 3 to 5 feet and widths of 3 to 8 feet so that the porous topsoil and underlying lenses of sand, buried logs, and the like could be removed. The void was filled with more compact material to retard seepage. Where soft materials extended to some depth, as happened at infilled sloughs, sheet piling was driven to reduce seepage. Both the muck ditch and plowed surface of the right-of-way were expected to afford good bonding and minimal seepage between the levee foundation and the structure. This foundation work preceded by several hundred feet the amassing of dirt from the borrow pit into the levee by dredge or dragline.[21]

Commonly, levee construction and ditch excavation began near the bluffs, at the upstream, or higher, end of the proposed systems. The large levee-building dredges were delivered on flatcars to railway sidings or freight yards near the assembly and work sites. A number were delivered to riverside towns such as Meredosia, Beardstown, and Havana, where it was relatively easy to marshal mechanics and carpenters, lumber and hardware supplies, teams and wagons, and the stationary or wheeled steam or gasoline engines used for hauling and winching heavy objects aboard the hull. Once in the water, dredges usually were moved to work sites by towboat. While positioning a dredge at the work site might await a high stage of the river, it could cut its way across the floodplain. Exceptionally, a large dredge was hauled overland by wagon to a reassembly site. For instance, one contractor moved a 2 1/2-cubic-yard dredge from the Sny River bottoms of the Mississippi River to Florence, on the Illinois River, where it was reassembled. The move required two and a half days of transit time for about 100 wagonloads of freight. The dredge's boom, which weighed about 30,000 pounds and was 72 feet long, required ten days to be eased through the 45-mile trip by a pair of wagons and a traction engine.[22] The small dredges used to cut ditches arrived in knocked-down condition at the railway siding nearest the work site or were towed to the prospective district. In the former case, the boilers were skidded and the machinery and lumber hauled to the assembly point, where either a creek was dammed or holes were dug to receive the hull. Sometimes, the hull was floated with water diverted from a tributary. The machinery was skidded aboard. Ideally, the dredge and ditcher assembly was undertaken in the winter, when the ground was firm.

A good deal of labor and specialized equipment was required to clear the right-of-way for a levee or ditch system. At least once, when the local and regional labor supply was short, a contractor recruited blacks in Alabama. As a rule, laborers and work stock were accommodated in camps near the work areas. Dredge and ditcher crews were housed aboard the machines or in quarters boats, which were shoe box–like craft with galley, mess, and sleeping quarters. Such assemblages of working men and stock, equipment and machinery, and the spectacle of blasting stumps provided entertainment for gatherings of the local populace. Danger was ever present. During the levee clearing operation along the river opposite Meredosia, for an extreme example, the foreman of the stump removal crew and the tree behind which he took cover were felled by a plummeting stump. A second man lost a leg to the sweep of a horse-powered stump-lifting device that released suddenly as a chain broke. Yet another man was injured seriously while loading timbers onto a barge, and a man drowned while escaping an exploding gasoline launch that ferried men between the camp and the work site.[23]

Commonly, dredges and ditchers worked day and night, which required that second shifts of operators be used and that the machines be fitted forward with large carbide or electric spotlights. Alternatively, and with dredges having very long booms, a roustabout with lanterns marked the discharge point on the distant levee.

Contractors usually committed one to three dipper dredges to a project, one or two larger ones primarily for levee building and a smaller one for ditching. Sometimes, one of the levee-building machines also excavated the main drainage canal, which indicated that the contractor had successfully bid on both tasks. The number of dredges that worked at a district reflected the volume of dirt to be dug and the timetable for completion, which was at least a season of ten months. In that period, a single circuit usually was made by the levee-building machines. It is evident at the Coal Creek and Nutwood districts, and probably elsewhere, that the machines made at least two circuits along the levee at riverside. The two to three passes required of the big levee-building dredge at the Coal Creek district were due to running and slumping by soft material. The first circuit by the dredge at the Nutwood district raised the structure to about 9 feet, enough to withstand a heavy flow of the river. The merit of making two such circuits was that the prospective levee core could drain and consolidate before being brought up to grade. When two dredges were employed on a levee system, the machine that raised the northerly flank levee then followed the circuit of the machine that started just below the intersection of that creek with the river.

Whatever the procedure, amassing material into levees resulted in a water-filled borrow trench located 10 to 15 feet beyond the toe of the levee perimeter. These borrow trenches were 4 to 8 feet deep and about 60 to 70 feet across.[24]

The recommended practice in building levees along the river with dipper dredges, at least by 1905, was to align the structure landward of the low-water bank by 50 or 150 feet (Pekin and La Marsh district) and sometimes by as much as 200 or 300 feet (Spring Lake and Banner Special districts). The residual strip of wooded bank left between the borrow trench and river formed a barrier against waves and ice at all river stages. Gaps in the natural breakwater were apt to be filled by dredge and then planted to willows. Such an auxiliary levee at the Chautauqua district required 63,000 cubic yards of spoil. Sometimes, wave protection was obtained by mooring timber rafts along the levee. A particularly vulnerable southern sector of the Spring Lake district's levee system was faced with concrete to prevent wave damage, and riprap was used at the East Peoria and Keach districts.[25] However, such costly armoring is not known to have been adopted elsewhere.

Although placing the river levee 100 to 300 feet back from the low-water bank and leaving a borrow trench adjacent facilitated the flow of flood stages, the accommodation was incidental to achieving security against wind-whipped wave wash. Moreover, to impede the development of erosive currents, the continuity of borrow trenches sometimes was broken every few hundred feet by leaving residual segments of natural levee in place or by depositing transverse barriers of earth, as was done at the Big Prairie and Thompson Lake districts.[26] Such barriers facilitated trapping alluvium in the borrow trenches, stockpiling (as it were) material for levee dressing.

The rule of thumb in determining the height of a proposed levee system was to have the freeboard 2 to 4 feet above record flood stages, the margin being expected to withstand the effect of waves or lost floodplain area. The early levees were from 8 to 18 feet higher than adjacent land surfaces and averaged 12 to 15 feet high. The early machine-made levees were as little as 30 to 50 or 60 to 80 feet across the base (fig. 2.1). Nevertheless, levees of massive proportions were built as early as 1909, at which time the Nutwood district's system was from 198 to 231 feet across the base at riverside, 82 to 165 feet across at the flanks, and about 16 feet higher than the land adjacent. The object in levee building was to have enough width or mass at prospective water levels that a plane of saturation would not extend through to the inner slope of the structure, which could result in a break if the unsaturated overlying material slumped along the plane of saturation.[27]

To achieve desired mass at prospective water levels, the levees were built with riverside slopes of 2:1 or 3:1 (i.e., 2 or 3 ft. horizontal to 1 ft. vertical) and landward slopes of 1½:1 or 2:1. Levee crowns and flanks built by dipper dredge were rough and irregular (fig. 2.2). As the material consolidated and was worked by scraper, crowns of 3 to 8 feet

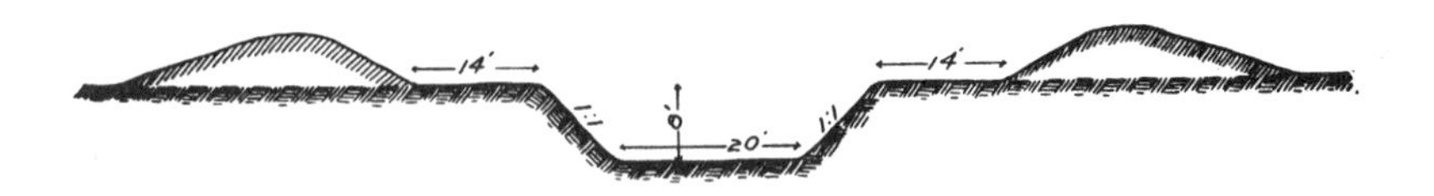


**Fig. 2.1.** Section of levee and borrow pit, Coal Creek Drainage and Levee District, 1904. From Elliott, "Drainage Investigations," 673.

evolved. In general, the alluvium was a good levee-making material. It compacted well, and weathered surfaces were resistant to erosion by overbank flow; it was susceptible to wave attack and sloughing, however. Although it was recommended that levee slopes be smoothed and planted in grass, many districts allowed rank volunteer cover to develop. Consequently, brush and timber predominated on the levees at most districts in 1926.[28]

In specifying the amount of material to be amassed into levees, the general contractual practice was to factor in 15 or 20 percent for shrinkage when power excavators were used, although as little as 10 percent was added at the Scott County district. As a rule, 10 percent shrinkage was agreed to where teams and scrapers were used. The excess material, besides compensating for prospective settling in the spoil, was insurance against the difficulty that operators had in maintaining prescribed levee grades and profiles with dipper dredges. On the other hand, sometimes levees had exceptional height and mass because big dredges had to deepen borrow ditches to maintain flotation at times of low water. While some inexperienced levee builders, such as those at the Coal Creek district, may have emplaced the equivalent of twice the volume called for, there is evidence that other levee systems were not built to grade from the outset, which suggests negligence all around. Now and again, suits were threatened or brought, such as the one by the East Peoria district, to force contractors to complete projects to grade. There was a tendency to allow completed levee systems to slip below grade adjacent pumping stations and where roads crossed them. The sags were tolerated because they eased transit for men and wheelbarrows as coal was taken from barges to storage sheds, in the former case, and because transit by wagons was facilitated, in the latter case. Presumably, it was expected that these accessible sags would be raised to grade when high water threatened. On the other hand, a general settling of up to 2 feet along lengthy stretches of levee could be unrecognized until major flood stages, as noted by Pickels after the flood of 1926–27.[29]

The problems posed by runoff and sediment loads from bluff areas appear to have been understood at the Willow Creek, Coon Run, Lost Creek, and Otter Creek districts, where overflow by creeks motivated the formation of the districts. However, there were early districts whose principals assumed that runoff and the flow of seasonal creeks would be promptly disposed of by sluice and pump capacities. The Pekin and La Marsh

**Fig. 2.2.** Embankment across state-built canal into Spring Lake, ca. 1909. The vegetated embankments flanking the artificial canal were raised with spoil from the ditch. The seal across the canal shows serrated crest typical of work by dipper dredge. The fresh material probably was added to compensate for settling in spoil emplaced a season or two earlier. Illinois Submerged and Shore Lands Investigating Committee; courtesy of Illinois State Museum.

and the Coal Creek districts for example, were flooded by heavy rain and runoff several times after the works of reclamation were completed in the 1890s. The frequency of crop damage and pump house immobilization diminished after 1907, as the engineers devised means to contain problems outside the perimeter of cropland.[30]

The immediate response to a flood at the Coal Creek district in 1898 was to divert the creek through a ditch skirting the foot of the bluff and rounding the north end of the district's western flank levee, as may be seen in maps 2.3 and 2.6. The ditch, which was embanked on the district side, intercepted Coal Creek and all runoff from the bluff area to the west. Nevertheless, as late as 1906, runoff from the remainder of the bluff area and unruly Coal Creek continued to be considered a greater hindrance to reclamation than the river. The Pekin and La Marsh district undertook a lesser diversion project in 1899, as well. The lessons were being learned yet in 1916–17, when the Big Lake district excavated a diversion ditch along part of the bluff margin two years after original plans for reclamation were completed.[31]

Thus, at some drainage and levee districts, the embanked bluff ditch became as

**Map 2.6.** Drainage system, Coal Creek Drainage and Levee District, 1915. From Woodward, *Drainage by Pumps,* 15.

integral a part of the works of drainage as were the levee systems, the interior drainage systems and pump houses, and the rectified flank creeks. As may will be discussed in chapter 3, some districts receive no runoff from the bluffs, and at least 12 districts receive runoff from no more than 1,000 acres of upland. Excluding runoff with diversion ditches enabled district engineers to design internal ditch systems and pump houses of lower capacity and lower cost than would have been required otherwise. Also, the incidence of bank erosion and deposition in the drainage system was reduced.

The major difficulty with bluff ditches and flank creeks was a tendency to fill quickly with sediment, raising beds substantially above the land protected by the embankments. At the Coal Creek district, for instance, it was evident by 1904 that the intercepting ditch filled so quickly with sediment that more water entered the district than the pumps could remove in a timely fashion. Such channel alluviation was especially acute for districts that were organized with a major creek in their midst, such as the one adjacent Coon Run that broke across the levees at least nine times between 1902 and 1918. In 1918, the 6.6 miles of dredged and leveed channel was all but filled in to about levee height. At the Otter Creek district, the channel was excavated and the levees set back (fig. 2.3) to accommodate flood flows during any stage of the Illinois River. The expectation that scouring would cleanse the channel as the river receded was not realized. No way was found to prevent the accumulation of sediment in channels and diversion ditches where the "hill water" gradient was broken upon reaching the floor of the valley, least of all when the backwater of a river flood stage was encountered. Channels had to be cleaned and embankments enlarged with some frequency. The frequency of such remedial work was reduced after 1909 at the Coal Creek district by excavating settling basins of 4 to 20 acres opposite the mouths of watercourses that entered a new bluff ditch. Some districts with bluff areas that could only be drained directly into the tract incorporated embanked settling basins in the internal ditch system,[32] as happened at the Nutwood and Valley City districts. The filled settling basins were returned to the landowners for cropping, and new basins were shaped nearby. Only the Banner Special district is known to have abandoned the sediment basin concept in favor of directing a troublesome seasonal watercourse to the river.[33]

Documentation of the rate of sediment deposition in ditch systems is elusive in district archives. Indicative of the problem, however, are data from a handful of districts. The Big Swan district performed piecemeal ditch cleaning before engaging a contractor to remove 278,000 cubic yards of dirt from the system between 1911 and 1913, and 176,000 cubic yards more in 1920. The dirt deposited in 15 miles of ditches came from 3,260 acres of upland, the district's fields, and ditch banks. The known volume removed was equivalent to 80 percent of the volume excavated to make the original ditch system. The general ponding condition that prevailed in 1918 and 1919 at the McGee Creek district was partially alleviated by removing 156,000 cubic yards of material from 16.5 miles of ditch system. The volume is estimated to have exceeded 90 percent of the material excavated to create the original ditch system in 1908. At

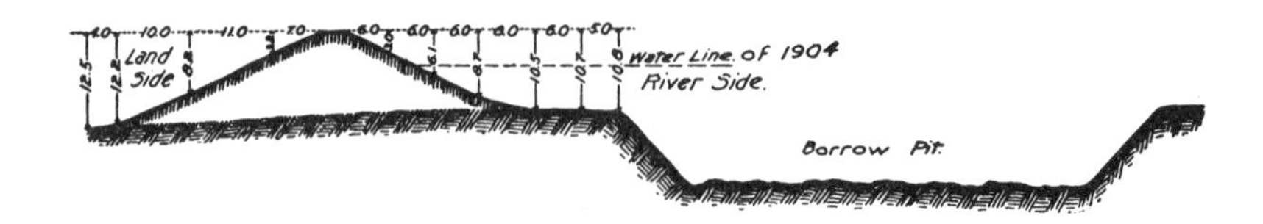


**Fig. 2.3.** Section across Otter Creek, Otter Creek Drainage and Levee District, 1904. From Elliott, "Drainage Investigations," 704.

the Hillview district, about 60,000 cubic yards of sediment buried the uppermost half mile of a main ditch (40 ft. wide at bottom, 70 ft. at top, and 12 ft. deep) between 1906 and 1916. The debris stood higher than the land on either side. Apparently, the practice was to allow ditches to become heavily alluviated before engaging contractors to clean them out, which occurred at intervals of several years. Illustrative was the Coal Creek district, where major ditch cleaning jobs were done in 1908–9 and 1913. However, the district purchased a dragline in 1917, making the task a routine in-house maintenance activity. The Big Creek and McGee Creek districts purchased draglines for the same reason in 1916 and 1919, respectively. Thus, a new practice was being adopted.[34]

Although the procedure may have been adopted elsewhere, only the Little Creek district, in Brown County, can be documented to have diverted minor creeks so as to systematically spread their loads over property. The twenty-year process, begun about 1906, filled in part of Perkins Lake, which lay about a mile south of the La Grange Lock. The idea of building a dam within the bluff area to trap sediment was broached for the Nutwood district in 1915,[35] but the concept of effecting erosion control within the bluff areas is not know to have been acted upon until the 1930s.

## Development of Interior Ditch Systems

Interior drainage systems consisted of main, submain, and lateral open ditches and tile lines of large diameter. The systems afforded gravity drainage from the landowners' systems of field and tile drains and the district's internal sediment traps to the sumps from which water was pumped over the levee or, in a very few cases, allowed to exit the districts by gravity through concrete or steel sluiceways. The gravity flow option was not available generally because the level of the river maintained by dams and locks for navigation purposes was too high.

It was common practice to incorporate beheaded creek channels and sloughs into district drainage systems. The channels were straightened and enlarged by excavators. Nevertheless, most of the aggregate length of main and submain ditches was newly excavated by small dipper dredge or dryland machine. Teams and scrapers were used to cut most small ditches until about 1911.[36] As a rule, the rights-of-way were cleared and grubbed by the contractor prior to excavation, which usually began in the northerly and upslope corner of the district. The artificial canals might conform with the slope of the land, following troughs or channels in the floodplain, or they would be excavated across the grain of the land (map 2.6). Whatever the case, they tended to follow property boundaries. The systems either terminated near the downstream corner of the district or at some more central point adjacent the riverside levee, whichever was determined to be less costly to construct. In some instances, notably at the Pekin and La Marsh and the McGee Creek districts, the original drainage mains were lacking in sufficient fall and width to function satisfactorily, which resulted in ponding and crop damage. The problem was especially costly to correct at the McGee Creek district, where in 1914 the drainage system had to be divided into two segments, each with a pump house.[37] Otherwise, deficient systems were reexcavated to the single pump house. The floodplain material and local relief made ditching relatively easy, even to the extent of enabling the Lacey, Langellier, West Matanzas, and Kerton Valley districts in 1930 to integrate separate ditch systems into one served by a single pump house, as described in the chapter following.

The mains and submains of open ditch systems, it was agreed generally, should be about 6 to 8 feet deep. As a rule, the steep-sided open ditches were at least 12 to 16 feet across the bottom. They cost about half of what tile lines of equivalent capacity cost. Smaller ditches were cut, but by 1915 it was general practice to substitute large-diameter tile for the open ditches that would have been up to 12 feet across earlier. Ditches were given about 10 feet of berm between their edges and the ridge of excavated spoil. Dipper machines and draglines dumped to both sides. The latter machines replaced the former for ditch cleaning work because, among other reasons, they could cast spoil more widely, facilitating subsequent spreading by landowners or farmers into adjacent fields. Thus, embankments of spoil tend to be absent, which minimizes the loss of cropland and facilitates access to the ditches for maintenance. By and large, the interval between ditches allowed water tables to be drawn down to at least 3 feet below the soil surface, which fostered root development as well as providing storage capacity in the soil and in the ditches to accommodate heavy rains.[38] Such reservoir capacities and the rate of water conveyance to the pump house had a bearing on pumping capacities required by districts.

## Construction and Operating Costs

The district engineer's estimate of the cost of achieving security from overflow for a drainage and levee district was complicated by numerous variables. To begin with, there were estimates for the rights-of-way on which levee and ditch systems were to

**Table 2.1.** Allocation of initial reclamation costs, as percentage of total, selected districts

| | | | | Allocation of Costs (%) | | | | |
|---|---|---|---|---|---|---|---|---|
| District | Year of construction | Assessed acres | Total cost ($) | Rts.-of-way | Levees | Ditches & creeks | Pumping station | Admin., ct., eng., etc. |
| Big Prairie | 1916 | 1,700 | 120,388 | ? | 67 | 14 | 6 | 13 |
| Big Swan | 1904–10 | 11,850 | 172,221 | 10 | 67 | [a] | 12 | 11 |
| Coal Cr. | 1897 | 6,290 | 72,395 | ? | 69 | 22 | 8 | ? |
| E. Liverpool | 1916 | 2,765 | 191,991 | 3 | 55 | 11 | 20 | 11 |
| E. Peoria | 1907 | 736 | 45,525 | 2 | 54 | 17 | 18 | 9 |
| Hillview | 1905 | 12,318 | 226,250 | 5 | 51 | 33 | 6 | 5 |
| Kelly L. | 1918–20 | 985 | 70,103 | 5 | 68 | [a] | 24 | 4 |
| Lacey | 1896 | 2,987 | 54,886 | 6 | 42 | 26 | 9 | 17 |
| Liverpool | 1915 | 3,024 | 177,966 | 11 | 46 | 17 | 11 | 15 |
| Scott Co. | 1911–14 | 9,200 | 218,278 | 7 | 52 | 18 | 17 | 6 |
| Thompson L. | 1919–21 | 5,600 | 300,192 | 17 | 55 | [a] | 7 | 21 |

[a]Included under levee costs.

be constructed; the cubic yards of earth to be moved in shaping the systems, modifying adjacent streams, and controlling hill drainage; the pumping facility and appurtenances; and such overhead costs as engineering, attorney, administrative, and court services. However, as work proceeded, unforeseen vagaries of rainfall, creek and river stages, subsoil conditions, contractor and machinery performance, litigation, and inflation came into play. It was the rare project that did not require a supplemental plan and budget within a year or so of completion of the initial plan. Nevertheless, generalizations about the allocation of construction costs may be made using comparable data for 11 of the 36 most closely studied districts (table 2.1).[39]

On average, roughly 70 percent of the total investment in the first construction phase was spent for the levee and ditch systems; the pumping stations required 15 percent. Acquisition and clearing of rights-of-way for the works accounted for 5 to 10 percent of initial budgets. The engineering, legal, court, and administrative costs, plus contingencies, were about 10 percent.

Once the districts were reclaimed, the major operating cost was for the energy consumed by the pumping stations. Quite secondary were the salaries of operating personnel at the stations, which annually aggregated about the same as the sum for levee and ditch maintenance and for administration. An indication of operating costs for the Coal Creek district during the period January 1914 to July 1923 appears in table 2.2.[40] Annual reports for a number of other districts show that energy costs were the major operating expense everywhere, although those costs varied with the year and district.

Only for the Big Swan Drainage and Levee District was documentation found for the long-term cost of borrowing money relative to other costs of reclamation (table 2.3).[41] The record shows fund allocations for the start-up phase of reclamation and for the entire span of operations between June 1, 1904, and June 30, 1931. At the Big Swan district, the aggregate cost of borrowing money over the span of operations was 41 percent of all expenditures, whereas construction costs represented 29 percent of the outlay. Operation and repairs at the pumping station were 21 percent, of which 16 percent were energy costs. Engineering, court, and administrative costs were a little over 4 percent, whereas the purchase of rights-of-way approached 2 percent. Unspecified expenditures represented 3 percent of the long-term total. The outlay averaged $3.87 per acre per annum for the 27 years, a cost of doing business which upland farmers were spared.

In the long run, to judge from the 1904 to 1931 record at the Big Swan district, construction costs were well exceeded by the cost of borrowed money. Not greatly different was an informed estimate made in 1937 that borrowed money generally cost the districts one-third again the aggregate construction costs.[42]

**Table 2.2.** Average annual operating costs, Coal Creek Drainage and Levee District, January 7, 1914, to July 1, 1923

| | Average annual costs ($) | | |
|---|---|---|---|
| | Total | Per acre | Percent of total outlay |
| Pumping plant | 10,612.05 | 1.641 | 83.9 |
| Electric power | 9,120.10 | 1.410 | 72.2 |
| Operating labor | 1,248.81 | 0.193 | 9.8 |
| Heating coal | 116.02 | 0.018 | .9 |
| Supplies, repairs | 103.00 | 0.016 | .8 |
| Insurance | 24.12 | 0.004 | .2 |
| Levees and ditches | 907.70 | 0.140 | 7.2 |
| Maintenance | 755.05 | 0.117 | 6.0 |
| Bounty on rodents | 152.65 | 0.023 | 1.2 |
| Other maintenance | 103.63 | 0.016 | .8 |
| Bridges | 68.98 | 0.011 | .5 |
| Buildings | 34.65 | 0.005 | .3 |
| Administrative | 629.11 | 0.096 | 5.0 |
| Office and legal | 313.18 | 0.048 | 2.5 |
| Commissioners | 215.91 | 0.033 | 1.7 |
| Engineer services | 100.02 | 0.015 | .8 |
| Interest | 386.16 | 0.060 | 3.1 |
| Total | 12,638.65 | 1.953 | 100.0 |

## Assessment of the Work

It was the rare drainage and levee district where the original reclamation plan was completed as budgeted. Although Pickels noted in 1921 for Illinois generally that commissioners and their engineers might understate the first costs to allay opposition to the commencement of work,[43] such criticism was not evident in the record of court hearings or annual reports of the districts studied. More commonly, shortfalls were attributed to unanticipated physical problems in construction, flood-induced construction delays and rehabilitation costs, inflation, errors in design and construction, or defaults among contractors. At least two or three districts were affected by each type of encumbrance. Critics at one district focused on the extravagance of commissioners when the problem may have been undue permissiveness with the designer of excavators whom they recruited as contractor, whose machinery did not perform to expectation. A fresh slate of commissioners terminated the contract.[44]

**Table 2.3.** Summary of costs to develop, operate, and maintain the Big Swan Drainage and Levee District, June 1, 1904, to August 18, 1910, and to June 30, 1931

| | Development phase, 1904–10 | | All phases, 1904–31 | |
|---|---|---|---|---|
| Expenditure | $ | % | $ | % |
| Right-of-way purchase | 18,015.70 | 7 | 19,269.70 | 2 |
| Construction | 135,831.23 | 54 | 365,285.19 | 29 |
| Pumping plant operations | 16,721.11 | 7 | 258,813.30 | 21 |
| coal/oil | 8,894.45 | | 42,851.81 | |
| electricity | ... | | 150,821.66 | |
| lubricants, etc. | 322.96 | | 2,212.84 | |
| repairs | 1,783.65 | | 29,264.99 | |
| insurance | 271.25 | | 3,109.65 | |
| engineer | 4,474.54 | | 26,193.67 | |
| fireman | 974.26 | | 4,358.68 | |
| Diverse | 8,726.56 | 3 | 35,811.57 | 3 |
| Matured bonds, notes | 14,807.40 | 6 | 291,054.61 | 23 |
| Interest | 46,932.09 | 19 | 181,576.05 | 15 |
| Loans, certificates of deposit | ... | | 32,200.00 | 3 |
| Engineer's fees, expenses | 2,996.26 | 1 | 22,210.47 | 2 |
| Commissioners' fees | 3,548.03 | 1 | 15,788.48 | 1 |
| Treasurer's salary | 254.64 | | 1,903.24 | |
| Attorneys' fees, court costs | 2,849.00 | 1 | 11,139.83 | 1 |
| Ill. Drainage Assoc. dues | ... | | 2,876.46 | |
| Total | 250,682.02 | 99 | 1,237,928.90 | 100 |

The need for greater proficiency among drainage engineers, noted by Elliott in 1908, was cast more clearly later in Pickels's observation that local surveyors, rather than trained civil engineers, often were selected to design and construct works for the districts.[45] In 1913, the attorney for the Meredosia Lake district, whose first engineer made a very costly mistake, described those who masqueraded as engineers as "evils" outranking incompetent or indifferent district commissioners, because the costly work had to be done a second time.[46] Other costly errors were attributed to hacks at the Liverpool Drainage and Levee District in 1922, and still others were avoided at the Scott County district in 1911, when a flawed plan was recognized in time to secure the services of new commissioners and a credentialed civil engineer. The errors made in levee design and construction at the Pekin and La Marsh and the Coal Creek districts in the 1890s may have been due to inexperience, but the later problems reflected greater effectiveness in bidding on contracts than in technical expertise.[47]

At the time that reclamation was undertaken along the river, drainage work was not considered suitable for attention by better civil engineers, at least in the Midwest. Indicative of the low esteem for drainage engineering was the assessment by a civil engineering student in 1912 at the University of Illinois. He observed that texts and reference works were scarce, that professional articles were so general as to be useless in the preparation of a thesis, and that relatively little serious investigation had demonstrated the usefulness of the theories advanced. That it was "generally conceded" that "scientific investigation has not kept pace with the expenditures of money for drainage purposes" was consistent with Pickels's observation that the University of Illinois had begun instruction in drainage engineering only in the preceding ten years;[48] baccalaureate theses in civil engineering related to the subject began to appear in 1905 and 1906.[49]

The belief held by young engineers and their mentors at Urbana that drainage engineering was little evolved as a science suggests that a 200-year legacy of levee building and drainage of agricultural land along the lower Mississippi River was little known or appreciated in the Midwest; yet, by 1916 there were 3.5 million acres of drained farmland behind hundreds of miles of levees on and near the Father of Waters. Between 1882 and 1916, local entities and the Southern states in question expended $90 million on works that protected land from overflow and that removed unwanted water from cropland, and Congress had appropriated another $29 million for levee work by the Mississippi River Commission, which was organized in 1879.[50] Whether the annual reports of the agency and the articles on levee building in *Engineering News* were seen but found wanting for circumstances in the Midwest cannot be documented in the work of Pickels or of his students. It is curious, too, that the work of the contemporary dean of drainage engineers, Charles G. Elliott, whose degrees were obtained at the University of Illinois (B.S., 1877; C.E., 1895), was not referred to.[51] Whatever the case, it is indeed shown by the record along the Illinois River that drainage engineers had a good deal to learn.

Not only did the Midwestern drainage engineer who planned to realign streams, build levees, and lay out ditch systems on a large scale in the first decade of this century have "comparatively little reliable data to guide him," but many farmers approached him "with the idea that he has some supernatural, intuitive power which will enable him to look over the land and 'squint' through his instrument a few times to tell all about it; when an engineer does not measure up to this ideal, and insists that he needs a carefully made topographical survey of the lands which are to be drained in order for him to form a proper judgment as to the best location and the necessary size and slope of the various ditches required to drain the lands, he immediately loses his prestige and very frequently loses the job."[52] For the conscientious district engineer, there was no more useful reference than the topographic maps of the floodplain with 1-foot contours that, under the supervision of J. W. Woermann, were surveyed in 1902–4 by the U.S. Army Corps of Engineers for the office of the District Engineer in Chicago. The maps and monuments were used universally to locate levee lines and drainage systems at the districts, thereby sparing time and some of the $15.20 per acre cost of thorough surveys (1916).[53]

The presumed shortcomings in drainage expertise highlighted by flood events in the lower valley in 1903 and 1904 caught the attention of the then congressman Henry T. Rainey. He and other interested legislators persuaded the Secretary of Agriculture that enabling legislation for studies of irrigation and alkali problems in the Great Plains and the West should be modified to permit drainage investigations in the Midwest and the South. As a result, investigations of drainage along the Wabash, Illinois, Mississippi, and upper Missouri river basins and in Florida were added to a larger series of studies of irrigation projects in the subhumid and arid West. Although Rainey asserted in 1928 that the resulting report of a federal agency was a guide for land drainers of the Illinois River area, his motivation was to provide grounds to extend the jurisdiction of the Mississippi River Commission to the Illinois Valley. References to a seminal role for the report prepared by Charles G. Elliott were not found in the engineering and contracting literature nor in the reports of the district engineers and commissioners of the valley. Nevertheless, the report must have been read, because so many local and regional examples were used to illustrate guiding principles of levee building. Many of the ideas appear among the specifications for drainage contracts and in Pickels's work on drainage engineering, as they did earlier in *Engineering News.* In sum, the work by Elliott was a digest of costly errors made over the years in the Midwest by engineers, contractors, and commissioners. This pioneer drainage engineer published on the topic beginning in 1882 and prepared the first of the U.S. Department of Agriculture's bulletins on the subject in 1895; Elliott knew whereof he wrote.[54]

It is evident that the people who first undertook large-scale land drainage projects along the lower Illinois River paid little heed to the fine points of levee building on floodplains that were described in contemporary professional literature. Perhaps this was attributable to the fact that most local experience was gained in ditching the wet prairie uplands (where levees were not important); to the sense among farmers and other investors that anyone could lay out a ditch or throw up an embankment; and, therefore, to the reluctance to seek out and pay for professional engineering advice that could be found.[55] Whatever the case, the drainage works that began to be shaped in 1890–91 at the Pekin and La Marsh Drainage and Levee District were deficient in the extreme. The area largely remained under water between 1902 and 1907 as a result. Among the errors were the construction of the levee on the unmodified natural surface and the incorporation into the structure of logs, standing trees, and stumps that occupied the right-of-way. By the same token, the collaborating railway company used timber, trash, and earth to fill the trestled sectors of embankment that became the northern levee. Also, in the process of raising the grade, laborers simply buried old ties in place. The pipe into which the pump discharged was placed through the levee without any baffling designed to prevent avenues of seepage from developing; the sluiceway placed in the bed of the slough at the lower end of the district (map 2.4) was defective, as well. As woody material disintegrated and muskrats burrowed, and as pipes vibrated in use, the porosity of the levee became so great that crayfish entered in the flows. Moreover, the borrow ditches (30–40 ft. by 3–6 ft.) lay moatlike immediately along both sides of the levee. Seepage was a persistent early problem at the Coal Creek and McGee district levees, as well. At the McGee district, the cause is not documented; but at Coal Creek, organic matter was left beneath the levee, although clearing and grubbing the foundation and borrow pit were supposed to have been done in 1897. Perhaps dredge operators substituted timber for spoil here and there as an economy measure. Contemporaneous work was done properly at the Lacey district, where thorough plowing of the right-of-way occurred and a short section of muck ditch was dug.[56]

## The Reclamation Achievement

Between 1890 and 1928, according to Harman et al., about 187,000 acres of assessable land were reclaimed from the 280,000 acres of floodplain opposite and downstream of Peoria. Pickels and Leonard estimated that about 196,500 acres were organized or were being organized at the end of 1927, which was about 1,800 acres less than all of the land that Harman et al. associated with the reclaimed entities. About 178,700 acres of assessed land are represented in the districts researched here (table 2.4). They were reclaimed from tracts aggregating about 192,700 acres of bottoms. The discrepancy between the aggregate assessed area and earlier figures is accounted for largely by a number of small private reclamations. The districts with 178,700 acres of assessed land constituted about 64 percent of the floodplain (table 2.5); another 7 percent was occupied by levees, borrow ditches, and unimproved bottoms associated with the reclaimed tracts.[57]

The progression of reclamation is best appreciated by relating it to three sectors of the floodplain, as outlined in table 2.5 and in accompanying maps 2.7, 2.8, 2.9, and 2.10. Respectively, the three sectors lie between Peoria and Havana, Havana and Beardstown, and Beardstown and Grafton. The maps consolidate data from table

**Table 2.4.** District reclamation sequence, by periods and areas, 1890–1924

| District | Year of reclam. | Assessed land (ac.) | Average assess. (ac.) | Total area (ac.) | Aggregate (ac.) |
|---|---|---|---|---|---|
| 1890–94 | | | 3,443.5 | 6,887 | 6,887 |
| Pekin and La Marsh | ('91) | 2,674 | | | |
| Willow Creek[a] | ('94) | 4,213 | | | |
| 1895–99 | | | 6,290 | 6,290 | 13,177 |
| Coal Creek | ('96) | 6,290 | | | |
| 1900–04 | | | 3,975 | 7,950 | 21,127 |
| Coon Run | ('02) | 4,250 | | | |
| Meredosia L. | ('04) | 3,700 | | | |
| 1905–09 | | | 8,340 | 66,723 | 87,850 |
| Lacey | ('06) | 2,987 | | | |
| Langellier | ('06) | 1,978 | | | |
| Hartwell | ('06) | 8,650 | | | |
| Big Swan | ('07) | 11,850 | | | |
| Hillview | ('09) | 12,318 | | | |
| Keach (F'banks) | ('09) | 8,000 | | | |
| McGee Cr. | ('09) | 10,080 | | | |
| Nutwood | ('09) | 10,860 | | | |
| 1910–14 | | | 5,439 | 48,955 | 136,805 |
| Spring L. | ('10) | 12,000 | | | |
| Lost Cr.[a] | ('10) | 2,400 | | | |
| E. Peoria | ('11) | 736 | | | |
| Crane Cr. | ('11) | 5,050 | | | |
| Eldred | ('11) | 8,460 | | | |
| Scott Co. | ('12) | 9,200 | | | |
| Banner Sp. | ('14) | 4,279 | | | |
| Otter Cr.[a] | ('14) | 3,680 | | | |
| Big L. | ('14) | 3,150 | | | |
| 1915–19 | | | 2,910 | 23,276 | 160,081 |
| Mauvaise T.[a] | ('15) | 3,980 | | | |
| W. Matanzas | ('16) | 2,678 | | | |
| S. Beardstown | ('16) | 7,313 | | | |
| Valley | ('16) | 3,030 | | | |

*(continued)* **Table 2.4.** District reclamation sequence, by periods and areas, 1890–1924

| District | Year of reclam. | Assessed land (ac.) | Average assess. (ac.) | Total area (ac.) | Aggregate (ac.) |
|---|---|---|---|---|---|
| Kelly L. | ('18) | 985 | | | |
| Big Prairie | ('18) | 1,700 | | | |
| E. Liverpool | ('19) | 2,765 | | | |
| Seahorn[b] | ('19) | (1,820) | | | |
| Spankey | ('19) | 825 | | | |
| 1920–24 | | | 3,715 | 18,574 | 178,655 |
| Little Cr.[a] | ('20) | 1,700 | | | |
| Liverpool | ('21) | 3,024 | | | |
| Chautauqua | ('21) | 3,500 | | | |
| Thompson L. | ('21) | 5,600 | | | |
| Kerton V.[b] | ('21) | (1,741) | | | |
| Valley City | ('22) | 4,750 | | | |

[a]Originally organized as drainage district.
[b]Successor to Otter Creek district; acreage not counted.

2.5 into seven five-year periods between 1890 and 1924, by which time the development of new drainage and levee districts ended.[58] These periods have been adopted because they highlight areal variations in reclamation; moreover, in the absence of a clear break in the chronology of reclamations, it is useful to adopt periods that end with the major flood challenges of 1904 and 1913 and in the 1920s, a decade in which three catastrophic floods and a crash in commodity prices ended the expansive era of land drainage.

The three-sector grouping of floodplain reclamations between Peoria and Grafton derives from a traditional orientation to the reaches of the river between major towns, rather than coinciding with areas where distinctive hydrography and topography were relevant to land drainage. The northernmost of these three sectors, between Peoria and Havana, is a little over 42 miles long, while the distance from Havana to Beardstown is just over 31 miles; it is almost 89 miles between Beardstown and Grafton. These three sectors, from north to south, share 18, 22, and 59 percent of the floodplain's 280,090 acres. The extent of floodplain potentially available for reclamation per mile of river varied from 1,151 acres between Peoria and Havana, to 2,017 acres between Havana and Beardstown, to 1,875 acres between Beardstown and Grafton.

**Table 2.5.** Progression of floodplain reclamation by organized districts in sectors of the lower Illinois valley, 1890–1924

| Sector | Floodplain acres | Assessed area 1890–99 (ac.) | 1900–1904 (ac.) | Subtotal area (ac.) | 1890–1904 % of sector |
|---|---|---|---|---|---|
| Peoria–Havana | 50,860 | 2,674 | . . . | 2,674 | 5 |
| Havana–Beardstown | 62,740 | . . . | . . . | . . . | . . . |
| Beardstown–Grafton | 166,490 | 10,503 | 7,950 | 18,453 | 11 |
| Total | 280,090 | 13,177 | 7,950 | 21,127 | 8 |

| Sector | Assessed area 1905–9 (ac.) | 1910–14 (ac.) | Subtotal area (ac.) | 1905–14 % of sector | Total area (ac.) | 1890–1914 % of sector |
|---|---|---|---|---|---|---|
| Peoria–Havana | . . . | 17,015 | 17,015 | 33 | 19,689 | 39 |
| Havana–Beardstown | 4,965 | 6,830 | 11,795 | 19 | 11,795 | 19 |
| Beardstown–Grafton | 61,758 | 25,110 | 86,868 | 52 | 105,321 | 63 |
| Total | 66,723 | 48,955 | 115,678 | 41 | 136,805 | 49 |

| Sector | Assessed area 1915–19 (ac.) | 1920–24 (ac.) | Subtotal area (ac.) | 1915–24 % of sector | Total area (ac.) | 1890–1924 % of sector |
|---|---|---|---|---|---|---|
| Peoria–Havana | 2,765 | 12,124 | 14,889 | 29 | 34,578 | 68 |
| Havana–Beardstown | 3,663 | . . . | 3,663 | 6 | 15,458 | 25 |
| Beardstown–Grafton | 16,848 | 6,450 | 23,298 | 14 | 128,619 | 77 |
| Total | 23,276 | 18,574 | 41,850 | 15 | 178,655 | 64 |

*Note:* Includes drainage and levee districts and selected drainage districts.

Because the river flowed close to the western edge of the valley in the lowest sector, notably below Naples, the bulk of the 1,875 acres per mile potentially available for drainage lay in a broad swath east of the river. Northward of Naples, the river tended to be more central to the lowlands, which made possible the development of drainage and levee districts on either side. The bottomland areas were relatively wide to the west of the river between Naples and Beardstown (or Frederick), less so to the east because of the extent of Pleistocene terraces. These relationships of bottomland to the river and the easterly terraces persisted upstream to about Pekin, resulting in leveed tracts that tended to be narrower than those to the south.

By 1900, an aggregate of about 13,000 acres of arable land lay within three reclaimed districts; they occupied about 5 percent of the lower valley's floodplain. Although the districts were dispersed (map 2.7), they were alike in having good access to trade, service, and shipping centers—Pekin (Pekin and La Marsh district), Beardstown (Coal Creek), and Meredosia (Willow Creek). The districts opposite Pekin and Beardstown were largely composed of bottomland, land subject to overflow from the river and the creeks that issued from the bluffs. The reclaimers of the Willow Creek district, however, were concerned with reducing the incidence of overflow by a creek from the bluff, not overflow by the river. Much of their relatively sandy land occupied somewhat elevated Pleistocene outwash terraces and aggrading fans. Thus, from the outset, district formation was motivated by the hope of countering overflow from creeks that headed beyond the bluffs and of countering overflow from the river.

Between 1900 and 1904, the Coon Run and Meredosia Lake districts were reclaimed (table 2.4 and map 2.8) adjacent the older (1894) Willow Creek tract, increasing collaboratively reclaimed areas to over 21,000 acres, or 8 percent of the floodplain below Peoria. The landowners of the somewhat elevated Coon Run district wished to control unruly and heavily alluviating watercourses off the bluff and to drain hummocky land, while landowners in the vicinity of Meredosia Lake sought protection from rising riverine backwater, as well as from creeks.

Contemporaneous with the development of the drainage and levee districts near Meredosia was the completion of about 18 miles of levees around nearly 10,900 acres of valley floor near Hillview and Spankey, in Greene County. These systems of narrow and low levees, noted earlier, were emplaced at the direction of the owners of large properties then known as the Hartwell, Kaser (or Roberts), and Keach ranches and by one or two farmers of the Spankey vicinity. Before 1904, both the Hartwell and Kaser ranches had sluice gates and pumps at the lower end of simple drainage systems, as did the 525-acre Lowenstein Ranch, just northwest of Hillview. The large ranches were incorporated into the Hartwell Drainage and Levee District in 1906, and the Hillview district was formed two years later. Floods in 1903 and 1904 demonstrated that existing levee systems and pumps were deficient. The relatively high Spankey land

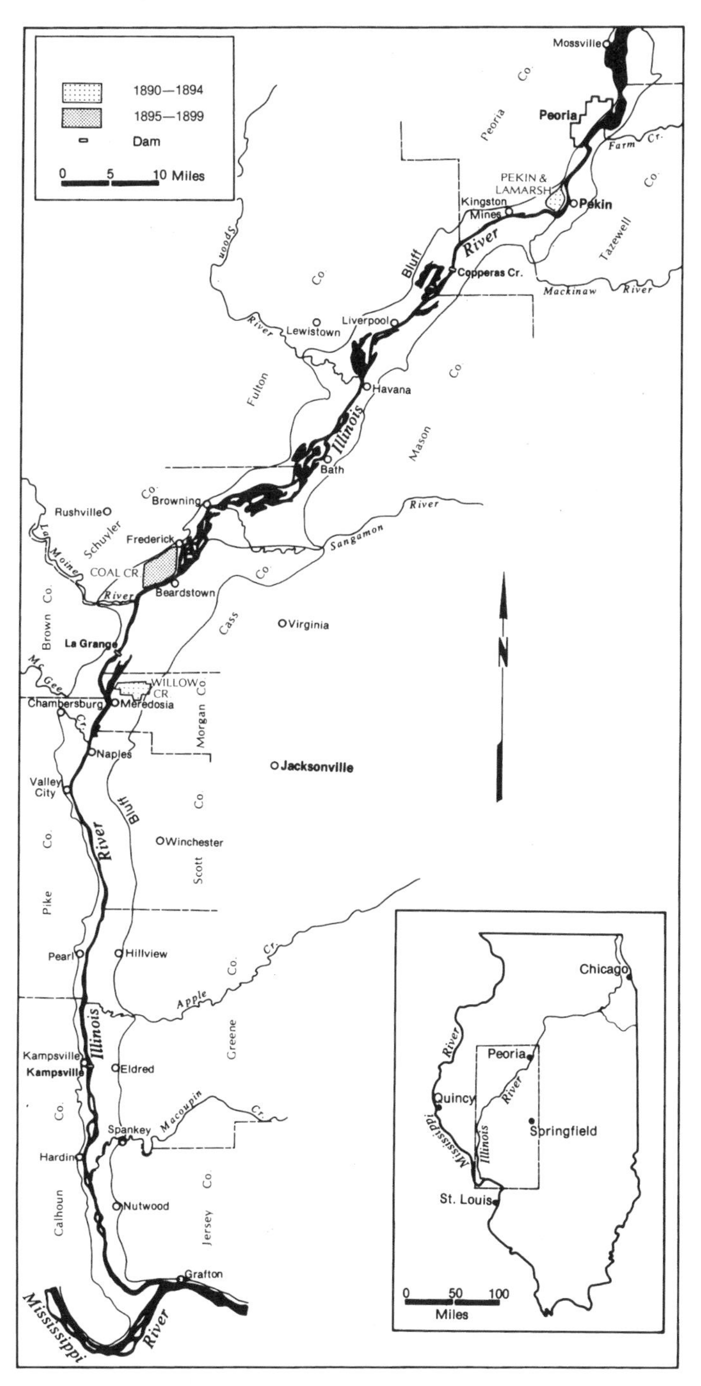


**Map 2.7.** Drainage and levee district formation, 1890–1894, 1895–1899

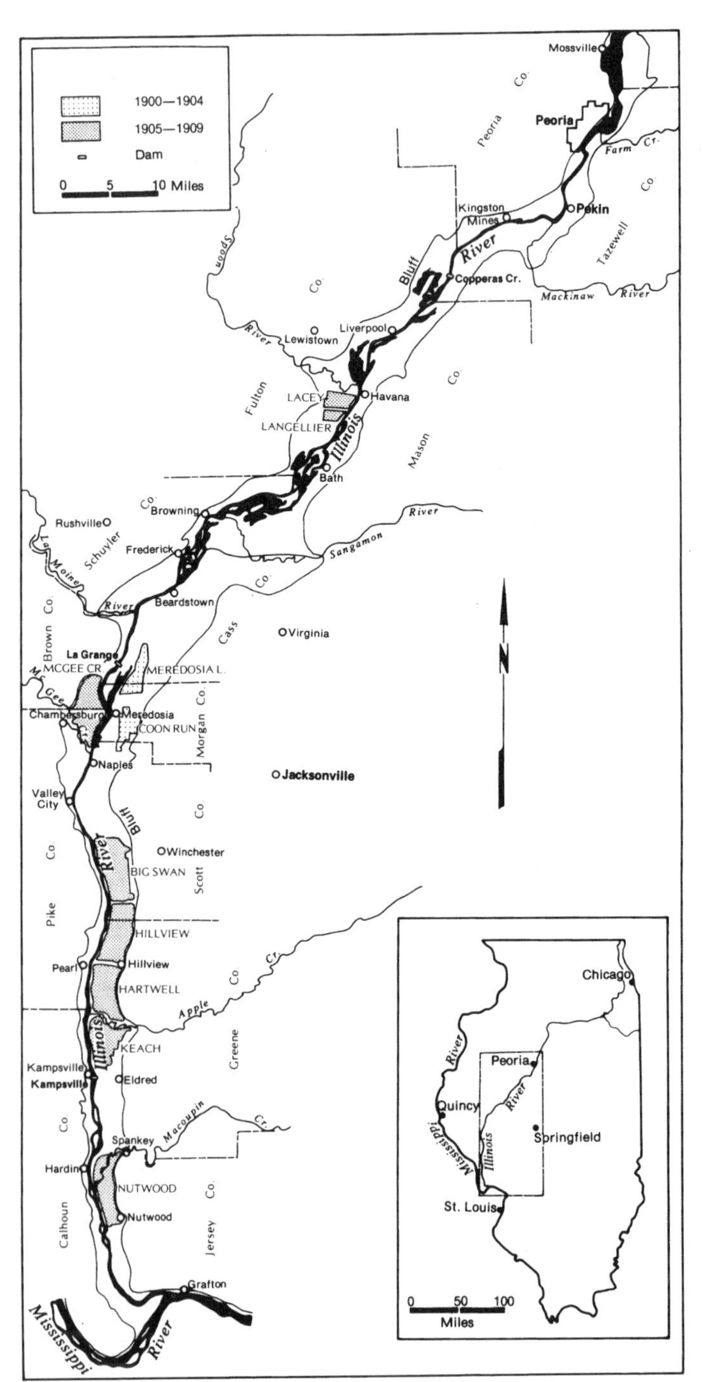


**Map 2.8.** Drainage and levee district formation, 1900–1904, 1905–1909

became a district in 1917. Elsewhere, scattered through the valley, were a very few small areas with levees around them. In general, the flood event of 1904 demonstrated that levee patrolling and the bolstering of weak places "were ineffectual because the levees were of such small cross sections that every trivial injury to the original embankments produced a serious weakness and there was not sufficient foundation nor material at hand to work with."[59]

Whereas only about 32,000 acres of land were enclosed with levee systems between 1891 and 1904 (21,100 acres in districts), the collaborative reclamation achievement soared in 1905–9 and remained at a high level in 1910–14, as table 2.5 shows. Over 66,700 acres were enclosed with levees between 1905 and 1909—about 24 percent of the floodplain area; and over 48,900 acres were enclosed with levee systems between 1910 and 1914—17 percent of the floodplain. In sum, 41 percent of the floodplain was leveed in the decade. It represented over three-fourths of the acreage of land, lakes, and ponds reclaimed between 1891 and 1924.

The reclamations made during 1905–9 (map 2.8) were most evident opposite and below Meredosia, over 90 percent of the newly leveed land being at the McGee Creek, Big Swan, Hillview, Hartwell, Fairbanks Ranch (Keach), and Nutwood districts. The remainder of the tracts, Lacey and Langellier, were opposite Havana and downstream of the confluence of the Spoon River with the Illinois River. These two west bank districts, like the McGee Creek district, consisted almost entirely of bottomland, whereas districts to the east of the river embraced areas of bottomland, terrace, and alluvial fan.

Levee and drainage district completion continued vigorously in the southern sector of the valley between 1910 and 1914, and more land was leveed then in the northern and central sectors than in any other 5-year period (map 2.9). In the southern sector, the Scott County and Eldred districts were added to the string of tracts below Meredosia, and the Lost Creek and Crane Creek districts were added near Beardstown. Somewhat less acreage was leveed in the northern sector of the study area, where the Spring Lake and Banner Special districts were reclaimed between Liverpool and Kingston Mines, and where the East Peoria district was located. Industrial, residential, and truck farming uses of the land set the East Peoria district apart from the remaining tracts. The Otter Creek and Big Lake districts were leveed in the middle sector in 1914. The reclaimers at most of the tracts were concerned with overflow and seepage from the river primarily, but the Scott County and Lost Creek districts were more susceptible to overflow from creeks.

After 1915, the removal of land from the floodplain into new drainage and levee districts affected about 41,900 acres, about 23 percent of all lands transformed between 1891 and 1922. The floodplain lost 15 percent of its area, most of the reclamations in 1915–19 occurring just above and below Havana and in a group between Beardstown and Valley City (map 2.9). Between 1920 and 1924, newly leveed bottomland districts

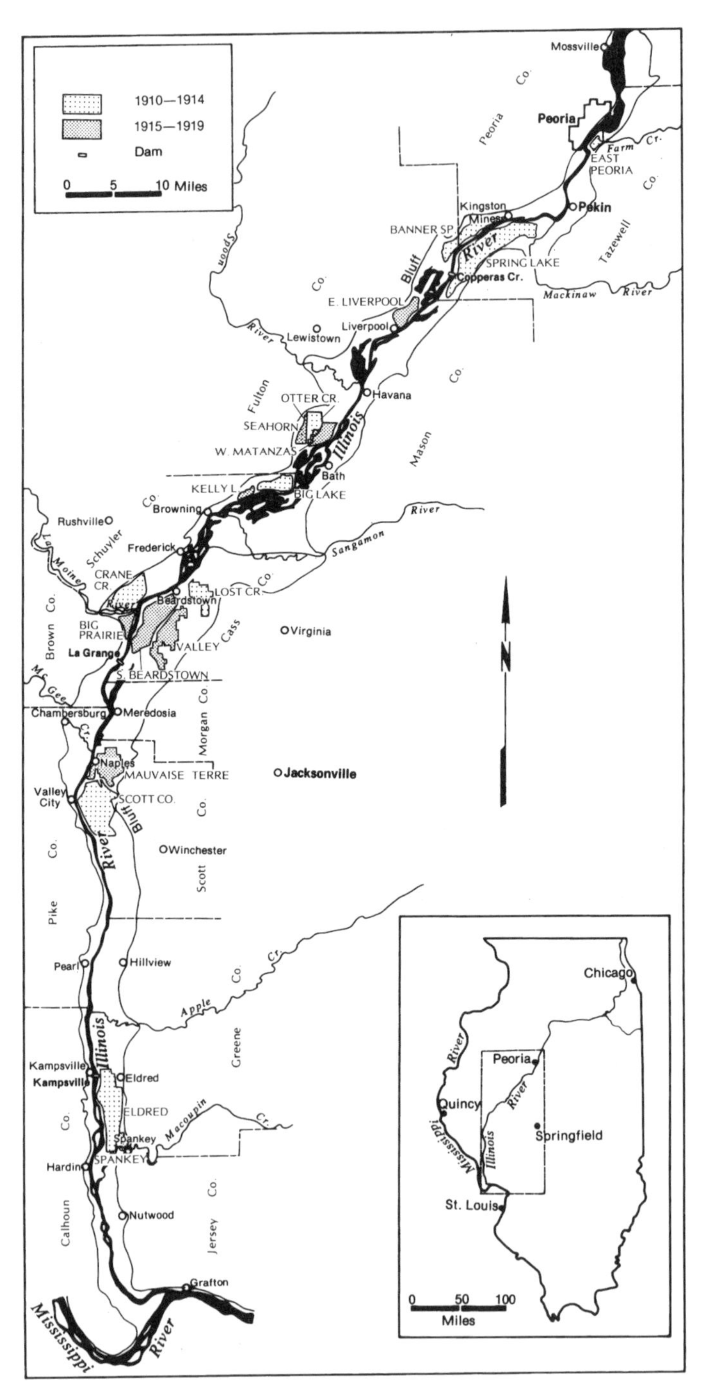


**Map 2.9.** Drainage and levee district formation, 1910–1914, 1915–1919

were either clustered near Havana or above and below Meredosia (map 2.10). With the exception of the tracts to the south of Beardstown, the districts were leveed primarily against the river.

With completion of drainage by the districts between Peoria and Grafton in 1924, about 178,700 acres of assessable land lay behind levees. And there were at least 4,600 acres of land associated with drained private properties. These 183,300 acres of arable land are essentially the same as what was reported by earlier investigators, although the inclusive tracts vary. The distribution of the districts studied was far from uniform (map 2.1). There were 21 tracts in the sector of floodplain below the Beardstown-Frederick vicinity, accounting for 72 percent of the drained wetlands. Another 9 percent of the drained land was occupied by the 7 districts that lay between the mouths of the Sangamon and Spoon rivers; and between Havana and Peoria, the 8 districts of the northern sector of the study area accounted for 19 percent of the drained wetlands.[60] The preponderance of drained tracts to the south of the Beardstown-Frederick vicinity included clusters on both sides of the river around Beardstown and Meredosia and were strung along the valley floor east of the river between Naples and Nutwood. Whereas the clustered districts tended to be of intermediate size (about 4,800 ac.), the districts to the south tended to be large (about 9,300 ac.). The small districts of the middle sector of the floodplain (about 2,500 ac.) lay in two right-bank clusters between the Sangamon and Spoon rivers. Upstream of Havana, most of the districts lay in two clusters astride the river. The group immediately above Havana tended to be of intermediate size (about 4,100 ac.), which was the case for two of the four districts above Liverpool, where the large Spring Lake (13,120 ac.) and small East Peoria (800 ac.) districts also lay.

Areas not reclaimed by the late 1920s consisted of relatively small strips of bottoms between Peoria and Havana; an extensive area adjacent the mouth of the heavily aggrading Sangamon River, which was a formidable deterrent to reclaimers; and large areas of water and islands above and below Meredosia, together with floodplain between the Nutwood district and the Mississippi River, where landowners chose not to challenge the elements with levees. Floodplain residuals between Peoria and Havana may have aggregated 14,000 to 16,000 acres, about half identified with an aborted reclamation undertaken a short distance above Liverpool. Over 43,000 acres of perennially wet and seasonally overflowed land lay adjacent the lower Sangamon River. About 5,000 acres of residual floodplain lay above and below Meredosia, and another 31,000 acres adjoined the confluence of the Illinois River with the Mississippi River. In the aggregate, possibly 35 percent remained of the wetlands that had supported one of the finest fish and game bird resources of the interior; but the residuals were hardly pristine. Together with areas of the valley between Peoria and Ottawa yet subject to overflow in 1931, the unreclaimed land amounted to about 134,500 acres.

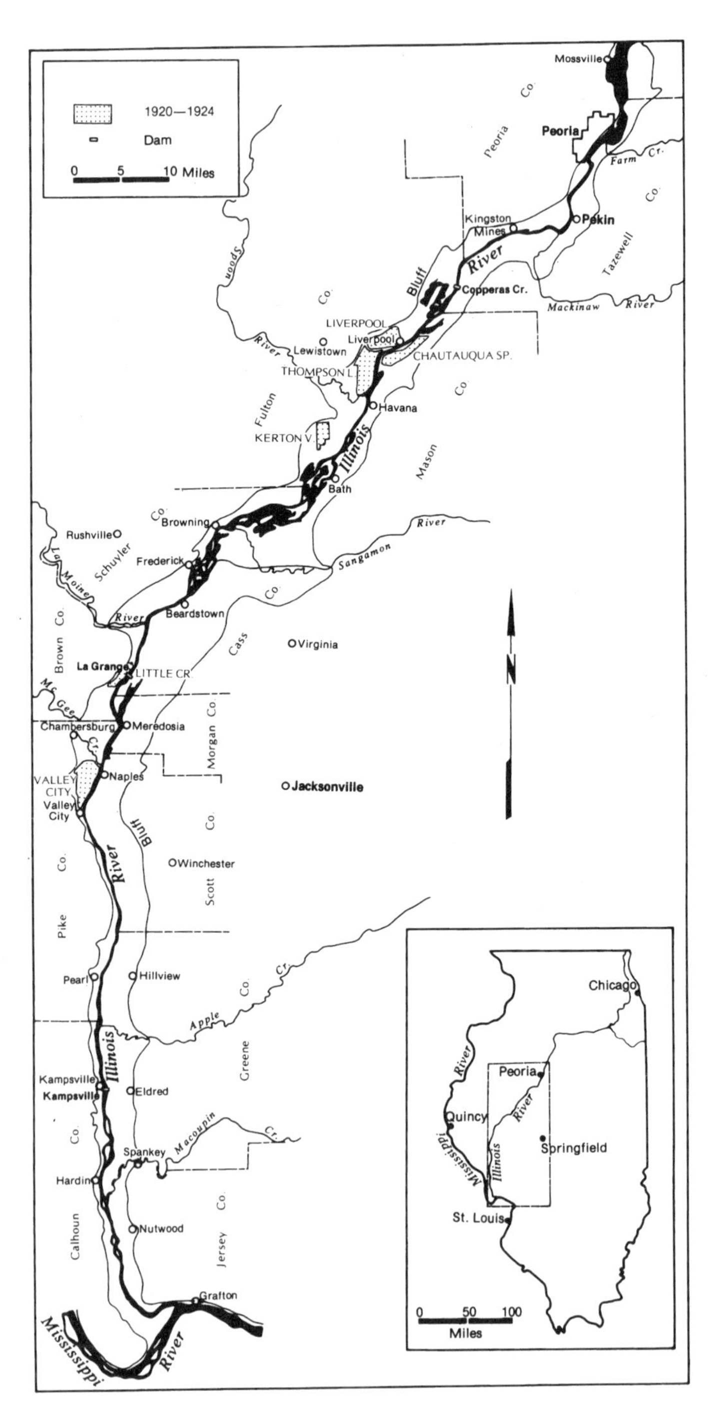


**Map 2.10.** Drainage and levee district formation, 1920–1924

Perhaps 46,000 acres were cultivated when conditions permitted, most of them (34,500 ac.) downstream of Peoria.[61]

The likelihood of reclaiming segments from the 102,000 acres of residual floodplain ended with the crash in commodity prices of 1921, when capital to finance new projects dried up. Moreover, the value of undrained wetlands to hunting and fishing clubs was growing. Persons of means in cities as distant as St. Louis, Indianapolis, and Chicago found the area quite accessible by rail and motor vehicle.[62]

Clearly, the reclamation on the floodplain increased flood stages along the lower Illinois River. The river was constricted along several reaches by levee systems, and about 65 percent of the floodplain that once accommodated overflow was behind levees. The constrictions were notable: (1) in a 35-mile stretch from opposite Nutwood to Naples, where the overbank section of the stream was just about eliminated; (2) in the 10 miles downstream of Beardstown, where levees of the South Beardstown district stood across from those of the Coal Creek, Crane Creek, and Big Prairie districts, and near the lower end of which was the dam and lock at La Grange; (3) a short distance north of Havana; (4) much of the 15 miles below Pekin, where the levees of the Spring Lake and Banner Special districts were most prominent; and (5) for the 1¹/4 miles that largely coincided with the levee of the East Peoria district. The major bottleneck was the 2,000-foot corridor between levees that ran for 10 miles downstream of Beardstown's bridges. Within the corridor, some levee sectors were as little as 1,100 to 1,200 feet apart. There were only 1,080 feet of open water at Beardstown's wagon bridge and about 1,750 feet at the railroad bridge. With abandonment of the Chautauqua district in 1927, its approach to within 910 feet of the levee of the Thompson Lake district ceased to attract attention. On the other hand, the constriction that narrowed to about 700 feet at Peoria remained a matter of concern downstream because it was believed to prolong flood stages by holding back water in the extensive reservoir area that was Peoria Lake.[63]

The investment in drainage works by 1930 at the 36 studied districts may be approximated.[64] Over 178,000 acres of assessable farmland were protected by levee and ditch systems and by pumping facilities, for which construction assessments of a good $12.5 million were made. Perhaps another 30 percent, or $3.75 million, represented annual operating assessments—an educated guess based on incomplete data. As to the cost of borrowed capital for construction of district facilities, the estimate of 1937 noted earlier was that interest costs were about one-third again of the aggregate construction assessments. The records of the Big Swan district showed the cost of borrowed money to have been 41 percent of all expenditures for construction and maintenance, which seems a reasonable indication of the long-run experience in the lower valley. Were one to assume that the Big Swan Drainage and Levee District was representative, the aggregate costs of achieving and maintaining reclamation for the districts in the study area was roughly 2.7 times the $12.5 million in construction assessments of record. While the burden may have seemed to the landowners and bondholders in 1930 like $33.75 million, the reality is that insufficient data are available to do more than suggest what the cost of borrowed money added to reclamation costs at the districts.

Assuming that known construction assessments were $12.5 million and that operating assessments were about $3.75 million, the works of drainage cost $16.25 million, plus the cost of borrowed money—much of which was discounted during the Depression.[65] The State of Illinois, as will be noted in chapter 6, committed $0.8 million to levee rehabilitation in the late 1920s. Thus, a reasonable estimate is that over $17 million went into the works of large-scale reclamations by about 1930.

The cost of drainage works in place by 1930 was about $96 per acre of assessed land, excluding the cost of borrowed money. This was over three times the cost to reclaim estimated by Alvord and Burdick in 1915, and a third above the average estimated in Harman et al. in 1929. Land breaking, on-farm tiling and ditching costs, and other individual farm improvement costs are unknown for the period ending in 1930. However, they were estimated at $5.3 million, or over $26 per acre, in 1923. The costs incurred by the state and the counties to build and maintain over 500 miles of roads in the reclaimed areas and to provide other services are unknown. Considering that the full value of the land was estimated to be $112 per acre in 1914,[66] estimates of districtwide average values of between $50 and $105 per acre in 1930[67] must have sobered even the speculators who began to invest in cheap, unreclaimed land a decade or two earlier. The mounting challenges to their interests and equanimity are described in subsequent chapters.

Meanwhile, organizers and developers of drainage and levee districts and of farm-sized drainage enterprises had accomplished a great deal. They created a high level of security from overflow, runoff, and elevated water tables for about 183,000 acres of highly productive, revenue producing farmland. The protective works included over 330 miles of levees, at least 370 miles of outlet canals and drainage ditches, and some 260 miles of tile lines.[68] There were as yet an estimated 34,500 acres of unreclaimed floodplain that were cultivated intermittently. Thus, the cutting edge of agricultural land use had modified close to 78 percent of the floodplain. Meanwhile, a sizable part of the remaining wetlands, pasture, and intermittently cropped land was subjected to prolonged high-standing water, another by-product of contemporaneous regional economic development. That facet of the valley's heritage is addressed in subsequent chapters.

There was a pride of achievement in the 1920s among the officers of the drainage and levee districts located between Peoria and Grafton, especially given the desultory results at some 50 contemporary districts located on floodplains in seven southern states. The districts along the lower Illinois River held about 64 percent of the floodplain area, most of it in cropland, whereas comparable districts to the south and southeast reclaimed but 26 percent of the acreage within their bounds. Nevertheless, the distant entities of larger areal extent than along the Illinois River accounted for about 900,000 acres of improved land by 1926.[69]

The relative success of large-scale drainage projects in the study area may be accounted for in several ways, among them an early start with power machinery. Implicit in the early start were relatively low land values and drainage costs and high agricultural commodity prices. Moreover, most of the reclaimed land was in production within a season or so after the levee systems were closed. By 1915, three-fourths of the ultimate acreage was behind levees, whereas the work of dredges and draglines on the timbered floodplain of the Missouri River and the Mississippi River below Cairo, in coastal Louisiana and in the Carolinas, barely had begun.[70] As a matter of fact, the drainage contractors who developed the largest operating plants in Illinois extended their businesses by securing contracts elsewhere, notably beginning in Missouri and to the south. The later development of contracting opportunities in land drainage elsewhere in the humid eastern half of the nation is reflected, too, in the destinations of new dipper dredges shipped between 1900 and 1925 by the four leading manufacturers of such machines for land drainage.[71]

Developers of large-scale drainage enterprises in the valley of the lower Illinois River faced a less formidable task than would-be reclaimers of the bottoms along the Mississippi River. The gradient and size of the former was less. Its comparatively narrow floodplain and the spacing and size of comparatively modest affluents provided challenges that were less daunting than along the Mississippi River. Moreover, the Illinois River was less prone to devour its banks and levees than was the actively meandering Father of Waters.

# 3

# The Role of Pumps in Land Drainage

In the era under investigation, it may be recalled, the floodplain of the Illinois River had very low relief, and sloughs, shallow ponds and lakes, and perennially wet areas of land were common. The floodplain waters were replenished by direct flow from the river, by precipitation, by runoff from the uplands, and by springs. Gravity drainage removed the general overflows of late winter and spring and the more localized overflows caused by heavy precipitation over the valley and its immediate tributaries. In general, the bottomland was most free of water between June and early winter. However, when dams and locks for navigation were built across the lower river

**Fig. 3.1.** View in upper slough of Thompson Lake, ca. 1909. This scene of an elevated water level in the Illinois River shows relationship of water to foot of trunks on mature trees, some of which are dead. Bluffs are distant. Illinois Submerged and Shore Lands Investigating Committee; courtesy of Illinois State Museum.

**Fig. 3.2.** Pumping station in lower Spring Lake district, ca. 1909. The facility and levee are on left. The expanse of water, cabin boats, and tent suggest that the pumping station is under construction and that the levee system has yet to be closed. Illinois Submerged and Shore Lands Investigating Committee; courtesy of Illinois State Museum.

at Copperas Creek (1877), La Grange (1889), and Kampsville (1893), the resulting navigation pools lifted low water levels and soil water tables, as is described in chapter 5. The effects predated organized land drainage activity, but the incentive to reclaim land mounted in 1900, when water diverted from Lake Michigan by the Sanitary District of Chicago entered the river in all seasons (fig. 3.1).

Once levee systems were emplaced on the floodplain to prevent overflow of the new drainage and levee districts, provision had to be made to remove water from within the perimeter of earthworks, especially during prolonged river stages that stood 6 to 10 feet above much of the bottoms. In a few cases, gravity flow could enter a low river through opened sluices and tile lines, as happened at the Fairbanks Ranch, or into leveed creeks adjacent Pleistocene terraces. Such drainage was enjoyed seasonally in 1914 at tracts located to the east of the river between the Valley and Nutwood districts, at the prospective Valley City district, and the Otter Creek, Langellier, and West Matanzas districts, located to the west of the river and downstream of Havana.[1] For most of the leveed districts, however, it was essential to install a pumping facility to remove unwanted water (fig. 3.2).

## Adoption of Pumps

The adoption of drainage pumps at the districts along the lower Illinois River followed by some decades the use of low-lift centrifugal pumps in land drainage, which began in 1850, but it did not lag more than a decade or so behind the introduction of greatly improved pumps to reclaimed tidal marshes in Louisiana and to the delta of the Sacramento and San Joaquin rivers in California. Nevertheless, the development was in the forefront of Midwestern adoptions of such technology in the late nineteenth century. After 1900, the installation of pumps for land drainage was widespread.[2]

Pumping stations, as they were called, were integral parts of the drainage and levee districts that first developed along the Illinois River in the 1890s. They were operating in 1891 at the Pekin and La Marsh district, above the Copperas Creek dam, and in 1898 at the Coal Creek district, upstream of the La Grange dam. Another 21 pumping stations were installed at drainage and levee districts between 1901 and 1919; and pumps were installed at three ranches in Greene County, where the modest levees and ditch systems built by individual landowners were antecedent to the works of the Hillview and Hartwell districts. Also, in 1915, 8-inch- or 12-inch-diameter pumps operated at three privately reclaimed tracts of 400 or 500 acres in Greene, Morgan, and Schuyler counties (fig. 3.3). By 1919, another five pumping facilities were functioning at districts on the floodplain of the Mississippi River in Illinois and at a handful of drainage districts in Iowa and Missouri. The development of pumping capabilities at districts and private holdings continued into the early 1920s; eight new pump houses were built along the lower Illinois River. There were over 35 districts in the valley with pumping facilities in 1930.[3]

Pumping stations were a major investment. By 1914, for example, five steam-powered facilities had an initial investment of about $100 per horsepower installed and roughly $3,300 per 1,000 acres of watershed served. Moreover, their operating costs, including fixed charges, averaged $1.38 per acre per annum.[4] The need for such outlays mounted with each transfer of acreage from floodplain function to farming. Moreover, floodplain areas that were submerged by the water of navigation pools were extended and deepened as the Sanitary District of Chicago diverted increasing volumes from Lake Michigan after 1900. The surfeit of water also increased because the drainage of upland wet prairie areas was extended and improved in central and northern Illinois and in northern Indiana, and the bottoms of tributary streams such as the Sangamon River were dredged, ditched, and leveed. The general effect of these activities was to raise the level of the river, increasing seepage through and below the levees, and to raise the head against which pumps had to discharge. Illustrative of the problem created by the dam at La Grange was a drainage engineer's assessment of the

**Fig. 3.3.** Representative early private works of drainage. The location and date of scene are unknown, but the size and configuration of levee, discharge pipe, pump, and gasoline engine suggest a small tract some distance from the river. The dredge-built levee has been dressed by scraper. Courtesy of George and Virginia Karl Collection, Havana.

Crane Creek area: "Prior to the time the dam was put in, a large part of the Crane Creek district was cultivated without pumping. In addition to increasing the amount of pumping, backwater from dams no doubt has caused a permanent rise in the groundwater table in affected districts, making impracticable the cultivation of certain areas which otherwise could be cultivated, and has necessitated the installation of additional drains for yet other cultivated lands."[5] As a consequence, too, drainage and levee districts commonly increased the capacities of their first pumping station.

## Pumping Facility Functions

The entrepreneurs and engineers who first undertook the development of drainage and levee districts along the lower Illinois River underestimated the magnitude of water removal needs. Invariably, the original pumping facilities had deficient capacities. Drainage engineers appear to have been finding their way rather than proposing economizing strategies for adoption by district commissioners. Data were lacking on local storm intensity and runoff from adjacent uplands. Seepage rates through and under levee systems were unpredictable, and pumps were installed without appreciation of their capabilities against the heads accompanying high-water stages, among other incompatibilities.

The shortcomings of ditch systems and installed pumping capacities were evident soon after the first districts began to operate in the 1890s. At the Pekin and La Marsh district, where the levee system was completed in 1891, the ditch system completed in 1894 lacked the capacity and fall to serve the pump properly. The district was occupied by large areas of standing water even when the levee system stood. Direct runoff from the bluffs, together with the freshets of La Marsh Creek, so overwhelmed the ditch system and pump that in 1899 a diversion ditch was cut westerly along the foot of the bluff into the strongly leveed flank creek. A more elaborate diversion ditch was excavated at the Coal Creek district (map 2.3), opposite Beardstown, in the winter of 1899–1900 to counter "hill water" that damaged crops during the preceding two years. In essence, Coal Creek was diverted westward 3.5 miles from the point where it exited the bluff area to a new channel that skirted about 2.5 miles of the district's western flank levee in reaching the Illinois River. The diversion truncated the segment of Coal Creek that was enlarged by dredging in 1897 to serve as the district's main canal, which ended at the pumping station in the southwestern corner of the reclaimed tract.[6] Opined a critic, "The folly of attempting to handle the Coal Creek flood with pumps has proved expensive and has kept the owners from realizing anything upon their investment."[7] He might have added that seepage through the levee system and underlying material was a contributory problem.

Among the lessons learned at the Pekin and La Marsh and the Coal Creek districts was the advisability of minimizing the extent to which upland areas drained into the tracts. Runoff from storms exceeded pumping capacities, resulting in ponding and crop damage. Also, the rivulets and creeks so filled the ditch systems with sediment that they had to be excavated anew. At the Pekin and La Marsh district, a first ditch cleaning was required in 1893, barely two years after the levee system was completed. The sedimentation problem was accentuated thereafter whenever embankments were breached, which happened frequently because of the rate of alluvium accumulation in the intercepting ditches and creeks. The bed of La Marsh Creek, for instance, stood 2 to 4 feet higher than adjacent fields in 1905, barely a decade after the channel was deepened to provide spoil for the levee that protected the fields. Coal Creek first burst through its embankment within two years of completion of the bluff ditch.[8]

Understandably, drainage engineers subsequently gave consideration to having diversion ditches reduce as much as practicable the runoff that might enter reclaimed tracts. As a result, a handful of districts today are free of runoff from higher land. Among the 25 districts for which data are available regarding upland watershed areas, only one-third had 10 percent or less of their total watershed in the uplands. At another third of the districts, the ditch systems and pumps coped with runoff from upland areas comprising between 10 and 29 percent of the watershed. The districts with the largest acreage tended to have 30 to 45 percent of their watersheds in the bluff areas, which is understandable given the extent of the tracts along the valley. The small East Peoria district was unusual in that about 48 percent of a 1,550 acre watershed was outside of the district proper. The data may be seen in table 3.1.[9] Having written of plans to divert runoff around the newly organized Eldred district (1909), in southwestern Greene County, a commissioner went on to inform a public more accustomed to farming on the uplands: "The great difficulty is to get rid of the water that falls on the land itself—the spring rainfall is a far greater problem than high water in the Illinois. Where the farmer on ordinary wet land has only to thoroughly tile his farm, that method is out of the question with the bottom farmer whose land is so much lower and so near the level of the river. His only solution of the problem is in powerful pumping plants."[10] While the removal of accumulations of water from reclaimed bottoms was required following passage of storms, there was a sustained need to lower ground water levels within the districts so that tillage could be done and the crops would have early and optimal growing conditions. By the same token, the artificial drainage systems could be used for subirrigation in times of drought, which practice occurred at the Fairbanks Ranch (Keach district) at least as early as 1912 and at a farm in the Hillview district in 1915. No earlier conscious adoptions of subirrigation are known than in these districts of Greene County.[11]

An ancillary benefit provided by pumping plants to the landowners and general public was that roads became passable at all seasons across the districts. In this respect, the *Carrollton Patriot* noted in June 1909: "At the present time, and usually for several months at this time of the year, transportation from Eldred to Columbiana is mainly by water. To be able to drive there at any time from February to July, without the necessary intervention of a gasoline launch or a row boat, will be an unheard-of benefit."[12] As noted earlier, the merchants and city officials of Beardstown, Havana, and Pekin recognized the advantage of roads across the floodplain since they subsidized road construction and maintenance in counties across the river before the formation of the Coal Creek, Lacey, and Pekin and La Marsh drainage and levee districts.[13]

## Trial and Error

To judge from the experiences of formally trained drainage engineers and of practitioners whose background was in surveying, a good deal was achieved in land drainage through trial and error. That was consistent with the perspectives of contemporary manuals of drainage engineering, which held that although guiding principles could be enunciated, there was less predictability in building land drainage systems than in "the construction of many other kinds of engineering work."[14] Each prospective reclaimed tract presented numerous variables of microtopography, composition of soil and underlying materials, and surface and underground water sources and outlets. Then, too, the quality of engineering advice and supervision, like the responsiveness of commissioners, varied. If there was collusion in understating project costs as a means of allaying opposition to projects, it was not aired by critics. On the other hand, a local critic in the engineering profession reported that errors "caused exceedingly heavy losses" in many districts.[15]

The errors made at a handful of pumping stations soon were evident. The facilities

**Table 3.1.** Pumping facilities and watershed areas, drainage and levee districts, lower Illinois valley

| District | Began operating | Watershed area (ac.) | | Size of pumps (inches diam.) | | Capacity in 1927 (in./ac. per 24hr.) | Power source[a] |
|---|---|---|---|---|---|---|---|
| | | Total | Upland | 1913–15 | 1927 | | |
| Banner Sp. | 1915 | 8,100 | 3,539 | ... | 20,36 | 0.366 | E |
| Big Lake | 1914 | 4,300 | 900 | 26,26 | 26,26 | 0.41 | E; D'20 |
| Big Prairie | 1918 | 1,800 | 80 | ... | 30 | ... | E |
| Big Swan | 1906 | 15,700 | 3,220 | 24,45 | 24,30,45 | 0.31 | S; E'12 |
| Chautauqua | 1921 | 4,413 | 613 | ... | 20,30 | ... | E |
| Coal Cr. | 1898 | 7,718 | 1,000 | 15,15,24 | 20,20,20 | 0.28 | S; E'13; D'40 |
| Coon Run | 1923 | 4,500 | 0 | ... | 18 | ... | D |
| Crane Cr. | 1910 | 6,200 | 700 | 36 | 36 | 0.255 | S |
| E. Liverpool | ca.1922 | 3,300 | ... | ... | 18,24 | ... | E |
| E. Peoria | 1911 | 1,550 | 750 | 15,18 | 18,24 | 0.44 | S; E'13 |
| Eldred | 1911 | 9,500 | 1,048 | 24,36 | 24,36,36 | 0.522 | S; D'20,'21,'29 |
| Hartwell | 1915 | 12,300 | 3,400 | 30,30,30 | 30,30,30,48 | 0.554 | E |
| Hillview | 1908 | 18,500 | 5,600 | 24,30,30,32 | 24,24,30,30,60 | 0.565 | S; E'14 |
| Keach (F'banks) | 1910 | 15,000 | 6,600 | 36,48 | 36,48 | 0.315 | S |
| Kelly L. | 1918 | 985 | ... | ... | 20,22 | ... | E |
| Kerton V.[b] | 1921 | 1,740 | ... | ... | ? | ... | E |
| Lacey[c] | 1901 | 3,200 | 300 | 12,20 | 12,20,20 | 0.50 | S; E'12 |
| Langellier[c] | 1905 | 2,200 | 0 | 18,24 | ... | 0.50 | S; E'16 |
| Little Cr. | ca.1921 | ... | ... | ... | ? | ... | ... |
| Liverpool | 1923 | ... | ... | ... | 20,20 | ... | D; E'30 |

*(continued)* **Table 3.1.** Pumping facilities and watershed areas, drainage and levee districts, lower Illinois valley

| District | Began operating | Watershed area (ac.) | | Size of pumps (inches diam.) | | Capacity in 1927 (in./ac. per 24hr.) | Power source[a] |
|---|---|---|---|---|---|---|---|
| | | Total | Upland | 1913–15 | 1927 | | |
| Lost Creek | 1921 | 2,740 | 300 | ... | 24 | ... | E |
| Mauvaise T. | 1925 | 7,180 | 1,400 | ... | 30 | 0.152 | D |
| McGee Cr. | 1909 | 16,500 | 5,700 | 32,32,36 | 30,30,30,30,36 | 0.401 | S; D'26 |
| Meredosia L. | 1906 | 5,000 | 1,000 | 24,24 | 36 | ... | S; E'16 |
| Nutwood | 1909 | 16,050 | 4,750 | 36,48 | 36,48 | 0.31 | S; D'34 |
| Otter Cr. | 1914 | 3,680 | ... | 15,20 | ... | ... | S |
| Pekin and La Marsh | 1892 | 3,200 | 190 | 12,26 | 18,22 | ... | S; S'02; E'09 |
| Seahorn[b] | 1918 | 1,820 | ... | ... | 20,36 | ... | D'20 |
| S. Beardstown | 1918 | 8,350 | ... | ... | 26,30,30,30 | 0.382 | E |
| Spankey | none | ... | ... | ... | ... | ... | ... |
| Spring L. | 1909 | 22,500 | 9,380 | 24,48,48 | 24,48,48 | 0.265 | S; E'14 |
| Thompson L. | 1922 | 6,000 | 0 | ... | 24,30 | ... | E; D'41 |
| Valley | ca.1916 | ... | ... | ... | 24 | ... | E |
| Valley City | 1923 | 7,160 | 2,160 | ... | 24,36 | 0.265 | D |
| W. Matanzas[c] | 1916 | 2,700 | 0 | 18,24 | 18,24 | 0.435 | E |
| Willow Cr. | none | ... | ... | ... | ... | ... | ... |

[a]Power source: S = steam; E = electricity; D = diesel or semidiesel. Number following symbol is year of subsequent installation.
[b]Successor to Otter Creek.
[c]Early pumping facilities consolidated into one in 1930 with 24-, 30-, and 36-in. electric-powered pumps.

were immobilized by settling foundations and by the breakdown of improperly installed linkages between power sources and pumps. At least one facility went into service with a direct drive arrangement that could not be adjusted to operate each pump separately. More common was the failure to provide pumps and power sources adapted to operate against the heads that developed. Some lift maxima reflected unnecessarily high levee crossings with the discharge pipes, and there were instances of pumps being furnished with intake and discharge pipes of incompatible diameters. On average, the vertical differences between water levels in the suction bay and the river were from 5 to 10 feet, but every few years a head of 20 feet might be sustained for a few days. Unfortunately, the largest heads to pump against occurred when the most pumping had to be done. Sometimes, the security of a pumping station was threatened by boils and caving around the concrete suction bay, and several facilities were built low enough that water trapped inside breached levees reached the machinery. In at least one instance, at the McGee Creek district, the first pumping station was enclosed within a separate levee. More commonly, machinery was reinstalled a little higher after being stilled by an inundation. Assuming the design and construction were sound, districts might suffer the sloth of commissioners whose pumping station was "out of fuel when it rains," and whose pumps were "invariably out of commission when they are needed." Costly mistakes and oversights occurred wherever pumping plants were built in the early days.[16]

Obviously, the determination of an optimal pumping capacity for a new drainage and levee district was a complex problem for the engineer. The area of the district and of the tributary bluff and uplands had to be calculated. The precipitation record, especially the frequency and intensity of storms during the growing season, had to be established. And the incidence and duration of high river stages had to be figured in for their bearing upon seepage and pumping requirements. Other variables included the rates at which soils and underlying sediment, tile lines, open feeder ditches, and outlet canals conveyed water to the pump station. The water storage capacity of residual sloughs and ponds, like the storage capacity of the tile systems, open ditches, and outlet canals, were important to know, as well.

The object of gathering such comprehensive information was to design a pumping facility capable of discharging unwanted water fast enough that levels of ground water would not unduly delay planting activity or damage growing crops, consistent with installation costs and economical and effective schedules of pumping. Since it took several years to accommodate field tile and ditch systems to farm needs and to adjust the district's water gathering and disposal capacity, and since annual precipitation and river regimes were variable, the achievement of uniformly effective drainage within a tract took time to approach. Nevertheless, it was clear by 1914 that roughly 60 to 75 percent of the annual work in discharging water occurred between mid-March and mid-June and that the rest of the work was spread fairly evenly through nine months.[17] Thus, optimal pumping capacities had to be determined for two seasons.

In the early days, a pumping facility with capacity to remove the equivalent of 0.25 inches of water over the entire watershed in 24 hours usually was considered the limit of economy. But the need for larger capacities was recognized shortly after operations began. At the McGee Creek district, for instance, two aspects of the problem showed up in a serious way between 1909 and 1917. The drainage system in the northern third of the tract had to be extended and rerouted to a second pumping station, which was built opposite Meredosia in 1914, because the low gradient of the original (1909) main ditch and distant southerly pumping facility could not drain the land quick enough to save crops after heavy rains. Then, in 1917, the ditch and pumping capacities were found wanting in the face of heavy seepage attributable to prolonged high stages on the Illinois River. Ultimately, the lower pump house was abandoned, and a revamped ditch system conveyed all water to the single pumping facility opposite Meredosia.[18]

By the late 1920s, favored runoff coefficients for the districts were between 0.28 and 0.44 inches, as may be noted in table 3.1. A 24-hour pumping capacity to remove 0.30 inch in depth over the drainage area seemed to be the most common. It was equivalent to discharging 5,650 gallons per minute per 1,000 acres. At such rates, two pumps of different size were prescribed for drainage areas of 3,000 to 8,000 acres. For drainage areas of 8,000 to 10,000 acres, there might be three pumps of the same size or a pair of pumps of different sizes.[19] The districts that had as many as five or six pumps, McGee Creek and Hillview, had extensive watersheds, and the landowners seem to have been willing to underwrite the high cost of having their pumps discharge in a day the equivalent of 0.48 and 0.56 inch of water per acre, respectively. The facilities approached a capacity that prominent drainage engineers would have preferred for all districts had it not been for the cost. The quest for sound drainage coefficients upon which engineers might perfect drainage plans did not mature locally until the late 1930s, when extensive streamflow studies were made in the Midwest by the Civilian Conservation Corps.[20]

Although the first pumping stations had only a single pump, the rule soon became to have two or three, as shown in figure 3.4. The most common unit was a horizontal double-suction centrifugal pump with closed-vane impellers, which was suited to high lift requirements and was reliable and simple to operate, although having to prime it to start was a disadvantage. By 1920, the units in service ranged from 12 to 60 inches in diameter. As may be noted in table 3.1, adoptions of larger pumps were common between 1913 and 1915 and in the late 1920s, which were very wet years. By and large, the small pump in a plant represented about one-third of the total capacity. It sufficed during most of the year, pumping steadily against a fairly constant head.

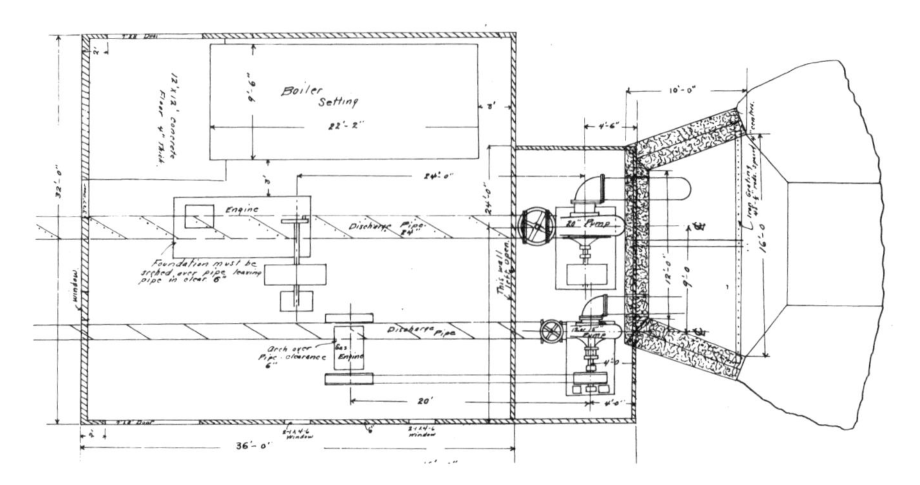

**Fig. 3.4.** Plan for east-side pumping station, Otter Creek district, 1912. The pumps discharged into a dredged channel with a gravity flow outlet. The relationship of the 100 h.p. boiler, 75 h.p. steam engine, and belt-driven 20-inch-diameter pump, like that of the gasoline engine and 15-inch pump, were conventional. Courtesy of Havana Public Library.

Otherwise, the operation of a steam plant required that water accumulate in the ditch system until the large-capacity machinery could be operated efficiently.[21] When steam engines were replaced by electric motors in the plants, the tendency was to have three pumps rather than two.

Ordinarily, pumps were operated no more than 60 to 90 full days per year. In some years, only 15 to 20 days of pumping were required. While those days tended to be between mid-March and June, the passage of downpours might require that the pumps be engaged for short runs in summer or winter. Although sluice gates were opened to allow gravity drainage in the summer at some districts, enough evaporation and transpiration generally occurred during the season that pumping could be reduced. As a rule, the pumps were shut down subsequent to the harvest.[22] The major exceptions to the general regime were the northerly Spring Lake and South Beardstown districts. The flow of springs at the former required that one or more pumps operate daily throughout the year; heavy, year-round seepage through sandy levees and subsoils required steady pumping at the South Beardstown district.[23]

## Pump House Location and Characteristics

Pump houses were of frame, brick, or cement block construction. They were set on concrete foundations that rested on piling. The structures housed the pumps, a priming device, some combination of boilers, engines, and motors, and the essential

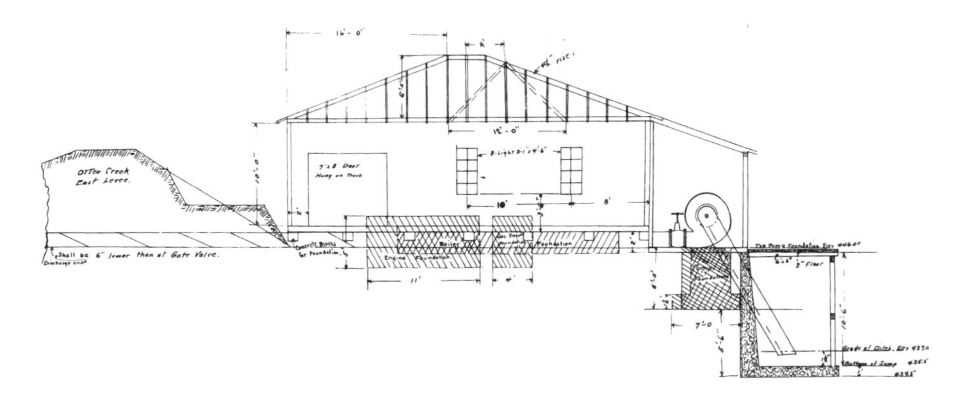


**Fig. 3.5.** Section for east-side pumping station, Otter Creek district, 1912. The relationship of about 40 cubic yards of concrete foundation for the machinery and sump, together with relationship of the facility to the levee, is shown. Debris screen is to right of sump. Courtesy of Havana Public Library.

ancillary equipment. The intake and discharge pipes tended to be made of riveted or welded steel. The discharge pipes were placed through or over the levee (fig. 3.5). Nearby, depending upon the energy source, was a coal storage bin, a transformer substation, or oil storage tanks. Small generators were installed for illumination in the steam-powered pump houses.

Pump houses tended to be located at the lowest point in a district, adjacent levees that faced the Illinois River (maps 2.5 and 2.6). About a third of the sites were in the downstream corner of districts. The siting of two-thirds of the pumping plants upstream of the lower corner by a half-mile to four miles either reflected surface drainage patterns or the economy of ditching to the more central site. The selection of a site opposite Pekin (map 2.4) for the pumping station in the Pekin and La Marsh district (1892) was made primarily for accessibility,[24] a matter of importance considering the prospective cost of hauling coal and of providing housing for a resident fireman at a distant pumping station. With few such exceptions, pumping plants received coal or oil by barge. To handle such barges, modest improvements were made in borrow ditches or sloughs, and several districts had conveyor devices installed to move the coal over the levee to a storage shed. Otherwise, men with wheelbarrows did the work. It was slow going. For instance, 10.5 man-days (at $2 per man per day) were required to unload, weigh, and store over 215 tons of coal at the Scott County district pumping station in 1917. Deliveries were especially large in the fall and early winter, before the river froze. Otherwise, fuel for the heavy pumping season of late winter and spring might have to arrive by wagon from the nearest railway siding or coal yard, which was costly. The same consideration favored barging pumps, boilers, and the like to the sites during construction.[25]

At first, the most economical energy source for pumping stations was bituminous coal, very commonly delivered from Kingston Mines, southwest of Pekin, by the Lancaster Landing Coal and Transfer Company. Prior to 1915, virtually all pumping stations had fire-tube boilers and Corliss steam engines, but no new plants were driven by steam thereafter (table 3.1). In 1914, the owners of one or two privately reclaimed pieces of 400 or 500 acres adopted kerosene or gasoline engines or electric motors to run the pumps. However, electric power was adopted as the initial energy source for most of the pump houses built between 1914 and 1922, when diesel engines were installed at the new Valley City district. Meanwhile, the oldest districts gradually adopted electric power sources after 1909. The switch from coal to electricity at the pumping stations was marked following the extension by competing power companies of transmission lines to both sides of the river at several points in 1913 through 1915. The utilities were expanding rapidly, replacing acquired small power plants with a few strategically placed large generating facilities and linking these to an expanding grid of transmission lines. To such firms as the Central Illinois Public Service Company and the Canton Gas and Electric Company, pumping stations represented a large potential demand and the possibility of diurnal and seasonal load factors to complement urban markets and the use by coal mines in the region, which were then being electrified. Understandably, the success of the utility companies was reflected in a decline of coal barging to the districts along the river.[26]

The installation of transformer stations by the utility companies and of constant-speed induction motors by the districts occurred as pumping stations underwent first and second cycles of equipment replacement. The attractiveness of electric power over steam to the districts was related to a lower initial investment, the substantially reduced operating manpower requirements, the greater ease of starting up and adjusting pumping capacity to needs, and the generally longer life and lower maintenance needs for the machinery. Still, the cost of electricity was high. District officers viewed the standby charges for electric power as onerous, given the cyclical and short-term needs for power in pumping. After 1920, the diesel engine became the preferred replacement power source. Nevertheless, at the start of the 1930s, about half of the pump houses among about 40 districts along the Illinois River and among 35 districts along the Mississippi River in Iowa and Illinois were using electric power. Steam boilers were yet in use at 35 percent of the pump houses, and the remainder had diesel engines. Commonly, the older steam-powered units were kept on standby for emergency use. This retention of steam-powered units probably was related to the proximity of inexpensive coal supplies, as was the case for all pumping districts in Illinois. Districts in the state were slower to give up boilers than elsewhere in the nation; capacity yet attributable to steam was 13 percent in 1930 (vs. 9.8 percent nationwide).

By then, dependence upon electric power in the districts of Illinois was declining sharply, being replaced with power generated by diesel engines.[27]

Given that the bulk of the drainage and levee districts took shape in an era when steam power was familiar, and the use of electric motors and internal combustion engines with drainage pumps was relatively new, it is probable that the design of drainage systems was constrained by the need to channel water to one point. It was economically impracticable to build, operate, and maintain more than a single steam pump house.[28] Had the other power sources and automatic controls been more in vogue, it is conceivable that the configuration of drainage systems and the dimensions of main canals would be different today.

The districts varied in their policies toward hiring operators for the pump houses, perhaps because of the differing sizes in facilities and assessment areas. In some districts, the operator or stationary engineer was hired for the season of greatest activity. In others, operators were employed throughout the year and were provided with a frame residence near the pumping station. The man operated and maintained the equipment and pump house, doing minor repairs and keeping records of fuel deliveries and use and of the running time on the pumps; he supervised one or two firemen and saw to the removal of woody trash from the screens located across the main ditch near the intake bay. Also, the longer-term employee inspected levees regularly, seeing to the mowing of grass and weeds and the destruction of rodents and their runs. For such work, the Scott County district held onto its operator with annual contracts for $75 per month (1916) and $100 per month (1918), plus housing. When the districts began to purchase draglines and other power machinery, the experienced equipment operator became an attractive candidate for the pump house position.[29] Diverse skills and reliability remain the prime assets of such employees.

In developing drainage and levee districts along the Illinois River, the initial outlay for a pumping station usually represented from 6 to 20 percent of the first construction assessment (table 2.1) but might exceed a third of such costs. The coal or electric power for the pumps became the principal cost in annual operations of the district, as noted earlier.

There is but one instance of the integrating of the systems of internal drainage ditches at adjoining districts into one system served by a single pumping plant, and that was done by the Lacey, Langellier, and West Matanzas districts in the wake of the floods of 1926 and 1927. The electrified facility, located about five miles downstream of Havana, was operative in March 1930 with 24-inch, 30-inch, and 36-inch pumps. The districts prorated the cost of the new pumping plant in accordance with the areas benefited. In the aggregate, 8,700 acres were served. The Kerton Valley district (2,024 acres) joined the system subsequently.[30] No other districts are so well suited by configuration of land and streams to such consolidation.

## Retrospect

The 34 pumping districts of the lower Illinois Valley that are focal to this study represented about half of the drainage and levee districts that operated pumps in 1930 in Illinois, Missouri, and Iowa upstream of the confluence of the Illinois and Mississippi rivers. The districts in the study area served about 55 percent of all farmland in the state's drainage and levee districts. As a result of wet years and catastrophic flood events in the 1920s, the engine and motor capacities along the Illinois River were increased from 9,850 to 12,135 horsepower, or from about 54 to 65 percent of the capacity installed in all of the state's drainage and levee districts. The local pumping capacity rose from 1,466,000 to 1,974,500 gallons per minute.[31]

The effectiveness of the drainage systems along the lower river was reflected in higher land values than were found generally in pumping districts of the Midwest. The average pumping plant represented an investment of $53,600. Annual pumping costs per assessed acre (including fixed charges, ca. 1915) at ten districts where coal was used averaged $1.00; it was $1.92 at five districts where pumps were driven by electricity.[32] This outlay for pumping was essential to the success of agriculture on the bottoms. In good measure, the need was induced by water levels that were elevated because of human modification of land and water relationships in the watershed. While diversion of Lake Michigan water into the Illinois River system for sanitation and navigation purposes accounted for much of the problem, the earlier creation of slack water pools for navigation and the sustained, large-scale removal of bottoms from their natural function were important factors. Also contributory was the increased runoff from intensified land drainage in the wet prairie uplands of the watershed.

While the land reclaimed in the districts on the floodplain was as productive as the best upland farming areas, it cost more to drain and to maintain and was priced somewhat lower. Comparative drainage costs may be derived from the Census of 1920, using the counties of Brown, Fulton, Greene, Schuyler, and Scott to represent areas where pumping had to be done, and Champaign, Ford, and Piatt counties, where gravity drainage prevailed. In the counties where pumping was done on over 84 percent of the former wetlands, the reported cost of drainage averaged $39.33 per acre (from $27.58 to $58.91), whereas in the area dependent upon gravity flow, the average was $6.43 per acre (from $5.73 to $7.07). On the face of it, the data are comparable, but as may be recalled from chapter 2, drainage costs on the bottoms of the lower Illinois Valley tended to be understated in public documents. Adding to the price differential on the land was the cost of pumping and the larger maintenance requirements in the levee districts. Moreover, the threat of overflow from the bluffs and the Illinois River constrained investment in houses, barns, and the like on the reclaimed floodplain.[33]

The determination of pumping capacities and types and arrays of pumps and power sources suited to drainage needs in the districts along the lower river took some time to work out. Because the machinery was expensive, and because deficiencies and malfunctions could result in crop damage or loss, the learning experience was costly. Then, too, pump makers and drainage engineers were learning principles of efficient design. There was less uncertainty in determining the relative cost of energy sources.

The productive acreage in the pumping districts of the lower Illinois Valley represented almost 10 percent of the area in pumping entities nationally. The cost by 1920 to reclaim the districts in the state was about twice the average cost of pumping districts located elsewhere in the Midwest and South but only 61 percent of the cost per acre arising in districts located in California's Central Valley, where irrigation and drainage pumps were required. The degree to which higher drainage costs in the Illinois Valley and Central Valley represented idiosyncrasies of river regimes and intrusions of human modification probably is unknowable, but the intrusions were important. In any case, pumping districts in the lower valley employed 15 percent of the horsepower represented by all engines and motors in the nation's drainage pumping stations, and they achieved over 9 percent of the capacity (in gallons per minute) of all plants in the country's pumping districts. This experience of the pumping districts in the Illinois Valley was a noteworthy reflection of the enterprise nationally.[34]

# 4

# Shapers of the Drainage Landscape

To reclaim swamp and overflowed land from the floodplain of the lower Illinois River required the excavation of a large volume of alluvial material. Creek beds were diverted, straightened, and enlarged; miles of levees were raised; and extensive ditch systems were cut within the systems of levees. Moving prodigious volumes of dirt quickly and at relatively low cost usually involved the service of drainage contractors, who responded to advertisements placed by district commissioners in local newspapers and trade journals. These endeavors of contractors and district engineers on behalf of private investors preceded any involvement by agencies of the State of Illinois or the United States government in levee construction and stream rectification for flood control purposes.

Men with teams, scrapers, wagons, and spades had constructed numerous discontinuous ditches and embankments here and there on the bottoms prior to the development of drainage and levee districts. A few miles of low levees enclosed farms and ranches in a couple of areas. Also, most of the earth moving subsequently performed at a handful of early districts was done with scrapers. Once power machinery was adopted, however, the scraper work was confined to shaping diversion ditches and embankments near the bluffs, to cutting shallow lateral ditches, and to dressing or smoothing the rough levees that were raised with power machinery. Teams worked where the ground was firm.

The bulk of the dirt amassed into well over 300 miles of levees and excavated from at least 370 miles of open ditches was handled by power machinery, notably the dipper dredge. Dragline and other dryland excavators, a couple of clamshell dredges, and hydraulic dredges equipped with cutterheads were used, as well. Sometimes the men, teams, and machines that worked at a district were marshaled by a single contracting firm, but it was more common for one contractor to build the levee system and another to do the ditching. Presumably, they accomplished the work with the mix of team-drawn and power excavators that they judged was best suited to the tasks for which low bids were made. Now and again, district commissioners purchased machinery and engaged operators, teamsters, and the like to build the original levee and ditch systems under the supervision of the district engineer.

As a rule, the drainage contractors did not bid on the construction of pumping plants, which was done by a separate group of contractors, some of whom represented the firms that manufactured pumps and related machinery. These contractors were based in cities like St. Louis, Chicago, and Toledo, whereas the drainage contractors tended to operate from provincial towns or Chicago. By the same token, most of the engineers retained by the districts to design and supervise construction of drainage works had offices in Beardstown, Jacksonville, and Peoria; engineering firms in Chicago or Bloomington were retained sometimes.

In this chapter, the contribution to the land drainage process by the contracting sector, together with a description of the excavating machinery, will be examined. Considering the extent to which these firms and their tools modified the landscape, in the process destroying the wetlands, altering some stream courses, replacing others with ditch systems, and creating the dominant immediate relief features on the floodplain, it is appropriate to give them their due. Working with the landowners and the drainage engineers, drainage contractors shaped a new physical environment for agriculture—the ridged and ditched artificial landscape of the floodplain.

## Land Drainage Precedents in Central Illinois

The precedent for the application of power machinery to reclamation in the valley of the Illinois River was established in drainage work on the wet prairie uplands when "the drainage craze which has taken possession of the people" arose in the 1870s and 1880s. Most of the work by steam excavators was directed at the enlargement and extension of outlet ditches that served drainage districts where farmers' tile and ditch systems had existed for a few years. For the most part, the antecedent systems of open ditches and tile ditches were excavated by spade or with plows and drag scrapers. Sometimes, road graders were used to excavate shallow and broad ditches, whereas capstan plows were "employed whenever possible" to make ditches about 2½ to 3½ feet deep having sides that sloped from about a foot across the bottom to 4 to 6 feet across the top.[1]

A few steam dredges and "drag boats" were introduced in the mid to late 1880s and 1890s to excavate main and tributary drainage canals. The dipper, or scooplike digging device (figs. 4.1a and 4.1b), on the dredges was supported by a boom and A-frame. The dipper was cycled from a forward thrust into the water and dirt through a rising lateral swing to a point of discharge on either side and back again—in about a minute. The elongated and shallow wooden hull that bore the excavating device, operating machinery, and boiler was held in digging position with two large, retractable vertical spuds forward and a trailing spud aft. Before long, bank spuds were substituted for the vertical spuds on ditching dredges. The jacklike devices rested on the ground surface outboard of the A-frame instead of on the ditch bottom. The bank spuds made it possible to carry an equivalent excavating capacity on a narrower hull, thereby reducing the cubic yardage (and cost) of ditching. The stresses and strains of the digging cycle were transmitted to the ground through the A-frame and spuds, sparing the hull. The craft was moved ahead to a fresh digging position in a process akin to winching; the dipper, extended and grounded ahead, became a deadman, or anchor, towards which the hull (with raised forward spuds) was drawn with the operating chain or cable that retracted the dipper. Large versions of these early "boom dredges" had dippers of 1¾ cubic yards capacity, 60-foot booms, and "boats" that were 36 by 80 feet. These dredges, powered with 32-horsepower steam engines, could dig 600 to 1,200 cubic yards in the course of the usual 10-hour working day. They could cut ditches as much as 50 feet across the top. Ditches as little as 16 feet across and 4 to 6 feet deep were dug with the smaller excavators. These smaller drag boats or skid ditchers were mounted on heavy hull-like timber frames that rested on the bottom and flanks of the ditch. Some very early ditchers may have lacked bank spuds, but the retractable jacklike devices soon became standard equipment. The skid ditcher functioned best on bottoms of firm clay, being moved ahead by tensing a pair of steel cables that were anchored to deadmen well forward of the device, as shown in figure 4.1a.[2] It was a small-capacity (½ to ¾ cubic yard), relatively fast excavator but was short of reach. As a rule, the skid ditcher moved up-grade from the outlet of a ditch, whereas the dipper dredge began at the headward end, floating on accumulated water.[3] The boiler fireboxes in both machines could be adapted to burn firewood or coal.

Already in 1886, the contractors who operated steam dredges received between 11 and 17 cents per cubic yard, which compared favorably with the 8 to 12 cents, and sometimes 16 cents, per cubic yard for scraper work. Capstan ditchers excavated for rates equivalent to 4 or 5 cents per cubic yard, but they and the drag scrapers and road graders were not suitable for excavating outlet ditch systems that might be as much as 6 to 8 feet deep, 30 or 40 feet across the top, and 10 or 15 miles long.[4]

By the early 1900s, the floating dipper dredge (fig. 4.1b) largely had displaced the drag boat because of its greater capacity and reach. Also, relatively fast and reliable mechanisms were designed for raising and lowering the spuds. In this respect, it should be noted that improved, large versions of the dipper dredge were being developed steadily for harbor and waterway contracting.[5]

Whereas landowners of the prairie uplands favored broad and shallow ditches that did not interfere with livestock movement, the owners of bottomland were concerned with both drainage and the prevention of overflow, which required deep ditches and levees. The dipper dredge was well suited to the tasks. It was durable and simple to operate and maintain. The machine was unexcelled in digging indurated material and could be operated in places where water on and in the soil prevented men and horses

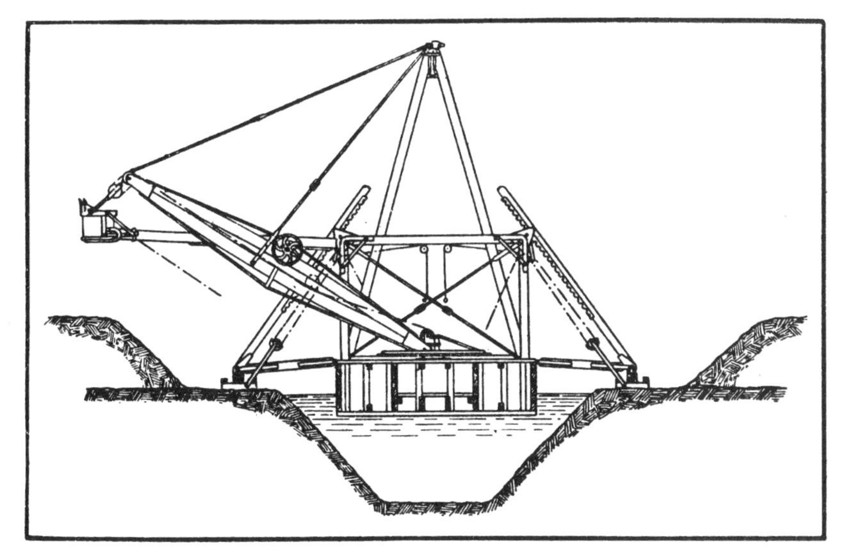

**Fig. 4.1a.** Mud boat or drag boat with bank spuds. From Wright, *Excavating Machinery*, 17.

**Fig. 4.1b.** Dipper dredge with bank spuds. From Wright, *Excavating Machinery*, 9.

from working. Although felled timber and stumps generally were removed from the rights-of-way of prospective levees and ditches before dredging began, the dredge was able to hoist and lay aside such heavy objects. With its boom and extendable dipper arm, the dredge could deposit excavated material some distance from and above the area from whence it was removed. Moreover, the reach was such that a berm could be kept between the embankment of fresh spoil and the edge of the ditch, which reduced caving and the return of spoil to the ditch through tumbling and erosion. Such berms of material in situ were desirable also to preserve levee stability, to reduce seepage, and to protect the levee base from stream erosion.

## Excavators on the Bottoms

The adoption of dipper dredges for land drainage began in the Midwest in 1884–85 in an area located some 12 to 24 miles northeast of Havana, where a large tract of wet prairie lay in Mason and Tazewell counties. The improvements were undertaken for the Mason and Tazewell Special Drainage District by a contractor who employed five dipper dredges, the first manufactured by the Bucyrus Foundry and Manufacturing Company of Bucyrus, Ohio. Three were floating machines with vertical spuds and dippers of 1-yard capacity; two probably were smaller skid ditchers. The project was remarkably successful, improving about 38,800 acres and inspiring other drainage districts in central Illinois to let contracts for the dredging of broad and lengthy systems of outlet ditches.[6]

The earliest subsequent dredge adoptions in areas near the river were purchased by drainage districts whose commissioners believed that the work could be accomplished at lower cost using their own machines and local crews than by engaging contractors. The Hager Slough Drainage District, located just northeast of Beardstown and adjacent the Sangamon River, purchased a 1-yard dredge from the Marion Steam Shovel Company in 1886 in order to cut about five miles of gravity-flow outlet ditch to the river. Three years later, the Hickory Grove Drainage District (now inoperative) put a new 1-yard Marion dredge to work some 25 to 30 miles northeast of Havana—adjacent the Mason and Tazewell Special Drainage District. The machine at the Hager Slough district remained operative there until 1897, but the dredge used by the Hickory Grove district was sold upon completion of its task.[7]

It is possible that dredges or skid ditchers were used on the bottoms of the Illinois Valley before 1890, but this has not been documented prior to 1891, when a 1-yard Marion skid ditcher was delivered to the Hartwell Ranch, in Greene County. By 1902, some 8 miles of main ditch system were in place at the Hartwell Ranch, and it is evident that creek modification and levee building had occurred along nearby Hurricane and Apple creeks where they crossed the valley to the Illinois River.[8]

Thereafter, dipper dredges were used to excavate ditches and raise levees at several drainage and levee districts adjacent and near the river, among them the Willow Creek district (1894), just northeast of Meredosia; the Coal Creek district (1896), opposite Beardstown; the Lacey district (1899), west of Havana; and in 1902, in the Otter Creek, Coon Run, and Pekin and La Marsh districts. Few, if any, drainage projects were completed between 1900 and 1921 without using dipper dredges.[9] Dredges with dipper capacities of 1½ cubic yards or less were used as ditchers, while those with capacities of 2 to 3½ cubic yards were used to build and repair levees. A dredge with a 2½-cubic-yard capacity could move 100 to 200 cubic yards of dirt per hour. Assuming a boom length of 50 feet, the dredge could discharge as far as 37 feet beyond the side of the hull and up to 22 feet above the water. Ditching dredges with a 1½-cubic-yard capacity and a 50-foot boom could excavate to 16 feet below the surface and deposit spoil as much as 33 feet away and 22 feet above the level of the water at a rate of 40 to 80 cubic yards per hour.[10] It is doubtful, however, that ditches were excavated deeper than 8 feet.

The dipper dredge and skid ditcher were the excavators most commonly used to raise levees and to cut systems of outlet ditches on the floodplain of the lower Illinois River, as they were in most of the agricultural land drainage work in the humid eastern half of the United States. However, two other types of dredges were employed in the study area—a pair of long-boom clamshell dredges and at least one 12-inch cutterhead hydraulic dredge, all made by the Marion Steam Shovel Company on order to the Edward Gillen Dock, Dredge and Construction Company of Beardstown (fig. 4.2). The first long-boom clamshell machine and the hydraulic dredge were assembled at Beardstown in 1915 to reclaim the South Beardstown Drainage and Levee District, where Gillen put a pair of 2-yard draglines to work as well. It was an assemblage of substantial excavating capacity. The clamshell dredge, which carried a 5-yard bucket at the end of a 110-foot steel boom, could build a broader and higher levee than any dipper dredge, moving an average of 370 cubic yards per hour in the process. The hydraulic dredge, rated at 125 cubic yards per hour, excavated the major drainage ditches, pumping the slurrylike material into low areas.[11]

A second clamshell dredge with a 5-yard capacity and 105 feet of boom was introduced by Gillen in 1920; it was placed on a hull made of timber, unlike the first all-steel machine. Like the earlier clamshell machine, it was designed by Gillen and Marion engineers along lines akin to a craft that had evolved in California.[12] These rigs could deposit spoil a good 80 feet from the side of the dredge, which facilitated setting levees back from the river after the floods of 1926–27, when the State of Illinois underwrote restorative work within the framework of a first general flood-control plan for the Illinois River. The clamshell dredges continued in levee building service well into the 1940s and were employed on the Ohio and Mississippi rivers, also. While the first hydraulic dredge is not known to have been used to excavate drainage systems elsewhere, it is evident that hydraulic dredges were used to restore levees and refill the borrow ditches that builders of early levees excavated just inside their levees at riverside. This filling added to levee security and reduced seepage from the river. Land filling of industrial sites and railway yards as well as levee building were done by hydraulic dredge at Peoria and East Peoria, however.[13]

**Fig. 4.2.** Gillen-Marion long-boom clamshell dredge, 1922. The dredge is restoring the levee at the Thompson Lake district. The machine is the second of its class to be assembled (1920) at Beardstown. Courtesy of G. and V. Karl Collection, Havana.

Ownership of the dredges used on the Illinois River bottoms took one of three forms. Either they were owned by the drainage contractors themselves, as described below; or they were purchased by the districts, as noted for the Hager Slough and Hickory Grove districts; or they were purchased by landowners who controlled virtually all of a tract and answered only to themselves, as was the case at the Hartwell Ranch, the Coal Creek and Langellier districts, and the Fairbanks Ranch (Keach district). The brothers Fairbanks owned and operated three dredges between 1905 and 1910, whereas A. L. Langellier owned but one. The 2-cubic-yard dredge with 85 feet of boom used to build the levee for the Coal Creek district was owned by Christie and Lowe, whose engineering and contracting firm in Chicago (1889–1913) appears not to have been involved at Coal Creek, although its principals were the owners of most of the land in the district. The dredge lay in the artificial channel of Coal Creek until 1903, when it was rehabilitated. Although the machine was to work in the reclamation of the South Beardstown Drainage and Levee District, the project was not begun until some years later.[14] Dredges owned by the other districts or landowners probably were sold or scrapped sooner or later. There is no evidence that they were kept to clean out the ditch systems or to restore levees once the initial drainage project was completed. The dredges represented too much machine and labor costs to be used for maintenance work.

Most of the dipper dredges were owned by drainage contractors whose fleets, or plants, are described subsequently. Almost invariably, the dredges worked under contracts on which the owners were low bidders. However, there were instances in which shared ownership between a contractor and a landowner or the long-term leasing of a dredge occurred. The former occurred at the Big Swan district in 1904–7, where a dredge was purchased by the firm of Wills and Woodall with money advanced by the district. The same machine, apparently owned by Woodall alone, was leased and operated subsequently by the Hillview Drainage and Levee District for eighteen months.[15] In 1910 and 1911, the machine completed a ditching job that was abandoned by the Quincy Dredging and Towing Company, which circumstance is related later. A similar contractual arrangement was made between the Hartwell Drainage and Levee District and C. C. Jacobs of Amboy, Illinois, in 1904. Two of Jacobs's dryland, steam-powered template excavators were purchased for $20,000, with the proviso that the designer and builder operate the machines. The arrangement was canceled at mid-point in the work because the machines failed to perform as promised.[16]

Although floating dipper dredges were the most common type of steam excavators employed in land drainage along the valley, it is apparent that several dryland excavators were used for ditching purposes as early as 1904, as noted above. Two land dredges in service at Spring Lake in 1907 could have been draglines or dipper machines mounted on skids or railroad trucks and portable tracks that straddled the excavation. Inasmuch as dryland dipper machines were not described again until one with an internal combustion engine was used for ditching at the Valley City district (1921–23), the early reference probably was to draglines. Other dryland machines used in ditching included the unsatisfactory Jacobs machines and the template excavators built by Frederick C. Austin of Chicago in 1910 for work at the Hartwell district. These two makes of template excavators drew buckets transversely across the line of a prospective canal, guiding the scraperlike bucket along a frame that spanned the ditch and beyond, where the bucket discharged onto a waste bank (figs. 4.3a and 4.3b). The Jacobs machine carried two buckets, each of which rose along one limb of the frame as it excavated; they dug alternately, each cutting half of the ditch from the center outward. The Austin excavator used one or two buckets to make a slice across the axis of the ditch. Both template excavators were supposed to shape the banks of excavations to about any desired angle, but Austin's was more successful. The machine rode on railway-type wheels and portable tracks or crawler-type treads. That the Jacobs machines were withdrawn from work and the Austin machines were reported to have been used but once suggests that they were both deficient, at least under local conditions, or too costly to use.[17]

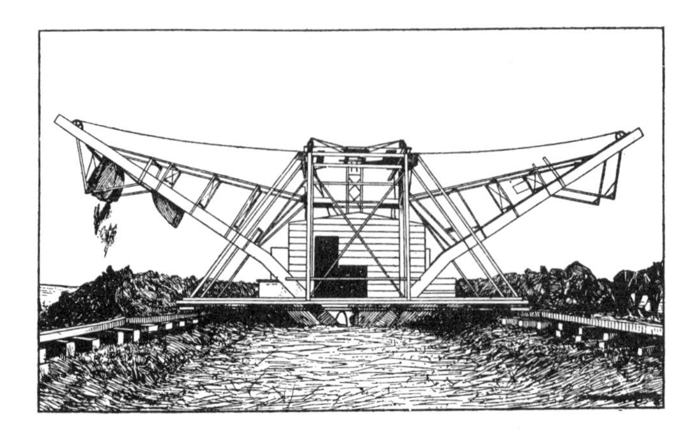

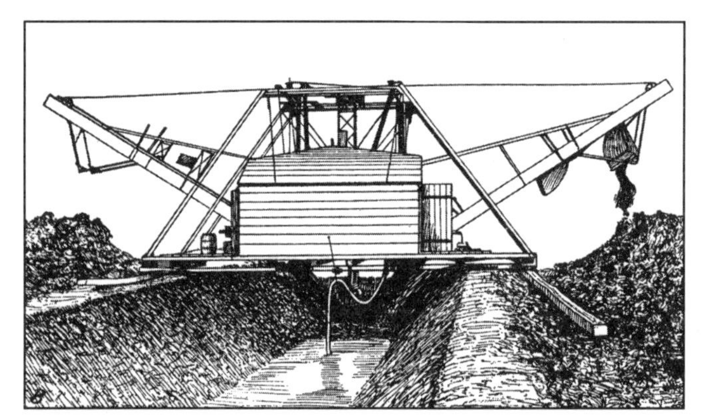

**Fig. 4.3a.** Template excavator *(front view).* From Wright, *Excavating Machinery,* 26.

**Fig. 4.3b.** Template excavator *(rear view).* From Wright, *Excavating Machinery,* 27.

Most common among the dryland excavators were draglines (fig. 4.4), first used by contractors for ditching and levee building; later, they were purchased and operated by the districts. Draglines were employed by contractors at least as early as 1907 at the Nutwood district, roughly three years after the device was developed and a year after its first application in cutting drainage ditches. Three draglines, along with three dipper dredges, worked in 1910 and 1911 at the Eldred district. A pair of 2½-cubic-yard draglines worked at the Scott County district between 1912 and 1914, where the plant included dipper dredges of 1½, 2, and 2½ cubic yards. In 1914, a dragline and some 90 teams and scrapers were used at the Hartwell district.[18] The large dragline that worked on the Meredosia Lake district in 1914 and 1915, according to "the general sentiment among people in the levee district . . . built a better and more permanent levee than any kind of machine heretofore used on similar work in the vicinity."[19] Whereas the levee built by dipper dredge was a rough structure of irregular crest, the dragline's levee "had the appearance of a railroad embankment being perfectly smooth and level

**Fig. 4.4.** Early dragline. From Wright, *Excavating Machinery,* 24.

on top and the side slopes neatly and accurately formed."[20] To judge from the inability of the commissioners of the Seahorn district to attract bids from operators of large draglines for such work in 1918, demand exceeded availability of such machines. On the other hand, a 2-yard dragline is known to have been withdrawn from levee building and ditching work by the contractor at the South Beardstown district in 1918 because its weight could not be supported on the saturated land. Such a 2-yard machine could move an average of 240 cubic yards per hour. It required an operator and a fireman and two or three hands to move the rollers and heavy planking on which it moved.[21]

The transformation of the early, large, and cumbersome draglines used by contractors into a small, single-operator machine that the districts could afford to buy and operate came about as a result of the adoption of such innovations as steel operating cable, the internal combustion engine, and walking or crawler-type traction systems.

Possibly the first dragline to be owned by a district was in operation at the Hillview district before 1913. It, like most early machines, was of unknown manufacture. However, a 1-yard-capacity walking dragline made by the Monighan Machine Company was purchased for $6,750 by the commissioners of the Eldred district in 1916. The Coal Creek district purchased a dragline in 1917, as did the McGee Creek district in 1919; and the Nutwood district purchased a Monighan in 1918 to replace the small suction dredge that it had used for ditch cleaning. Acquisition of such excavators continued through the 1920s, with purchases by the Scott County (1925), Big Swan (1926), West Matanzas (1926), and East Liverpool (1928) districts and major landowners at the Little Creek district (1921).[22] It is evident that the purchases were induced by the slowness of established drainage contractors to add small ditch-cleaning machines to their plant. However, it appears that in the 1920s, when the draglines with crawler-type traction appeared, a fresh group of contractors with a new generation of excavators became active in the valley. They tended to be from Beardstown and from county seats located near the valley.

A low-cost and fast alternative to using teams and scrapers for ditching in damp soil received attention during 1911 and 1912 in Pike and Scott counties, and probably elsewhere, because salesmen pressed for "the modern ditching" procedure. Their memorable demonstrations with dynamite may have resulted in some contracts with landowners. However, dynamite seems to have been used more by contractors in clearing rights-of-way and in excavating muck ditches.[23]

## Contract Advertising and Negotiation

In accordance with the Levee Act of May 29, 1879, the commissioners of drainage and levee districts advertised for sealed bids, offering a brief description of the components of the proposed work and advising prospective bidders where the detailed engineer's plans and specifications might be consulted. The resulting bids were opened at a specified place and time in the county seat where the district lay. As a rule, the low bid was accepted by the commissioners, but second rounds of bidding might result when the first bids were considered to be too high. Such second rounds evidently were preceded by negotiations between the commissioners, their engineer and attorney, and the contractors. Once in a while, the commissioners sought out additional contractors, and a very few groups of commissioners obtained their court's permission to negotiate directly with makers of machinery for work after disappointing first rounds of bidding. Such direct negotiations were more common for work on pumping plants than on earth-moving work. Sometimes, the commissioners were forced to step in because the contractor defaulted.

Either because the cost estimates by the district engineer were understated, or because inflation arose in the interim between project design and receiving bids, or because conditions required that the plans be modified after the work was let, it was common for projects to cost more than was estimated originally. There is no record of criticism that first estimates were deliberately understated by district commissioners and their engineer as a device to temper opposition. Nevertheless, second and third construction assessments frequently were called for to complete a reclamation project. These, like all other major transactions on behalf of the districts, had to be approved through hearings before the local circuit court. Contracts could not be let for amounts exceeding the approved assessment. On the other hand, the cost of a flood fight or of other pressing maintenance work might force commissioners to borrow money from local banks or individuals. Repayment was made by a subsequent assessment for construction.

The agreements between the commissioners and the contractors delineated project specifications and contained fairly uniform provisions as to standards of work, the payment schedule, and related responsibilities of the signatories. As a rule, the terms prescribed such matters as shrinkage allowances for spoil emplaced by one or another type of machine or by horses and scrapers; invariably, the fill emplaced in levees was required to be thawed and free of such woody waste as tree roots.[24] Sometimes, contracts specified the type of dredges that were to be used. Such instances appear to reflect an awareness that the low bidder had not performed to expectations earlier.[25]

The men retained as engineers by the district commissioners supervised construction. They and their aides placed the stakes, monitored the work done by contractors, and submitted to the commissioners the monthly progress reports upon which payments were authorized. At least a dozen engineers and surveyors served as district engineers. Most active among them were C. W. Brown of the Jacksonville Engineering Company (pre-1914), Jacob A. Harman of Peoria, Frank J. Traut of Beardstown, and John Goodell and C. H. Kreiling, who worked as individuals and partners out of Chandlerville and Beardstown (1909–13), from whence Kreiling moved in 1913 to Havana to work independently. R. W. Hunt and Company of Chicago worked briefly in the area. Although Harman and Traut were retained by districts throughout the lower valley, the other engineers tended to serve districts that were clustered. Jacksonville Engineering was most active below Meredosia, while Goodell and Kreiling worked most with districts between Meredosia and Pekin and in and adjacent the Sangamon River bottoms. Hunt and Company worked only at the districts adjacent Beardstown, where Jesse Lowe was the principal investor. Presumably, word-of-mouth endorsements and direct contact in the field produced the early patterns of service; job performance and longevity kept Goodell and Kreiling active well beyond 1930.[26]

## Comparative Costs

Although the dipper dredge usually excavated dirt at lower cost than did alternative excavating modes, there were recurring circumstances when other modes were adopted. Ditches of small cross section on the upslope edge of the districts, and laterals generally, were shaped by a dryland machine or by equipment drawn by either horses or steam traction engines. Also, where levees required relatively small volumes of dirt, as was the case near the margins of the valley, they could be shaped effectively by dryland machines and horse-drawn equipment. On the other hand, in at least one instance (Hillview district, 1906–8) after the introduction of power machinery, the successful bid to build a levee system was completed with as many as 400 men and 300 teams. Since the contractor was from St. Louis, it is surmised that the work force was drawn from there. The levee was restored in 1914 with 90 teams.[27]

Ditch maintenance became the task of small dryland machines because they were more mobile and better suited to the scale of work than were dredges. Moreover, obtaining the proper flotation of dredges might so elevate water levels in the ditch system that damage resulted to crops on low land. The dragline was especially useful because it was able to spread dirt into field margins, rather than heap it into an embankment; it moved yardage more cheaply than did teams with scrapers. There is no documentation of the use of trenching machines that were developed to facilitate the laying of field tile systems, although trenchers must have been used. By and large, the dipper dredge worked on levees at rates of 9 to 11 cents per cubic yard in the 1890s and early 1900s. Rates increased to about 11 to 16 cents per cubic yard between 1911 and 1916 and to about 17 to 32 cents per yard in the 1920s. In 1923, at the East Peoria district, dipper dredges worked for 32 cents per yard on levee repairs and 35 cents per yard in ditch cleaning. After 1910, dragline work cost about the same. Teams and scrapers worked for about 9 cents more per yard in 1912–16, and for about 3 cents more in 1928. The rates varied with the size and duration of the project, the degree to which rights-of-way were covered with timber, whether they were for original or restorative work, and (probably) the degree to which the contractor needed work.[28] It is evident that some contractors operated on thin margins; cases of work interruption due to nonpayment of dredge crews, of building levees and ditches to less than specified dimensions, and of suits for noncompliance with contracts may be found in the records of the districts and circuit courts.

## The Drainage Contractors

Drainage contractors were a migratory lot whose calling and achievements were all but ignored in local newspapers and in community or county histories. Moreover, the accounts of the contractors are elusive because incorporation, licensing, and performance monitoring were not required by the legislature.

The handful of drainage contractors who performed most of the work in the valley of the Illinois River had prior experience in public works contracting on the Great Lakes or the Mississippi River or in land drainage and road building on the prairie uplands of central Illinois, or they became contractors after some years of overseeing contract work for a drainage and levee district. One suspects that contractors with early starts along the river were advantaged by having local experience and contacts and by not having to factor into their bids the high towage or rail freight charges required to move machines to a new area of work. Although a few successful bidders on drainage contracts were from out of state, only one made a sustained contribution to work along the river. In the following section, the most active contractors are identified, and to the extent possible, their physical plants and the nature and places of their work are described. The record of projects along the river is indicative, as is the case for references to their projects elsewhere.

## A. V. Wills and Sons

Among the drainage contractors whose familiarity with land reclamation was developed through some years of experience in the management of affairs of a drainage district was Abner V. Wills, a second-generation breeder of cattle and farmer from the Pittsfield vicinity. He and a brother owned over 800 acres of bottoms on the Mississippi River in Pike County, and Wills is understood to have owned close to 1,500 acres of similar land along the lower Illinois River and in southern Indiana.[29]

The contracting firm of A. V. Wills and Sons began operating in the Sny River district in 1891, by which time Wills was approaching the midpoint in a 90-year lifetime. As was common among such family-owned firms, it was unincorporated, and family members formed the nucleus of machinery operators and field managers. At least three sons and Wills's brother were employed.[30] The company is not known to have used draglines on ditching and levee building projects in Illinois. Rather, the firm appears to have relied entirely on dredges made by the Marion Steam Shovel Company, of Marion, Ohio. Among the factors inducing strong product loyalty was the convenience of reusing compatible machinery and parts from retired or destroyed dredges. A. V. Wills and Sons operated dipper dredges of 1- to 4$^1/_2$-cubic-yard capacity.[31]

The firm was most active between 1899 and 1909 along the lower Illinois River, where Meredosia was used to tie up the machines when they were idle. The company owned a number of launches, quarters boats, and barges and is known to have employed teams and scrapers. For example, over 75 teams complemented the work of dredges in the Big Swan district project of 1904–7, as did an undocumented number of teams and scrapers at the McGee Creek district in 1906–9 and on the Crane Creek levee in 1909–11.[32]

The experience gained at the Sny River district and along the lower Illinois River led to work in Indiana, Ohio, Iowa, Missouri, and Arkansas. Ultimately, the dredges worked in such distant states as South Dakota, Mississippi, Louisiana, Florida, and North and South Carolina. The firm was remarkably successful, operating 7 dipper dredges by 1901 and as many as 8 such machines at one time. In all, at least 17 dredges were purchased between 1891 and 1914. The company's dredges did not exceed 2-cubic-yard capacity until after 1907; thereafter, 9 of 12 new machines were of 2½- and 3-cubic-yard capacity. A last 3-yard dipper dredge was purchased in 1920, by which time dragline and trenching machines were in the plant.[33] In sum, A. V. Wills and Sons operated one of the larger fleets of dipper dredges in land drainage work in the nation. When and how the firm went out of business is unknown. It worked at the Crane Creek district in 1922 or 1923, but it is doubtful that there was sufficient work to keep it very busy after 1925.

## Federal Contracting Company

Under oath in 1918, Frank J. Traut claimed to have built about 100 miles of levees and to have dug over 100 miles of ditches since his arrival in Beardstown in 1895. Some of the achievements were realized at the Coal Creek district, where Traut began work on a surveying crew in the employ of the Chicago contractors who purchased the bulk of the tract to have it reclaimed. Traut, who had a dozen years of experience with railroad and general engineering firms, soon became the farm manager and the sole drainage commissioner. The duties continued until 1909, when Jesse Lowe appears to have fired him over a matter of conflict of interest—Traut retained his own contracting firm for work at the district. Meanwhile, Traut managed the Goodell Ice and Fuel Company, its subsidiary fish-packing firms in Beardstown and Browning, and the Beardstown Stave and Lumber Company, of which he owned half.[34]

Traut organized the Federal Contracting Company in 1905, engaging a brother as superintendent. One is tempted to infer that Traut's entry into the contracting business resulted from a major and costly miscalculation as engineer for the Meredosia Lake district in 1904–5. Having underestimated by 50,000 cubic yards and $5,400 the needs of the main levee, he was replaced. Traut, the contractor, was backed financially by William T. Osner of Chicago, who assumed the presidency in 1911. The company was succeeded by the Traut and Osner Company in 1913, but the successor is not known to have worked in the 1920s, except at the Big Prairie district (1926). Traut's firms were active at numerous districts. Among the levee building tasks were those of the Nutwood, Eldred, and East Peoria districts between 1907 and 1912, the Scott County district in 1912–14, and the Big Prairie district in 1916 and 1917. A major ditch-cleaning job occupied Federal Contracting at the Big Swan district between 1911 and 1913. The firm worked for the short-lived Partridge Drainage and Levee District, near

Chillicothe, in 1907; and one of its dredges closed the outlet from the Spring Lake district in 1910. In this latter year, Traut was implicated in a legislative bribery charge related to blocking a bill deemed restrictive by the commercial fishermen of the Illinois River, which involvement is assumed to have resulted in Osner's larger role in the contracting firm, noted above. There were cash-flow problems in 1912.[35]

In addition to seeking contracts to design and execute drainage district plans, including the pumping plants, Traut is known to have operated as many as 5 Fairbanks dipper dredges between 1907 and 1912 and 2 draglines, as well as a towboat. Three of the dredges were 2-cubic-yard levee-building machines; and there were 1- and 1 1/2-cubic-yard ditchers. The towboat may have been the *City of Beardstown*, in which Traut apparently purchased some interest in 1918. The craft towed dredges and grain and coal barges. Much of the coal may have been sold by Traut for use in the pump houses of the drainage districts.[36] The man was quite an entrepreneur.

## Michael J. O'Meara and Son

The dredging firm of Michael J. O'Meara engaged in ditching work in Saline, Iroquois, and Livingston counties between 1909 and World War I. It obtained contracts from districts along the lower Illinois and Sangamon rivers from 1910 into the 1920s, by which time son Michael appears to have managed the firm. During most of the period, the family resided in Beardstown. Their machines worked in the Hager Slough, Lost Creek, Indian Creek, and Big Lake districts prior to 1915, on the ditches at the Big Prairie district in 1916, and on levee construction or repair projects at the West Matanzas, Crane Creek, East Liverpool, Lost Creek, Hillview, Big Swan, and Pekin and La Marsh districts between 1916 and 1929. Teams and scrapers were employed by O'Meara at the Spankey district in 1908.[37]

The firm performed ditching and levee work with dredges of 1 1/2-, 2 1/2-, and 3-cubic-yard capacity along the river. The 2 1/2-yard machine was a Fairbanks dredge; the other two were Marion dredges. The firm is known to have taken delivery of 3/4- and 1 1/2-cubic-yard Marion ditching machines for work elsewhere in the state. Its auxiliary craft included at least one coal barge and a large (18 ft. by 75 ft.) quarters boat for dredge crews. It is not known what machinery comprised the plant of the successor M. W. O'Meara Construction Company, which operated out of Quincy from about 1930 until World War II.[38]

## R. H. and G. A. McWilliams

The brothers G. A. and R. H. McWilliams of Chicago were involved in drainage contracting in the 1890s but did not become partners until 1910. The former operated a ditching dredge made by the Bucyrus Steam Shovel Company at the Coal Creek district in 1897 as junior partner to John J. Shea, a Chicago contractor. R. H. McWilliams excavated the ditches of the Otter Creek district in 1900 and the Willow Creek district in 1906–7. He used a large dragline in 1908 to elevate the road, now U.S. 136, which the City of Havana owned and maintained across the bottoms of southern Fulton County. Thus, both men were active from the outset of dredging on the bottoms of the Illinois Valley. Meanwhile, G. A. McWilliams had operated in east-central Illinois from about 1904. He was sufficiently well established by 1910 to own at least five Marion dipper dredges and to bid successfully on contracts for work in Arkansas and Louisiana.[39]

The brothers were competitors in bidding for the first contract to build the levee and ditch systems at the Eldred Drainage and Levee District in 1910, but neither had sufficient equipment available to complete the work in the time required by the commissioners. The shortcoming was overcome by forming a partnership and subcontracting to Traut, an unsuccessful bidder. Between them, the men assembled 5 dipper dredges, 3 draglines, and numerous teams with scrapers for the project. One of the draglines may have been the Monighan that the McWilliams brothers again used at the district in 1916. The brothers performed levee repair and ditch cleaning at the McGee Creek and Big Swan districts in 1914–15 and levee setback and creek-bed cleaning in 1917 at the Crane Creek district. The company built a 15-inch hydraulic dredge to restore and raise the levee system and to fill old borrow pits at the Hartwell Levee and Drainage District in 1923–24. The firm cleaned the heavily alluviated Coon Run in 1923 and was variously engaged in levee repair work, for instance at Naples, in the mid and late 1920s.[40]

The firm of R. H. and G. A. McWilliams was exceptionally successful in garnering contracts, reporting annual operations of $8 million in 1918. Their operating plant of 42 dredges, draglines, and other dryland excavators worked between the Great Lakes and the Gulf of Mexico. The firm, reorganized as McWilliams Dredging Company about 1910, maintained contracting offices in Chicago, Memphis, and New Orleans for a decade or so. Although R. H. McWilliams became active during the 1920s as a broker of municipal and drainage bonds in Memphis, G. A. McWilliams remained in contracting as McWilliams Company. His legacy is reflected by retention of the family name in successor companies located in the New Orleans vicinity.[41]

## Edward Gillen Dock, Dredge and Construction Company

After 1915, one of the most active operators of dredges in land drainage work along the river was the Edward Gillen Dock, Dredge and Construction Company, which theretofore engaged primarily in public works contracting on the Great Lakes. Although the parent firm remained in Wisconsin, the unit incorporated in Illinois in

1915 operated as a separate entity under Frank P. Gillen. The new venture, which accommodated a domestic breach, undertook the reclamation of the South Beardstown Drainage and Levee District and the abortive Rome View district, near Chillicothe. Gillen's sons set up an administrative and supply facility at Beardstown in support of about 100 men, plus dredges and draglines. George W. Gillen moved the base to Havana in 1918 to be close to major undertakings at the Chautauqua and Thompson Lake districts. Subsequently, major work was done for the East Liverpool and short-lived Wakonda districts and to excavate canals that circulated river water to cool a large thermoelectric power plant near Pekin. Lesser levee-dressing projects and one or two small navigation improvement jobs engaged the company's equipment, as well.[42]

The two long-boom clamshell dredges that Gillen obtained in 1915 and 1920 were better suited for levee enlargement tasks than dipper dredges because they could reach farther with a bucket of larger capacity than was the case with the accustomed machines. The advantages were appreciable, given that a lot of levee systems had to be restored and enlarged in the 1920s. Gillen's dredges, as noted earlier (fig. 4.2), carried a 5-yard bucket at the end of booms that, respectively, were 105 and 110 feet long, which enabled the machines to place up to about 6$^1/_2$ cubic yards of spoil in a digging cycle. The two machines, respectively costing over $52,000 and about $86,000,[43] must have paid for themselves several times over. Gillen's operating plant at the South Beardstown district was supported by three double-deck, 18-by-75-foot quarters boats, a pair of launches, tugs and barges, and an office boat. A plant of similar size—but without the cutterhead hydraulic dredge—operated out of Havana in the 1920s. The firm is believed to have operated for a couple of years after G. W. Gillen died in 1925. The surviving brothers sold the dredges and other assets.[44]

## Other Drainage Contractors

Among the drainage contractors whose work on channel modification and ditching is known to have brought them into or adjacent the bottomland of the Illinois River, but whose involvement seems to have been fleeting, were James S. Pollard, originally of Mason County but based for many years in Champaign; H. S. Brown, of Quincy; and C. L. Cook, of Pekin. The first, who learned the trade in northern Mason County on the first drainage project that employed dipper dredges in the Midwest, then organized Pollard, Goff, and Company, incorporated in 1899, and operated until 1906; Brown's Quincy Dredging and Towing Company was involved in land drainage work between 1897 and 1909;[45] and Cook was active from about 1910 to 1917. The latter two appear to have had no more than a pair of dipper dredges, but Pollard, Goff, and Company operated 8 or 10 dipper machines of various capacities.

At least another three or four small firms worked in the valley. Among them was one owned by John E. Rodgers, who learned the dredging trade in the middle 1880s in northern Mason County, and whose J. E. Rodgers and Company was particularly active in Douglas County for about 15 years, by which time he owned a pair of 1-cubic-yard Marion dredges. He was active in ditching until 1909, when a boiler explosion ended his career. Active in the mid-1920s was Pekin's E. M. Dirksen, who became a U.S. senator for Illinois.[46]

Pollard, Goff, and Company introduced dipper dredges to drainage work in east-central Illinois in 1885. Pollard died in 1906, but his sons carried on at least until 1907 as Pollard and Company. The firm's large 2½-cubic-yard dredges performed work in the lower Sangamon River and the Salt Fork between 1905 and 1909.[47] The dredges straightened and deepened the Sangamon River from the Oakford vicinity downstream. A substantial shortening of the stream channel reduced the overflow problem there, but the resulting acceleration of flow through the Sangamon River bottoms contributed to heightened crests, alluviation, and debris problems that came to be associated with the Sangamon River's course in the Illinois Valley.

In 1897, H. S. Brown's Quincy Dredging and Towing Company added drainage contracting to the accustomed work on federal navigation projects along the Mississippi and Illinois rivers, where he had worked at least since 1877. He started land drainage work with a Marion dipper dredge of 1½-cubic-yard capacity and purchased a used 1½-yard Marion dredge in 1906; and a new 1½-cubic-yard Marion ditcher was added to the plant in early 1907 for work at the Sny River district. Having lost one of the dredges to fire in the Sny River, Brown appears to have tried to recoup by low bidding on ditching work for the Hillview district in the winter of 1908–9. Faced with bankruptcy when the ditching dredge opened a subterranean flow of water and sand at the district on the Illinois River floodplain, Brown abandoned the task.[48] He is not known to have engaged in drainage contracting thereafter.

C. L. Cook is thought to have assembled his first of 2 dipper dredges and 3 dryland ditchers in Pekin in 1910, but the debt-ridden Cook-Devault Dredging Company is not known to have functioned beyond 1917. In the meantime, E. M. Dirksen was retained to run the firm (which he left to enter local politics). The company completed both the ditching and levee building at the Banner Special Drainage District in 1912 and 1913, and it appears to have been one of the contractors whose dredges built levees adjacent Thompson Lake at the short-lived Crabtree Drainage and Levee District in 1917.[49]

## End of the Dredge Contracting Era

By and large, the drainage contracting enterprises were family owned, some with participating partners, others with inactive partners. Most of them functioned so long as the founding partner and head of family remained active, but only the McWilliams firm is known to have remained active once family control ended. The McWilliams firm probably survived the demise of the land drainage era because it engaged in public works contracting in a large arena.

It is unlikely that the drainage contracting firms that owned one or a handful of dipper dredges survived beyond the 1920s. Certainly, there was no new work for them in the lower valley of the Illinois River. The levee and ditch systems that were in place by 1924 represented the culmination of the local reclamation process. Between the flood event of 1922 and the fallen commodity prices, interest in speculative land development for corn and wheat ended. Residual areas of the floodplain were too small and vulnerable to flooding to warrant reclaiming. The message was punctuated emphatically at the Chautauqua district, where the spades of fishermen were credited with the disaster of 1922, and where high water alone in 1926 and 1927 led to abandonment.[50]

The need to remove sediment and volunteer growth from drainage systems, to extend the ditches, and to maintain and restore levees continued; but the relatively large capacity of bulky dipper dredges, the expense of moving and manning them, their flotation needs, and the limitations of their reach did not compare favorably with draglines (fig. 4.5), least of all with the draglines owned by the districts. These machines of relatively small capacity but relatively long reach were mobile and had comparatively modest operating manpower needs. The dragline shaped sightly ditches and berms; the structures were broad, clean, and smooth, suiting the sensibilities of farmers and landowners. Moreover, a district's dragline could be rented to landowners or to other districts.

Evidently, most of the established local contracting firms were slow, if not unprepared, to add draglines to their steam-powered excavating plants. With such in mind, in 1919 the commissioners of the McGee district decided to purchase a dragline with crawler traction and lights plus a wagon for the operator's portable residence. Fully equipped, the machine was to cost $8,100 assembled at the district; and two years of operating costs were expected to be about $23,000. The option cost about $20,000 less than a contractor's estimate for equivalent work. An additional attraction for the commissioners was the expectation that the dragline would earn income for the district through rentals to landowners.[51] Consistent with the foregoing is the impression that familiar contractor names were replaced by new ones when ditch cleaning work was let to firms with draglines. Moreover, the economic circumstances of the 1920s favored the use of district-owned excavators and scrapers in levee and ditch maintenance work—when such work was done.

The men with steam dipper dredges dug themselves out of jobs when they transformed the wetlands of the lower Illinois Valley into surfaces whereon dryland machinery could operate. The contractors and their crews, along with local drainage

**Fig. 4.5.** Early dipper dredge and dragline raising levee. In this unusual view, the greater reach of the dragline is used to move dirt excavated by the Bucyrus-type ditching dredge. The dredge is about to discharge into the broad and shallow scraper bucket (est. 1.5 or 2 cu. yd.), which carried spoil to the levee crown. The dredge is borrowing from early levee and ditch. The costly "passing over" process probably took place ca. 1906 near Manito, northeast of Havana. Courtesy of G. and V. Karl Collection, Havana.

engineers and landowners, participated in a very early reclamation venture involving the application of dipper dredges for large-scale projects on a major floodplain. Their use of steam excavators began within a decade of the initial appearance of dredge boats on the upland wet prairies of the Midwest. The steam machines remained in service into the 1920s, when economic conditions ended the great era of speculation in land drainage nationwide. In the interim, speculation in wetlands drainage on a large scale developed in such diverse areas as the bottoms of the middle and upper Mississippi River, the wetlands of the upper Great Lakes states, in Florida, and elsewhere in the South.[52]

Two of the contracting firms that contributed significantly to the construction of drainage and levee districts in Illinois were early participants in land drainage work that developed elsewhere between Canada and the Gulf of Mexico. While the growing firms added draglines to their operating plants, they seemed to make a larger commitment with them in land drainage elsewhere than on maintenance jobs for districts along the lower Illinois River. Such maintenance work was being done by in-house operators at the drainage and levee districts and by a generation of local general contractors who used the mobile and versatile dragline with internal combustion engine and small crew.

The experience of the contractors who shaped the drained landscape in the lower Illinois Valley was representative of the larger experience in the nation.[53] By the same token, the compass of drainage engineering firms entered a new phase. Whereas firms usually remained small, serving the familiar locale and market, Peoria's most successful drainage engineer became associated with the retired senior drainage engineer of the U.S. Department of Agriculture in the Elliott and Harman Engineering Company, which functioned between 1916 and about 1928 with offices in Peoria, Washington, and Memphis.[54] The firm was positioned to participate in the work of the culminating years of the boom in land drainage in the interior, and it would have been very well positioned had Congress been persuaded to take an active role in furthering the land drainage and flood control measures that landed interests of the Mississippi River floodplain promoted.[55]

The people who altered the microtopography and drainage of the Illinois River floodplain with systems of artificial levees that repelled overflow, who diverted or rechanneled watercourses, and who integrated captive sloughs into networks of drainage ditches and canals that ended at pumping stations near the river comprised a small cadre of drainage engineers and surveyors, a larger group of drainage contractors and dredgermen, and numerous teamsters and laborers. Their principal tools were the axe and spade, team-drawn scraper, and the steam-powered dredge and dragline. The dipper and clamshell dredges moved much of the dirt; but the era of usefulness of such machines on the bottoms, like that of scrapers, ran its course in the three decades that ended in 1924. By then, the internal combustion engine mounted on tracked excavators, tractors, and trucks were being used to maintain, extend, or alter the essential physical works of the drained landscape.

# 5

# Challenges Engineered by Other Sectors

THE Illinois River system provided the most direct natural waterway between the Great Lakes and the Mississippi River system, although continuity was broken by a short and low divide near Chicago, arrested by rapids in the upper valley, and hindered by seasonally pronounced bars in the middle and lower valley. The divide between the Chicago and Des Plaines rivers was surmounted and the rapid water bypassed by the Illinois and Michigan Canal, which began to function in 1848. It was the first major achievement of the state's internal development endeavors. The usefulness of the canal between Chicago and La Salle was complemented between La Salle and the Mississippi River through snagging, dredging, and lock and dam construction funded by the state and United States governments.

The depth and breadth of the Illinois River and its relationship to adjacent bottoms was modified by navigation pools created by four dams and locks, three of which impacted the floodplain at Peoria and downstream—near the mouth of Copperas Creek (1877), at La Grange (1889), and at Kampsville (1893). The resulting year-round encroachment of water over adjacent bottoms affected areas that, for the most part, were not broken and cleared for crops. However, a substantial rise and spread of water resulted from the diversion of Lake Michigan water into the Illinois River system by the Chicago Sanitary District, starting in 1900. Water was drawn from the lake to flush the Chicago River and to convey the city's sewage and industrial waste through the Chicago Sanitary and Ship Canal into the Illinois River system. Enthusiasts who entwined the resolution of Chicago's sanitary problem with the concept of a ship canal envisioned the prospective engineering marvel as the first component of a grandiose Lakes-to-the-Gulf Deep Waterway with a 14-foot depth and a 300-foot bottom width. The proposed waterway to the Gulf of Mexico was expected to complement Chicago's primacy in the nation's railway system and to provide similar benefits for St. Louis and lesser commercial and industrial centers located on the waterways of the nation's interior. Although the paramount issue for Chicago was sanitation, proponents of the canal tried to persuade people in the nation's heartland that the waterway would counter deficient railway capacities and excessive rates, while providing access to the prospective Panama Canal that would counterbalance the advantage of ports on the East Coast. Chicago's boosters magnanimously proffered their Sanitary and Ship Canal to cities and states on the Mississippi River, and to the nation, as a catalyst to induce improvement of the remainder of the waterway. The existing discontent with the railways resulted in strengthening the regulatory role of the Interstate Commerce Commission, but the deep waterway proposal foundered in the Rivers and Harbors Committee of the House of Representatives, as it had in feasibility studies done by the Army Engineers.[1]

The idea of a Lakes-to-the-Gulf Deep Waterway was consistent with the thinking of engineers and other technocrats who favored the multipurpose river basin improvements concept that evolved during the administrations of President Theodore Roosevelt (1901–9). Supportive of the comprehensive approach to river basin development was Nevada congressman (1892–1903) and senator (1903–19) Francis G. Newlands, sponsor of the Reclamation Act (1902), proponent of national flood control legislation, and supporter of the conservation movement generally. In 1907, Newlands proposed the abortive National Waterways Commission through which Roosevelt hoped to centralize watershed planning and development and to bypass an unsympathetic Rivers and Harbors Committee. An ally of Newlands was Henry T. Rainey, congressman (1903–20; 1922–34) for the district that included the lower Illinois Valley. Improvement of the Illinois River waterway was a focal commitment for the Carrollton Democrat.[2]

Notwithstanding the opposition to a deep waterway by powerful congressmen and their advisers in the War Department, the Sanitary District of Chicago proceeded to construct a capacious canal in the 1890s. The grand scale of the undertaking, like the Columbian Exposition of 1892–93, caught the public's fancy. Visitors to the site marveled at the large and powerful steam shovels and spoil conveyors. The project was a proving ground for engineering methods and machines subsequently applied in surface mining, in railway regrading and realignment, and in the cutting of the Panama Canal.[3]

Chicago's proponents of the Sanitary and Ship Canal believed that it would provide sufficient water through the middle and lower Illinois River to warrant removal of the locks and dams. The idea was welcomed by landowners and prospective developers of the bottoms in the lower valley, who were persuaded that a 14-foot waterway would be the ultimate gravity-flow outlet canal for reclaimed lands. As it turned out, the dams and locks remained in place, causing widespread flooding of the bottoms as water diversions from Lake Michigan increased.

The injurious effects of existing and alternative approaches to navigation improvement on the Illinois River system drew the valley's land drainage sector into a high-stakes contest with regional and national ramifications, some of them noted above.

Local interests sought recourse against the Sanitary District of Chicago in state courts, the General Assembly, and the U.S. Congress. Meanwhile, the sanitary entity proposed by Chicago and endorsed by the State of Illinois sought approval of the requisite water diversions by the War Department and Congress, and it responded in the federal courts to complaints filed by Great Lakes states and the Dominion of Canada about the diversion from Lake Michigan.

This chapter describes the civil works that modified the Illinois River in the nineteenth and early twentieth centuries, focusing on their purpose, their effects on the floodplain of the lower valley, and the nature of landowner responses. Dams, locks, and navigation pools created in the lower valley between 1869 and 1899 are dealt with in a short early section. A more lengthy account that follows recounts the Sanitary District's ambitious and controversial diversion of water from Lake Michigan through the canal that opened in 1900. Resolution of the in-state contest, as will be seen, hinged upon outcomes of the larger political contests involving Chicago and the state with the governments of other states and the United States.

## Navigation Improvements, 1869–1899

During the last half of the nineteenth century, the national and state governments made navigation improvements on the Illinois River to complement the function of the Illinois and Michigan Canal. Congress began to make appropriations in 1852 for snag removal, wing dam construction, and dredging. The dredging overcame bars with no more than 1.5 to 2.0 feet of water at low-water stages and it improved access to landings for steamboats. However, the principal federal commitment was to a series of dams and locks that formed slack water pools for navigation. Plans for the first of these structures were prepared by engineers of the War Department in 1867 and 1868, but the dam and lock at Henry (about 33 miles upstream of Peoria) were built by the State of Illinois between 1869 and 1871. Foundations for the second dam and lock were begun at the mouth of Copperas Creek, about 25.5 miles downstream of Peoria, with funds appropriated by Congress in 1873. The structures were completed by the Illinois Canal Commission in 1877. Flash boards placed along the rims of the dams and supplemental dredging created 6 to 7 feet of navigable water as far upstream as the La Salle terminus for the Illinois and Michigan Canal.[4]

The system of slack water pools was extended downstream of Copperas Creek (fig. 5.1) by the United States government with dams and locks at La Grange (1889) and Kampsville (1893), creating 7-foot navigation pools about 59 and 46 miles long, respectively. Between Kampsville and the Mississippi River, the depth was maintained by dredging. Thus, working with the State of Illinois and at a cost of over $2.1 million in federal and state funds, the U.S. Army Corps of Engineers completed the system of

**Fig. 5.1.** View toward Copperas Creek Lock from below, 1903. The slack water pool above La Grange Lock and Dam is in foreground. First structure on left is cabin boat on blocks or piling. Illinois Submerged and Shore Lands Investigating Committee; courtesy of Illinois State Museum.

navigation pools begun in 1869 at Henry. The resulting waterway between the Mississippi River and La Salle was 225 miles long. Meanwhile, waterway proponents in Chicago and the Illinois Valley were buoyed by the General Assembly's endorsement (1889) of the Lakes-to-the-Gulf Deep Waterway. The State of Illinois was on record favoring a 14-foot channel and a diversion of water from Lake Michigan large enough to eliminate the need for dams and locks.[5]

Although the value of bottomland for agriculture began to appreciate about the time that the structures at La Grange and Kampsville were finished, they were not yet prized lands. Nevertheless, the artificially raised pools provided an incentive to build levees and install pumps after large volumes of water were turned into the Illinois River system from Lake Michigan. Permanent flooding of the bottoms, together with the perception among owners of the bottoms that the dams and locks trapped sediment and sludge in the navigation pools and tributaries—adding to the flood hazard—led to persistent and nearly universal demands that the structures be removed.[6]

## The Chicago Sanitary and Ship Canal

Among Chicago's major challenges in becoming a modern city were the development of systems of safe water supply and of sewage and industrial waste disposal.[7] The accustomed drawing of water for municipal needs from Lake Michigan a short distance from the mouth of the Chicago River, the city's principal sewer, became increasingly risky as population and industry grew in the latter half of the nineteenth century. To abate the risk of polluting the city's water source, the summit level of the Illinois and Michigan Canal was deepened (1866–71) so that waste water would flow by gravity to the Illinois River system. Since by 1891 some 200 tons of solid material entered the canal daily, the summit sector was dredged continually. Residents of Joliet and smaller communities along the Illinois and Michigan Canal suffered a nuisance that was almost intolerably foul. At least for Chicago, this mode of disposing of waste water was greatly improved upon in 1884 by the installing at Bridgeport of pumps that moved an average of 600 cubic feet per second (c.f.s.) from the Chicago River through the Illinois and Michigan Canal into the Illinois River system. The pumping arrangement continued until the Chicago Sanitary and Ship Canal became the conduit for the city's waste, as well as the first link in the proposed deep waterway to New Orleans.[8]

The proposal that a sanitary and ship canal be constructed for Chicago arose at a time (1885) when dilution, dispersal, and natural processes were the least costly and most familiar approach to breaking down and disposing of sewage and industrial waste. At the time, too, there was precedent for the General Assembly to form quasi-public corporate entities with taxing and bond issuing powers (like drainage and levee districts) to address regional sanitation problems. In May 1889, after much debate and amending, the General Assembly adopted a bill enabling the formation of the Sanitary District of Chicago. The proposed assessment district, which was soon determined to have an area of 185 square miles, was adopted by referendum later in 1889. Construction of the canal began in 1892. The work involved altering the Chicago River, relocating 13 miles of the Des Plaines River, and building a waterway that dwarfed the adjacent Illinois and Michigan Canal. The cross section of the Sanitary and Ship Canal was designed to convey up to 14,000 c.f.s. of water from Lake Michigan. A permit to connect the modified Chicago River with the new canal was granted by the Secretary of War in May 1899. In 1903, the General Assembly authorized the Sanitary District to nearly double its service area and to construct facilities through which to flush the reversed Calumet River and the prospective Sag Canal, as well as the north and south branches of the Chicago River. While work began on the northeasterly trending Calumet–Sag Canal in 1907, litigation delayed completion until 1922. In essence, the Secretary of War would not authorize diversions exceeding the 4,167 c.f.s. permitted in 1901.[9]

A fundamental difficulty with the diversion of water from Lake Michigan was the very large discrepancy between the volume authorized by the General Assembly in the Sanitary District enabling act and the volume permitted by the Secretary of War. The volume authorized by the Secretary of War was less than half the flow required by the state's formula for dilution of sewage in 1918–19, which was 3.3 c.f.s. per 1,000 population served. Even had the diversion exceeded 10,000 c.f.s., it would have been inadequate for sanitation needs. Already, the level of pollution in the middle Illinois Valley dismayed the area's residents. The deplorable conditions were not anticipated by the urban commercial interests and landowners along the Illinois River, whose main concern had to do with water volume. The early worries about the scale of proposed diversions were assuaged by provisions in the enabling legislation that required the Sanitary District to remove the state's dams at Henry and Copperas Creek and to pay for damages done by the water.[10] Congressman Guy L. Shaw, a single-term (1920–22) Republican from Beardstown, recalled in 1922: "We were driven to [land drainage] very largely because of the construction of the dams in the river and the flow from Lake Michigan through the Sanitary District Canal. The evidence shows that when the Sanitary District was created the people of the valley were led to believe that any and all damages which might accrue as a result of the flow from Lake Michigan would be paid by the Sanitary District of Chicago. Had it not been for such provisions the people of the valley would not in the beginning have stood for the creation of the Chicago Sanitary District."[11]

Unfortunately for the landowners, the U.S. Army District Engineer for Chicago was committed to retention of the system of dams designed to provide navigation depths of 7 feet in the Illinois River. The engineers, who had determined at least as early as 1887 that Chicago's sanitary canal was not needed for the foreseeable navigation interests of the United States, were sure that the permitted diversion of 4,167 c.f.s. could not maintain the prescribed navigation depths on the Illinois River without the dams. Furthermore, the Illinois Supreme Court ruled in 1900, in a suit brought against the Sanitary District by the Illinois and Michigan Canal Commissioners, that removal of the state's dams was not mandatory. So, the dams remained in place for over a quarter century. Meanwhile, petitions seeking authorization of larger water diversions from Lake Michigan were carried by the Sanitary District of Chicago to successive Secretaries of War and to Congress, where concerns over the threat to navigation and to power generating capacity at Niagara Falls were heard from other Great Lakes states and Canada. Ultimately, in 1922, the State of Wisconsin brought suit in federal court to end the threat from Illinois. Supporting the Wisconsin case at district court level and before the Supreme Court in Washington (to which the Sanitary District appealed) were other Great Lakes states, the U.S. Attorney General, and the Dominion of Canada. Eventually, the plaintiffs prevailed, learning in April 1930 that the Supreme Court endorsed its Special Master's decision that the Sanitary District's diversions were for sanitation purposes, not for navigation, and that existing diversions averaging 8,500 c.f.s. or more had to be reduced over a period of eight years to 1,500 c.f.s. The court's order did not affect water drawn for the city's water distribution system. Although the Sanitary District of Chicago began to investigate the viability of sewage treatment facilities in 1909, progress seemed to be dilatory to the state and federal agencies that urged construction. By 1925, about 4 percent of the waste entering the entire system was being processed in sewage treatment plants. The Sanitary District remained committed to the sewage dilution program until the diversion issue was adjudicated, much to the detriment of the owners of bottomland and to summer tourism in the Illinois Valley. The magnitude of the pollution of the Illinois River was unprecedented in the nation. As a dismayed victim of "Chicago's Cesspool" put it in 1924, bathers at Havana risked skin rashes; and depending on whether windows were closed or open, people suffered either oppressive heat or foul smells in summer. The situation depressed residential real estate prices. Nevertheless, it must be recognized that the Sanitary District's deplorable financial condition in the 1930s impeded construction of the requisite sewage treatment facilities.[12]

Between 1900 and 1930, the Sanitary District diverted an average of about 8,000 c.f.s. into the Illinois River system. It was about half of the average flow past Peoria (1913) and about six times the river's natural low-water volume there. Moreover, the discharge from the sanitary canal tended to be greater than average at times of flood in the Illinois River system. The diversion increased flood hazards and the cost of flood protection in the valley; the onset of flood stages was hastened, and the periods and range of overflows were extended. The appreciable increase in water levels during the period 1900 to 1914, as compared with the 1890s, was estimated to have raised flood stages 1.5 feet at Peoria and 1 foot at Beardstown. Damages ascribed to the higher flood stages were less detrimental to wetlands than the effect of Lake Michigan water at low river stage. The low-stage level was elevated by about 6 feet at Peoria and 3 feet at Beardstown when the diversion from Lake Michigan was about 9,000 c.f.s., as it was in 1924. At Thompson Lake, just upstream of Havana, Forbes and Richardson noted that the water was deepened about 3.6 feet in summer, which increased the lake's area from 1,943 to 5,072 acres. Officers of drainage and levee districts adjacent the pool created by the Kampsville dam attributed a rise of about 4 feet to Chicago in that levee-constricted part of the valley.[13] Long-time residents who knew the Beardstown, Sangamon Bottoms, and Crane Creek areas observed that the slack water pool above the La Grange dam ended cropping on some land and drowned out timbered areas where pecans, hickory nuts, and berries had been gathered in quantities. Pasture and volunteer hay lands were lost to water encroachment, reducing the area available to local stockmen for running cattle, hogs, and idle work stock and for cutting hay. Similar losses occurred above the dams at Copperas Creek and Kampsville. In 1912, owners of leveed land claimed that lost production on destroyed land exceeded the combined value of the public works, the freight carried through them, and the annual waterway maintenance costs. By then, snag removal work was practically continuous for the Corps of Engineers because of fallen timber that was killed largely by the spread of water attributable to the diversion from Lake Michigan.[14]

The people of the lower valley were no less annoyed to be victimized by Chicago than were residents of riverside communities upstream of Peoria, where mounting evidence of pollution from Chicago threatened assets that had begun to draw summer and weekend visitors from Peoria and Chicago and had fostered recreational and commercial fishing and hunting.[15] The apprehensions of people in the middle valley were captured at a mass meeting at Chillicothe in 1902, where it was resolved that the dams be torn out, "because by filling all the adjacent low lands with dead back water which by them is held on those lands and becomes stagnant and filthy, and filled with dead and decaying vegetable and animal matter, poisons the air with malaria and fills it with germs of contagious diseases."[16] Already by 1902, there was sufficient water (4,302 c.f.s.) moving into the Illinois River that a navigation channel of 6 feet could have been maintained with the dams removed. By 1910, the mean volume (6,833 c.f.s.) approached one-half of the canal's design capacity, sufficient to maintain a 7-foot channel for steamboats and barges. The water all but concealed the dams and locks a good part of the year.[17]

Although there was friction between the commercial fishing, recreational hunting, and agricultural land development sectors of the lower Illinois Valley over competing uses of the floodplain, all seemed to agree that the bulk of the damage done by higher water levels was caused by the Sanitary District of Chicago. Indicative of the difficulties, operators of hunting areas below Liverpool lamented the transformation of marsh, wet prairie, and timber into "just a lake,"[18] as was the case at adjacent Thompson Lake. The drainage canal was blamed for the immersion of low-lying cropland in the Spoon River Bottom in four out of five years after 1900 and the drowning of haying lands adjacent Muscooten Bay, near Beardstown.[19] An engineer for levee districts near and below Beardstown claimed that some were compelled to pump five or six months per year because of the diverted water. The litany was heard in public meetings at the river towns and county seats, in assemblies of representatives of drainage and levee districts, before committees of Congress, and at the October 1928 hearings of the Illinois Valley Flood Control Commission. The essence of the complaint was captured in Washington by former congressman Guy L. Shaw, in 1924: "They had overflowed the land, they had killed thousands of acres of timber, and they had ruined many acres of land from an agricultural standpoint. They have almost destroyed some of the villages and cities along the river, and the supply of fish and animal life in the river has been destroyed; the conditions have become intolerable."[20] Shaw's former constituents in the Illinois Valley had experienced two decades of augmented flows, capped by the devastating flood of 1922. They were feeling the effects of the collapse in commodity prices; and landowners among them were deeply frustrated by the outcome of suits for damages brought against the Sanitary District of Chicago.

In an effort to recover losses attributed to the Sanitary District, landowners in the lower Illinois Valley filed over 170 suits in local and Cook County courts between 1903 and mid-1919; most suits were filed by 1905. The suits arising in the lower valley represented about 38 percent of all suits filed against the Sanitary District to that time by owners of land along the Des Plaines and Illinois rivers. The damages sought by landowners of the lower Illinois Valley were close to $2.8 million of the $8.9 millions sought in the Sanitary District's "overflow cases." While neither the records of the trials nor a summary of their outcome by 1930 have been found, it is evident that the plaintiffs obtained little satisfaction; reversals of decisions appealed by the Sanitary District to the Illinois Supreme Court in Springfield were especially galling because the losers paid all court costs. At least 125 of the suits initiated over wetlands in the lower valley were lost (over 73 percent), and to judge from incomplete summaries of the outcome of the class of suits, successful plaintiffs recovered less than 15 percent of the alleged damages.[21]

Rather than seek accommodation, the management of the Sanitary District of Chicago committed substantial funds to preparing and conducting the defense against complainant landowners. The Sanitary District retained prominent downstate attorneys, and at least 50 surveyors, engineers, and support staff were hired to prepare detailed surveys of property, soil, land use, and river behavior throughout the lower valley between 1906 and 1909.[22] In view of such preparations, the *Peoria Herald-Transcript* cautioned in August 1906 that farmers "who have counted upon getting large damages . . . , will have to be more careful with their facts than they evidently were in recent suits heard in the Peoria circuit court."[23] A Pekin correspondent captured the game succinctly after interviewing a field-party supervisor: the Sanitary District was gathering "information and data with which to sand bag all land owners who start damage suits."[24] Such was evident in the costly 38-day trial in Peoria to which the newspaper referred in August 1906.[25] The trial was vituperatively fought, allegedly bribery-tainted (though unsubstantiated), and inconsequentially satisfying to the victors—the awarded damages being $750, instead of the $55,000 claimed. It was evident that landowners assumed a formidable and costly task in challenging the Sanitary District in court, even when plaintiffs had the advantage of appearing before local juries. It was evident, too, that the Sanitary District's thorough preparations and legal tactics were saving a very large amount of money. Their success appears to have been marked after the Supreme Court reversed several lower court decisions in 1912 and 1913. Thereafter, some of the plaintiff's attorneys settled for whatever the Sanitary District would concede. The disparity between the actual damages awarded to plaintiffs or conceded to their negotiators and the amount of the original claims suggests that some of these claims were overblown. At least one attorney had agents solicit the landowners, who agreed to share half of whatever money was recovered. This attorney appears to have brought 143 suits against the Sanitary District.[26]

It was evident to contemporaries that damage claims were inflated[27] and that the Sanitary District adopted a hard line rather than compromise. Clearly, the Sanitary District had more resources, better on-site research, and better legal staffs than did the plaintiffs. While the general outcome of the "overflow cases" supports the Sanitary District position, it is evident that the landowners in the valley believed that they were deeply wronged. The reversals in the Illinois Supreme Court contributed much to the view.[28] "In the first place the damages are of such an irreparable character that the Sanitary District would be bankrupt if it paid in full, and notwithstanding years of turmoil and litigation, practically no relief whatever has resulted to the persons actually damaged."[29] Moreover, "to get upon the Statute books any legislation that affects the Sanitary District of Chicago, especially legislation they think is unfriendly to them, is a job that is a man's task. If you will look at the public records of the Sanitary District, you will find that they have had for years on their payroll, some of the ablest legislators in this state, and they pay them salaries far in excess of their salaries as legislators, so you can judge for yourself where their interests lie, when it comes to legislation affecting the Sanitary District."[30] And: "The question of the Sanitary District of Chicago is the greatest crime ever perpetrated on any part of this country, permitting them to ruin a people as they have done to the landowners throughout the Illinois River Valley. And it is a big question whether or not the owners of nearly 400,000 acres of reclaimed land along the Illinois River can survive what has already happened to them, caused wholly by the Sanitary District of Chicago. To date we have no court nor any department of our Government which has made the district pay for the loss and damages caused by it to the Illinois Valley nor live up to the laws under which the sanitary district was created."[31] Enough innuendo was raised by the disenchanted and suspicious landowners to suggest that they believed that more than Chicago's water was foul smelling. The tenor of the thinking was confirmed with the unveiling in 1928 of a monumental and pervasive payroll scandal within the Sanitary District of Chicago. Legislative investigation confirmed that 53 assemblymen and 9 senators had received money from the Sanitary District.[32]

When owners of land in the lower valley realized that the Sanitary District was strongly opposed to recompensing them for damages attributable to water from Lake Michigan and that the interests of the Chicago area agency prevailed in the General Assembly, and for the most part, in the courts, they formed the Association of Drainage and Levee Districts of Illinois in 1910–11.[33] District engineers and attorneys joined, as well; but the organization attracted little participation from drainage and levee districts located elsewhere in Illinois. By the same token, high water and pollution attributed to Chicago do not appear to have heightened general downstate antipathy for Chicago.[34]

The Association of Drainage and Levee Districts of Illinois was more effective in Congress than in the General Assembly. Among proponents of the association were individuals who were active in organizing the first National Drainage Congress, which met in Chicago under the sponsorship of the Nineteenth National Irrigation Congress (December 6–9, 1911). The charter of the new National Drainage Congress called for the conservation of water and land, for the improvement of land drainage, public health, and inland navigation, and for the facilitation of public meetings and representations before state and federal agencies on behalf of the objectives—particularly as they related to land reclamation for agriculture. Before long, the Missouri Drainage and Levee Association, organized in October 1912, vigorously championed the movement, as did Missouri's congressmen. The growing network of partisans seeking direct federal involvement in planning and funding land drainage was helpful, but the key partisan in Illinois was Congressman Rainey. Rainey, who ultimately served as Speaker of the House in Franklin D. Roosevelt's first presidential term, believed that protecting floodplains from overflow should be added to the Corps of Engineer's traditional role in navigation improvement. Consistent with favoring federal management of riverine conditions was the opposition to privatization of water-power sites. Thus, the Sanitary District's revenue-driven proposal (1907) to extend its bounds downstream of Joliet in order to have a private utility develop the site was but another ill-received action.[35]

To induce tractability in Chicago's Sanitary District, the Association of Drainage and Levee Districts of Illinois opposed federal authorization of the Sanitary District's de facto water diversions. Understandably, the opposition was not taken well by governors and others who supported Chicago's aspirations for the sanitary and ship canal. Consequently, the land drainage interests, like the Sanitary District of Chicago, became supplicants before the Rivers and Harbors Committee.

## Resolving the In-State Conflict

Attempts were made now and again in Congress and the General Assembly to address the problem of high water in the Illinois Valley, but these bore superficial results. Nevertheless, the groundwork for ultimate change was being laid. Congressman Rainey, for instance, persuaded the Secretary of Agriculture in 1904 to include research on land drainage along the lower Illinois Valley with studies of irrigation in the West. Rainey later construed the resulting report to imply a federal responsibility for the levees along the river. More directly beneficial was a Rainey-sponsored congressional resolution (1904) authorizing the Sanitary District of Chicago to remove 2 feet from the crests of the dams at Kampsville and La Grange, in the lower valley. This project was completed by the Sanitary District at Kampsville but abandoned at La Grange in 1906, when the agency decided that it was less costly to litigate landowner suits. The passage of a constitutional amendment referendum (1908) enabled the State of Illinois to issue $20 million in bonds for public power development and construction of locks and dams in the 9-foot deep Illinois Waterway project of the 1920s. The referendum was supported by Chicago, the State of Illinois, and Rainey, but controversies over how to proceed delayed project initiation for years. Congressman Rainey's agenda to ultimately involve the Corps of Engineers with problems of drainage in the Illinois Valley was incidental to such larger issues as public versus private power generating and federal approval, if not participation, in the waterway project.[36]

The work begun by the state on the 60-mile Illinois Waterway project in the upper valley (Lockport to Utica) in 1920 was completed by the Corps of Engineers in the early 1930s because the state lacked the resources to continue. This involvement of the central government's agency was extended in 1927 by Congress, which authorized deepening of the Illinois River navigation channel from 7 to 9 feet. In the process, the state's dams at Henry and Copperas Creek were removed in 1929. Meanwhile, a continuing diversion of 8,500 c.f.s. into the system from Lake Michigan preserved a low-water navigation depth of 8 feet, except at Kampsville, where there were 6.5 feet of water at the gate sill. The federal project was completed in the late 1930s with the replacement of the remaining two fixed dams by wicket dams at Peoria, La Grange, and Alton.[37] Provisions were made to compensate the drainage and levee districts for damages resulting from seepage attributable to the raised slack water pools.

Meanwhile, the Association of Drainage and Levee Districts of Illinois petitioned the Secretary of War in 1912 to deny the Sanitary District's request for approval of a 10,000 c.f.s. diversion. The landowners' position was reiterated when representatives of the Sanitary District succeeded in having the matter come before the House Committee on Rivers and Harbors in 1924. By then, the Sanitary District was amenable to resolving the in-state issue. The Chicago area agency was burdened enough responding to opposition from the Great Lakes states and the Secretary of War in congressional hearings and in the federal court system. Rather than have the in-state conflict over claims resolved by a commission designated by Congress, the trustees of the Sanitary District petitioned (1925) the 55th General Assembly to designate an Illinois Valley Claims Commission to work within the Sanitary District of Chicago. Meanwhile, the Sanitary District began negotiating settlements in 1924. These were reported in 1928 to have approached about $1 million per year.[38]

## Retrospect

The owners and users of floodplain lands in the lower valley of the Illinois River claimed to have been induced to build and improve works of reclamation by water levels that were raised by dams and locks for navigation and subsequently were deepened and spread by water from Lake Michigan. The problem was exacerbated as levee systems enclosed increasing areas of wetlands during the first quarter of this century. The residual segments of unreclaimed floodplain suffered deeper and more prolonged seasons of flooding, and land drainage works faced more sustained seepage and more severe challenges from flood stages. Anticipation of the problem, it may be recalled, was reflected in the enabling legislation for the Sanitary District of Chicago, which was to remove the navigation structures at Copperas Creek and Henry and to pay for damages. However, the General Assembly may have been uneasy about the ultimate responsibility for costs incurred by the policies and works pursued by its creature, the controversial Sanitary District of Chicago. The General Assembly may have been hamstrung by suborned members; and it either did not consider or ignored the limitations to its authority that were raised in the federal court system subsequently. It may be, too, that the owners of bottomland along the Illinois River, like the leadership in the major river towns, were insufficiently attentive to the political realities of diverting water from Lake Michigan for a sanitary canal touted for its commercial promise and strategic importance. In any case, powerful congressmen and the War Department opposed the removal of locks and dams from the middle and lower Illinois River. Moreover, the officers of the Sanitary District chose not to undertake the promised remedial work on locks and dams, while strongly combating damage claims in the courts. The district's authorities were as reluctant to compromise on the matters as they seemed to be in responding to urgings that sewage treatment plants be constructed to mitigate the public health hazard created downstate.[39]

The Sanitary District became conciliatory in 1924, when it was evident that the legal challenge raised by Wisconsin and other Great Lakes states was about to reduce the Chicago agency to whatever options Congress might provide—and in Washington, the congressman from the district that embraced much of the lower Illinois Valley had priorities to be addressed. Without delving deeply into the complex politics of the matter, it is apparent that this aspect of the Chicago versus downstate confrontation had to be muted. Contributing to a change of heart in Springfield[40] was the difficult financial circumstance into which the drainage and levee districts and landowners fell as a result of the crash in commodity prices and the flood of the early 1920s. When major floods in the fall of 1926 and the spring of 1927 followed, the General Assembly had to recognize that its earlier generosity in facilitating Chicago's development caused a good deal of harm along the Des Plaines and Illinois rivers.

The commitment of funds by the General Assembly to restore levees at the drainage and levee districts, and its provision of matching funds with the Corps of Engineers to rebuild the levees, may have been compensation for the past. By 1926–27, too, the extent of flood damage throughout the Mississippi River basin resulted in congressional authorization of flood control measures in the lower reaches of tributaries, including the lower Illinois River. These policy changes in Springfield and Washington are traced as flood event responses in chapter 6. In the context of the events described here, the larger federal presence was consistent with the aspirations of proponents of an entwined sanitary and ship canal that, so it was hoped, would presage an improved waterway to New Orleans. Although the Corps of Engineers did not develop the 14-foot-deep waterway, its work in the 1930s included a second generation of locks and dams on the lower Illinois River and the adoption and completion of the Illinois Waterway in the upper valley. The federal involvements were broadly consistent with Congressman Rainey's objective in collaborating with Senator Francis G. Newlands and others who responded to the call for action by the National Drainage Congress after floods devastated the Midwest in 1913. Their bill of 1913 was abortive, but the enlargement of the role of the United States government in flood control work, in public power generation,[41] and in navigation improvement and extension through the inland waters eventually occurred.

# 6

# Days of Reckoning for Reclamation

WHILE flood events in the first two decades of the twentieth century added incentives for the owners of bottomland to reclaim their properties, floods in the 1920s intensified the problems created by the dismaying slump in prices for agricultural commodities that began in 1921. The general flood of 1922 staggered a number of drainage and levee districts; the floods of 1926 and 1927 devastated more. Many landowners were unable or unwilling to pay their share of the costs of rehabilitating levee and ditch systems destroyed in 1922, let alone pay for recovery after 1926 and 1927. Comparative data on the memorable flood stages are shown in tables 6.1 and 6.2.[1]

The flood events in the lower valley were related to weather anomalies that afflicted a large part of the continental interior. However, the land drainage process throughout the watershed of the Illinois River, together with the water delivered into the system through Chicago's Sanitary and Ship Canal, was a contributing factor. Nevertheless, the responses by local and state interests were akin to the broader national experience that led to federal involvement in flood control matters. By the same token, the collapse in commodity prices that stunned local farmers, landowners, and the drainage entities and their backers was experienced nationwide. The flood events and market disaster of the 1920s altered perspectives in many quarters on the merit of direct federal relief for the rural private sector and the institutions of land reclamation. The transforming local events and their regional and national contexts are discussed in this chapter.

## Foretaste of Trouble

The flood events of 1900 to 1919 occurred as land drainage progressed quickly. Some of the floods inundated riverside communities, and several drove farmers from unleveed and leveed bottomland, destroying crops and some livestock. These overflows were off the bluffs and from the tributaries of the Illinois River, as well as from the swollen main stream. In general, the damage resulting from rain and local runoff preceded the destruction by the river. Among the more severe early floods of the century were those of 1904 and 1913, each carrying maximum stages exceeded only by the great

**Table 6.1.** Maximum stages, in feet, of notable floods on the lower Illinois River, 1904–27

| Station | Station elevation (ft.)[a] | Distance from Grafton (mi.) | Maximum stages (ft.) 1904 | 1913 | 1916 | 1922 | 1926 | 1927 |
|---|---|---|---|---|---|---|---|---|
| Peoria | 428 | 160.6 | 23.0 | 22.3 | 23.1 | 24.8 | 25.0 | 24.7 |
| Havana | 425 | 119.9 | 19.9 | 19.9 | 19.5 | 22.5 | 23.1 | 22.0 |
| Beardstown | 421 | 88.8 | 20.0 | 21.8 | 20.6 | 25.1 | 26.2 | 25.2 |
| Pearl | 412 | 43.2 | 19.4 | 20.1 | 19.8 | 23.0 | 22.1 | 22.7 |
| Grafton | 404 | 0 | 19.9 | 20.5 | 23.4 | 25.8 | 23.7 | 25.7 |

[a]Approximate mean sea level.

**Table 6.2.** Maximum discharge and frequency of notable floods at Beardstown, 1904–27

| | 1904 | 1913 | 1916 | 1922 | 1926 | 1927 |
|---|---|---|---|---|---|---|
| Max. discharge (c.f.s.) | 115,000 | 92,000 | 77,800[a] | 93,100 | 105,000 | 89,900 |
| Frequency (yrs.) | 38 | 10 | . . . | 11 | 20 | 8 |

[a]Ice jam.

flood of 1844. Although not listed in table 6.1, the lesser floods of mid-July 1902 and January 1907 bear comment because of their effect upon the lower valley and its drainage and levee districts.

It may be recalled that there were several incentives to reclaim the bottoms in the early 1900s. The land's unusual productivity was confirmed during the droughty 1890s and very early 1900s as cultivation was extended from near the bluffs to the river in several areas. Investment in bottomland was made attractive by the rising commodity prices, by the proximity of the new farmland to the relatively well-developed and populous prairie uplands of the state, and by the accessibility of the bottoms to investors and markets for grain, forage crops, and livestock. Commitments having been made to farming the bottomland, investors were induced to protect the land from overflow by a series of years more rainy than normal (1900–1910) and by the diversion of water from Lake Michigan into the watershed (1900). Mean diversions averaged 4,220 c.f.s. in 1900–1904, increasing in successive five-year periods to 5,401, 7,264, 8,417 and 8,674 c.f.s. During the period 1925 to 1928, the mean flow was almost 8,756 c.f.s. During the river's low stages, the Chicago Sanitary District's discharge of 8,000 c.f.s. raised the water 1.5 to 4 feet above accustomed levels.[2] The elevated water levels attributed to the diversion, it was averred at the formative Scott County Drainage and Levee District, made it "impossible" to drain lands that the driest seasons bared theretofore.[3] The condition was not unique; the commissioners of the Kerton Valley Drainage and Levee District claimed that they had organized in response to "Chicago Canal water," as did the officers of the Otter Creek and Chautauqua districts. The commissioners of the Lacey district attributed to the same cause a need to raise the original levee system.[4]

In the first decade of the century, the most challenging flood events occurred in mid-1902, March–April 1904, and January 1907. The first two experiences were alike in that heavy local precipitation and runoff had much of the bottoms awash to the south of Peoria. However, the flood event of 1904 was remarkable for conveying into the lower valley the greatest and most rapidly moving rush of water known to have passed Peoria. The Mackinaw and Spoon rivers, too, had prolonged high flows. During both events, the Pekin and La Marsh district was flooded, and there was loss of the corn crop through ponding across low areas of the Coal Creek district in 1902.[5] The damage to both districts in 1902 is not surprising considering that the La Moine River flowed so heavily that a tug and barges from Beardstown were reported to have ascended it about a dozen miles to pick up wheat.[6] The wonder is that the Lacey (and Langellier) district was not inundated in 1902, as it was in 1904. In the latter flood event, which broke records for river stages at Peoria and Havana (table 6.1), flooding occurred at leveed tracts aggregating about 22,835 acres, 11,850 of them associated with organized districts (map 6.1) and the remainder with the five privately reclaimed properties that became the Hartwell district, in Greene County. Stored grain and fields of winter wheat were destroyed, and considerable damage was done to district and farm improvements. Spring planting was delayed greatly, too. Although the levees at the surviving Coal Creek district and at Beardstown, opposite, constricted the channel, the overflow reservoir function of nearly all of the bottoms was operable.[7]

Levee systems were again tested in January 1907, when the Sangamon and Spoon river bottoms, then the bottoms of the Illinois River, were swept by flows that broke into the Pekin and La Marsh district and demolished the community of Liverpool.[8] Flood-fighting crews managed to hold the levees at other districts, while the pumps worked day and night to keep up with the water that runoff from the bluffs and seepage admitted.

The flood of late March and early April 1913 swept into 6 of 19 completed districts (map 6.2) and into 3 districts being reclaimed. Although the Hartwell district's levee system was breached by the creek to its south, which nearly broke into the Fairbanks

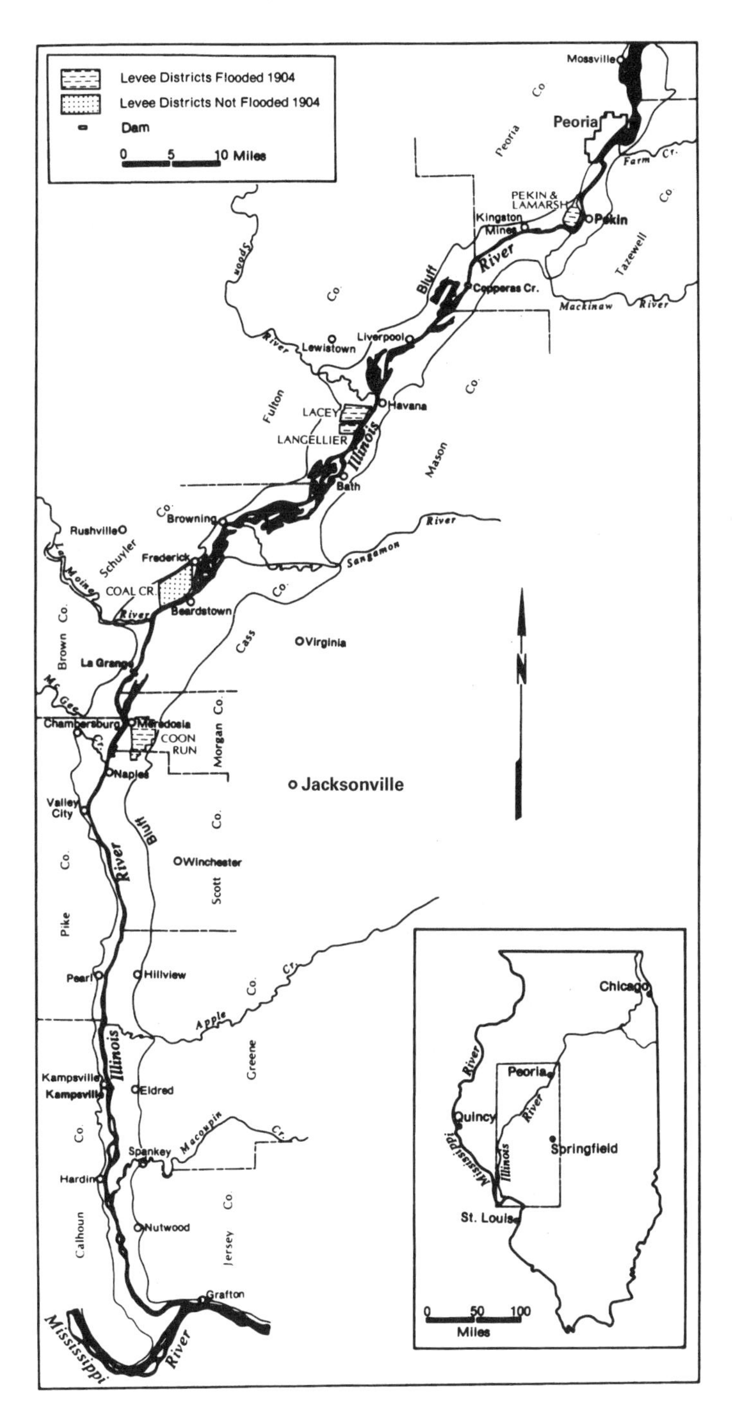


**Map 6.1.** Drainage and levee districts and the flood of 1904

(Keach) district, most of the damage was done by the river (map 6.2) at the Big Lake, Meredosia Lake, Willow Creek, Mauvaise Terre, and Scott County districts. Reclamation activity was interrupted at the Banner Special, South Beardstown, and Valley districts, which are not shown on map 6.2 because they were not completed. Since the Coal Creek and Coon Run districts were heavily ponded by precipitation and runoff, other districts with unbroken levees probably were, as well. The leveed communities of Browning and Naples were inundated, as were large areas of unleveed land and areas of privately leveed land. Because there was adequate forewarning, the residents of the bottoms had time to remove livestock and household effects to the bluffs, where tents were raised, or they took refuge in nearby churches and public halls.[9]

At the time of the 1913 flood, about 45 percent of the floodplain downstream of Peoria was leveed.[10] The breaking of river levees at 6 districts affected tracts having 28 percent of the 126,000 acres within fully developed districts. The degree to which the flooded districts were overwhelmed by direct runoff and rampaging flank creeks before the river entered is not documented. However, the circumstances at Coal Creek and Coon Run, and at the Willow Creek and Hartwell districts, suggest that the creeks and the bluff runoff inundated large areas. It probably was the same at the Scott County district, to judge from immediately subsequent land annexation and levee extension work toward the northeast.

The flood caused Governor Edward F. Dunne to call a meeting for March 14, 1914, to discuss the problems of the river with representatives of several state agencies, which meeting led to commissioning the study by Alvord and Burdick that was released in 1915.[11] This investigation revealed that at Peoria the crest discharge in 1913 was about 10 percent less than was the case in 1904 and that at Beardstown and below it was "substantially equal." However, the crest in 1913 exceeded the record of 1904, as shown in table 6.1. Although subsequent investigators noted that the Sangamon River and other tributaries delivered crests ahead of the crest of the Illinois River, a substantial share of the increase in water levels at and below Beardstown was attributed to the Sangamon River. The role of the Sangamon River is understandable; between 1903 and 1910, it was straightened and cleaned by dredges for a distance of 48 miles down to a point about 6 miles east of the confluence with the Illinois River.[12]

Alvord and Burdick described how, in the natural state, flood flows moved southward through the timber and brush of the lower third of the Illinois Valley as swaths of water averaging 3 miles across and 7 to 9 feet deep. Between 1904 and 1913, however, artificial levee systems below Beardstown had reduced the floodway to about 25 percent of the cross section available earlier. About 80 percent of the floodplain's reservoir area was sealed off by the levees. This modification, asserted the Corps of Engineers' investigators years later, was the reason why the 1913 flood stage between Beardstown and Valley City was 2.6 to 3.3 feet higher than it had been in 1904, although the volumes

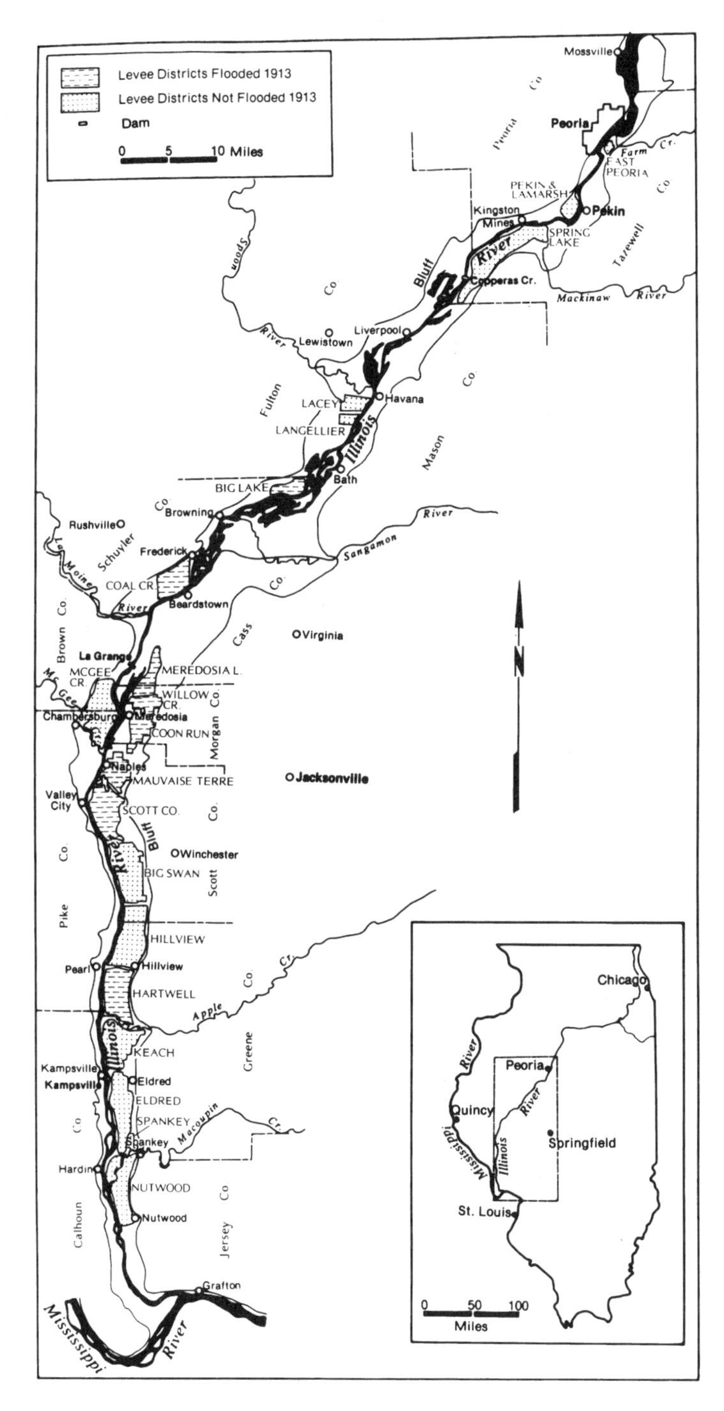


**Map 6.2.** Drainage and levee districts and the flood of 1913

of water were the same. To reduce the likelihood that levee systems would be overtopped in the future, Alvord and Burdick proposed that the floodway have at least 1,200 feet of cross section and that prospective districts be designed (and farming developed accordingly) to become emergency water-storage areas. The idea was to reduce the crests of great flows by 2.5 to 3.5 feet, depending upon the sector of the valley. The consultants' recommendation that levee systems be given a freeboard 3 feet above the flood level of 1844 would have required virtually all districts to add from 2 or 3 to 6 feet of height and commensurate cross section to their perimeters. Finally, the consulting engineers suggested that the state's Rivers and Lakes Commission be authorized to advise or regulate prospective levee building and modification by private interests. As it turned out, land reclamation proceeded apace with little heed to the advice of Alvord and Burdick, except in the Division of Waterways. Not until 1921 was the agency authorized to review plans for levee construction by the districts.[13]

Barely had the Alvord and Burdick prescription been published when, in January–February 1916, the river surged past Peoria, Havana, and Beardstown with crests comparable or higher than those of 1904 (table 6.1). Yet, the flow past Beardstown was calculated to have been only about 70 percent of the volume known in 1904. It was necessary to sandbag levees at the East Peoria and Spring Lake districts, and farmers took the precaution of removing stored grain from the Pekin and La Marsh district. Subsequently, flooding occurred at the Meredosia Lake, McGee Creek, Hartwell, and Eldred districts, and reclamation work was interrupted at the Banner Special and West Matanzas districts and at the privately reclaimed, short-lived Crabtree district. Already, runoff from the bluffs and such west-side affluents of the Illinois River as the Spoon River and Otter Creek had partially flooded a number of areas.[14]

The flood affected four functioning districts with 31,000 acres of improved land, a little less than in 1913. Engineers concluded that the difficulties resulted from defective levee design. Most of the levee systems were perfectly safe, concluded the Rivers and Lakes Commission. As a commissioner put it, "The reclamation of overflowed lands by levees and pumping has passed any experimental stage in the Illinois Valley."[15] The near loss at East Peoria was attributed to a defective levee; the Eldred district levee broke where persistent slumping had caused the contractor and the district to argue over culpability since construction; and the difficulty with excessive water inside the McGee Creek district was attributed to a chronically seeping river levee and successive levee overtoppings by the creek in 1915 and 1916. The collapse of the levee at the Hartwell district was induced by seepage along the outside of an abandoned and improperly sealed discharge line through the levee. Finally, loss of the Meredosia Lake district was considered to be a qualified victory for the river in 1916 because the defenses were not fully restored from the flood of 1913.

Enthusiasts believed in 1916 that levee systems would line both sides of the river between Peoria and Meredosia within a decade. The race to reclaim the bottoms was expected to result in higher and more substantial levee systems as developers vied to create the ultimate protection. The capability in engineering design was known; the newly proven long-boom clamshell dredge and the improving draglines made possible higher and broader levees; and the optimism of developers and investors was not dampened by worry about systemwide river considerations. Presumably, localized floods from unruly flank creeks and off the bluffs were to be endured, as happened at Crane Creek about June 6, 1917. The entire wheat and corn crops were lost, and a number of buildings were dislodged by water that exceeded a depth of 16 feet before an outlet was dynamited through the levee.[16]

## The Flood Event of 1922

Substantial progress was made in land drainage between 1916 and 1922, by which time virtually all of the drainage and levee districts downstream of Peoria had been organized. About 173,000 acres of improved land lay within completed levee systems, and nearly another 17,000 acres were being reclaimed. Additionally, over 6,000 acres lay in scattered, privately reclaimed properties. In the aggregate, improved land within levee systems represented about 64 percent of the floodplain, nearly 10 percent more than in 1913. The extent to which land was withdrawn from the floodplain was greatest downstream of Beardstown, where 95 percent of the bottomland was behind levees (map 6.3).[17]

The flood event of March and April 1922 surpassed all previous floods. As may be noted in table 6.1, the stage elevations marked in 1922 at Peoria, Havana, and Beardstown exceeded 1913 levels by 2.5 feet, 2.6 feet, and 3.3 feet, respectively. At Beardstown, the flow was calculated to have been 1,100 c.f.s. more than in the event of 1913 (93,100 c.f.s. vs. 92,000 c.f.s.). Although the 1922 flow past Beardstown carried but 81 percent of the cubic feet per second calculated for 1904, the high-water stage exceeded the 1904 high by over 5 feet. Thus, most of the havoc caused below Peoria was attributed to the loss of overflow area and floodway to the reclaimed districts. Also contributing to the flood was the mean 8,500 to 9,200 c.f.s. diverted into the drainage basin from Lake Michigan in March and April. While the water from the Sanitary District of Chicago accounted for about 14 percent of the total crest flow past Peoria and increased the flood height 18 inches there, the effect diminished at Beardstown to 7 percent and 6 inches. On the other hand, at Beardstown the water reportedly extended from bluff to bluff. Since the Sangamon River was over its banks from Springfield to the Illinois Valley, and the Spoon River flooded a good 10 miles of bottomland headward of its junction with the Illinois River, the assumption is that all the tributaries were booming.[18] The spectacle in the valley drew sightseers to the bluffs

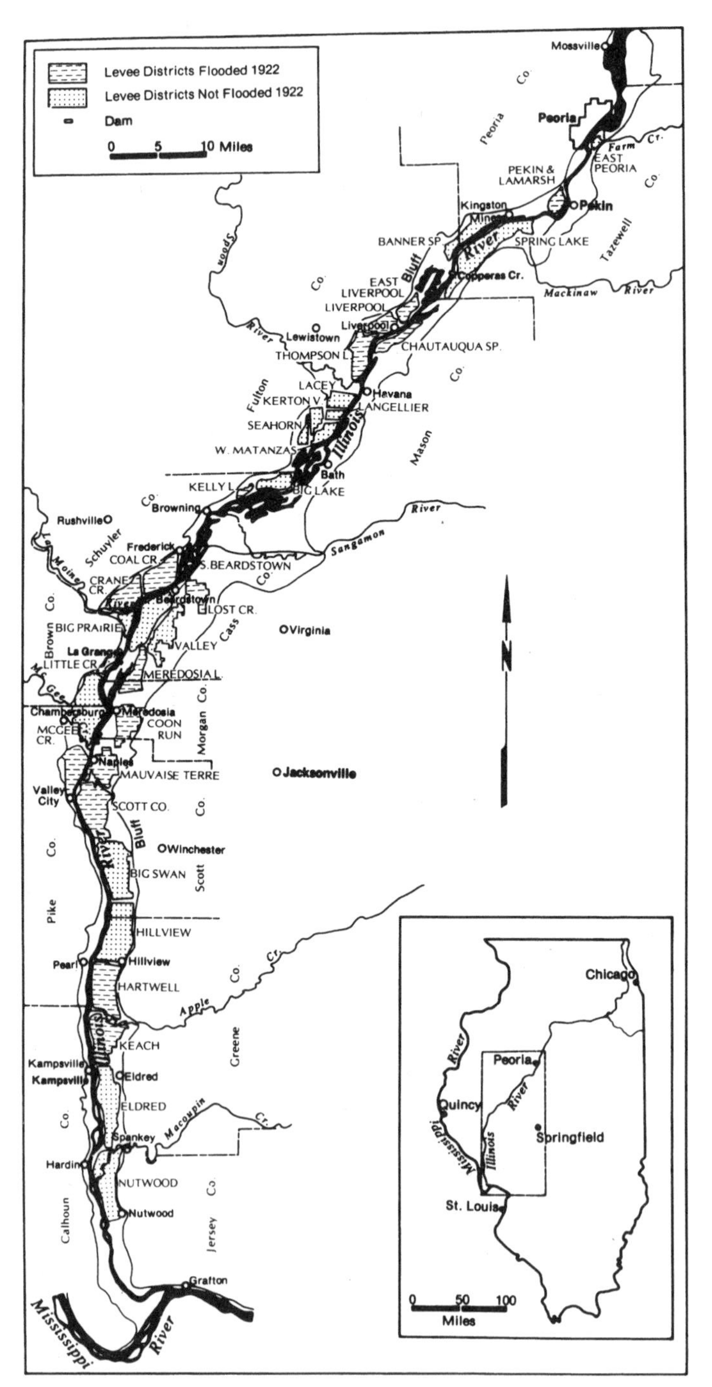


**Map 6.3.** Drainage and levee districts and the flood of 1922

**Table 6.3.** Estimated flood event losses, by class, lower Illinois valley, 1922, 1926, and 1927

| Class of loss | 1922 ($) | 1926 ($) | 1927 ($) |
|---|---|---|---|
| Farm losses in districts | 1,768,000 | 3,120,000 | 1,600,000 |
| Farm losses outside districts | 400,000 | 600,000 | 250,000 |
| Damage to levees | 848,000 | 1,560,000 | 800,000 |
| Urban damage | 1,300,000 | 1,500,000 | 800,000 |
| Industrial damage | 1,200,000 | 1,000,000 | 600,000 |
| Flood fight | 150,000 | 300,000 | 150,000 |
| Total | 5,666,000 | 8,080,000 | 4,200,000 |

from miles around. Considering road conditions, the convergence of 400 automobiles on the bluff grade into Frederick must have been a memorable spectacle, as well.[19]

The damages and flood fight costs to urban, industrial, and agricultural sectors were estimated conservatively at over $5.6 million (table 6.3) by the Corps of Engineers.[20] While no lives were lost, Beardstown and Naples were almost entirely flooded; Valley City, Frederick, and Browning were damaged badly; Meredosia was surrounded; and parts of Peoria, Pekin, and virtually every other riverside community were flooded. Almost all of the private reclamations and over half of the completed drainage and levee districts were broken into by the river or its tributaries.[21] Over the 20-day period of breachings, 17 districts were flooded; in approximate order, they included the Liverpool, East Liverpool, Chautauqua, Fairbanks (Keach) and Thompson Lake, Lost Creek, Valley City, Pekin and La Marsh, Kelly Lake, Coal Creek and Mauvaise Terre, Hartwell, Meredosia Lake and Scott County, Crane Creek, and Spankey districts. The Coon Run district flooded, as well. The levees at the Liverpool, Chautauqua, Thompson Lake, and Valley City districts were under construction at the time; at least another 6 districts suffered substantial crop loss or other damage from water accumulations within the levee systems. In sum, leveed districts representing just over 75,600 acres of improved land (44 percent of the total) suffered breachings; damage extended to land in districts representing another 23,000 acres and to private tracts. Most of the railroads and roads in the bottoms were awash.[22] The survival of the South Beardstown district may have been due to the deputizing of 40 members of the levee patrol, who carried arms because residents of Beardstown were apparently ready to cut the district's levee to ease the threat to the town.[23]

The estimated local expenditures on the flood fight and levee repairs and the losses

**Table 6.4.** Estimated flood costs, drainage and levee districts, March–April 1922

| Drainage and levee districts | Flood fight cost ($) | Levee break | Damages ($) | | | Total cost ($) |
|---|---|---|---|---|---|---|
| | | | Levee | Crops | Other | |
| Banner Sp. | 2,500 | | 18,686 | ... | 2,700 | 23,886 |
| Big L. | 3,500 | | 20,006 | ... | ... | 23,506 |
| Big Prairie | ... | | ... | ... | ... | ... |
| Big Swan | 5,000 | | 1,000 | 205,000 | 12,000 | 223,000 |
| Chautauqua[a] | ... | X | 100,000 | ... | ... | 100,000 |
| Coal Cr. | 4,000 | X | 125,000 | 209,000 | 135,000 | 473,000 |
| Coon Run | ... | X | ... | ... | ... | ... |
| Crane Cr. | 5,000 | X | 90,000 | 120,000 | 18,000 | 233,000 |
| E. Liverpool | 2,000 | X | 35,000 | ... | ... | 37,000 |
| E. Peoria | ... | | ... | ... | ... | ... |
| Eldred | ... | | ... | ... | ... | ... |
| Hartwell | ... | X | 200,000 | 300,000 | 50,000 | 550,000 |
| Hillview | 20,000 | | 15,000 | 75,000 | ... | 110,000 |
| Keach (F'banks) | ... | X | 150,000 | 250,000 | 50,000 | 450,000 |
| Kelly L. | ... | X | ... | ... | ... | 35,000 |
| Kerton V. | 400 | | ... | ... | ... | 400 |
| Lacey | 12,000 | | ... | ... | ... | 12,000 |
| Langellier | 15,000 | | 10,000 | ... | ... | 25,000 |
| Little Cr. | 2,000 | | 5,000 | 20,000 | 10,000 | 37,000 |
| Liverpool[a] | ... | X | ... | ... | ... | 7,000 |

*(continued)* **Table 6.4.** Estimated flood costs, drainage and levee districts, March–April 1922

| Drainage and levee districts | Flood fight cost ($) | Levee break | Damages ($) | | | Total cost ($) |
|---|---|---|---|---|---|---|
| | | | Levee | Crops | Other | |
| Lost Cr. | 5,000 | X | . . . | . . . | . . . | 45,000 |
| Mauvaise T. | . . . | X | . . . | 30,000 | 10,000 | 40,000 |
| McGee Cr. | 25,616 | | 20,000 | 125,000 | . . . | 170,616 |
| Meredosia L. | 10,000 | X | 35,000 | 62,300 | 15,000 | 122,300 |
| Nutwood | . . . | | . . . | . . . | . . . | . . . |
| Pekin & La Marsh | 5,000 | X | 115,000 | 70,000 | 10,000 | 200,000 |
| Scott Co. | 5,000 | X | 100,000 | 200,000 | 100,000 | 405,000 |
| Seahorn | 4,000 | | 2,500 | 5,000 | . . . | 11,500 |
| S. Beardstown | 1,000 | | 2,000 | 500 | 2,000 | 5,500 |
| Spankey | . . . | X | . . . | . . . | . . . | . . . |
| Spring L. | 16,000 | | 96,000 | . . . | . . . | 112,000 |
| Thompson L.[a] | . . . | X | . . . | . . . | . . . | 10,000 |
| Valley | . . . | | . . . | . . . | . . . | . . . |
| Valley City[a] | 7,650 | X | 9,500 | 5,000 | 15,000 | 37,150 |
| W. Matanzas | 10,000 | | 10,000 | . . . | . . . | 20,000 |
| Willow Cr. | . . . | ? | . . . | . . . | . . . | . . . |
| Private | 75 | X | 1,500 | 17,640 | 1,120 | 61,335 |
| Total | 160,741 | | 1,161,192 | 1,694,440 | 430,820 | 3,580,193 |

[a]Under construction.

in crops and improvements are shown in table 6.4.[24] Since damages were not reported for some districts, those losses have been estimated here at about $482,000, the assumption being that the losses per acre would have been equivalent to the losses reported for districts that did not have levee breaks. Whatever the damage, it is doubtful that the residents of many districts would have failed to prepare for the worst by moving as much as they could of livestock, implements, stored grain, and household effects to high ground. There was no saving the winter wheat, tame hay, or pasture, however; and farmers were prevented from planting their summer crops.[25]

Damage was so great, and the recrimination against the Chicago Sanitary District so strong, that the Division of Waterways undertook a study of the entire flood event. As indicated earlier, the study revealed that the enclosure of a large part of the floodplain with levee systems was a major causative factor. Such leveeing prevented or delayed the river from spreading across the bottoms. The encroachment of levee systems on the floodway was especially serious below Beardstown, where the distance between the earthen bulwarks was reduced to 1,130 feet at levee center lines and to 1,050 feet at levee toes. Investigators for the Corps of Engineers were appalled that "grasping or indifferent" landowners would have gone so far in reducing the floodway. Similar modification of floodplains tributary to the Illinois River affected flooding by accelerating water delivery into the main valley. A salient source of augmented flows arose in northern Indiana where extensive land drainage and the straightening and widening of the Kankakee River occurred. Channel rectification and land reclamation along the lower 40 miles of the Sangamon River between 1900 and 1910 also contributed to the seriousness of the flood. The accelerated descent of the Sangamon River into the Illinois Valley triggered heavy erosion, and large accumulations of sediment and woodland debris lodged in the lowest reaches of the tributary.[26]

To a substantial degree, the factors that contributed to the seriousness of the flood in the Illinois Valley were the product of the General Assembly's earlier responsiveness to sectoral and regional desires for economic and public health improvements. The tiling and ditching of the prairie uplands, the dredging of the bottoms in tributary streams, the reclamation of the floor of the Illinois Valley, and the diversion of water from Lake Michigan into the Illinois River system proceeded within frameworks shaped by the legislature. The General Assembly's responsiveness to local needs extended to making an appropriation in 1923 to reconstruct and improve the levees at Beardstown, a precedent having been set when Naples and the Ohio River towns of Shawneetown, Mound City, and Cairo were helped to recover from the flood of 1913. But the project designed for Beardstown by the Division of Waterways could not be implemented because the city failed to raise its share of funds. Although the State of Illinois made no financial commitment to assist the drainage and levee districts, future aid was foreshadowed. As Alvord and Burdick noted, the levees had to be made higher and more massive, the river's floodway had to be widened by setting back levee systems, and serious consideration had to be given to designing some districts to store excess flows in order to avert disasters. Alvord and Burdick believed that timber, brush, and islands should be removed from the floodway, and they suggested that reservoirs be built on tributary streams.[27] The ideas, except for reservoir development, were consistent with the policy of the Corps of Engineers.

Following the flood of 1922, a first Corps of Engineers flood control report on the Illinois River was authorized in the Rivers and Harbors Act approved September 22, 1922. Among the matters considered by the engineers was the suggestion of "a Congressman" that the work of flood control be done by the United States but with cost sharing by local interests on an equal basis. Congressman Rainey's idea of applying the cost-sharing formula newly adopted by Congress in the Flood Control Act of 1917 to address the needs along the lower Illinois River was endorsed by a large number of representatives of the drainage and levee districts. They undoubtedly recognized that the value of private investment in levee building to that time would count as their share of matching funds for prospective work. Notwithstanding local enthusiasm, there was no basis in the law to enable the Corps of Engineers to work beyond the course of the Mississippi River downstream of Rock Island. There was no benefit to navigation in the work proposed for the Illinois River. However, the Corps of Engineers conceded grudgingly that the flood control act approved by Congress on March 4, 1923, permitted work upstream to the lock and dam at La Grange. La Grange, located 77.5 miles above the confluence with the Mississippi River, then was the designated limit to which water backed from the Mississippi River at a projected peak flood stage—the criterion set by Congress for extending the jurisdiction of the Mississippi River Commission. At the same time, asserted the deferential chief investigator for the Corps of Engineers, the state's Division of Waterways had the authority and capability to undertake the work to La Grange and above. While the Corps of Engineers was prepared to offer advice to the state agency, it could not find justification under existing laws to fund work above La Grange.[28] Beyond that, there was no enthusiasm in the federal agency for any work outside navigation improvement. The perspective was shared by an important segment of Congress.

## The Floods of 1926 and 1927

New flood-stage records were achieved in September and October 1926, the high topping the crest of 1922 at Peoria by 0.2 feet, at Havana by 0.6 feet, and at Beardstown by 1.1 feet (table 6.1). Although the flow of the Sangamon River was unprecedented, the greatest volume of water carried by the Illinois River did not much exceed the record flows of 1913 and 1922. Thus, it was evident that the main causes of property destruction were the continued enclosure of Illinois Valley bottoms with levee systems and the encroachment of the levee systems upon the river's floodway. Encroachment was especially serious between Beardstown and Pearl, where only 1,100 to 2,000 feet separated the earthen bulwarks on either side of the river. Understandably, the river stages at Beardstown and vicinity had increased more than in other reaches of the river.[29]

In 1926, convergence upon the Illinois Valley of runoff from record-breaking general rains resulted in two particularly damaging flood periods, during the first eight days of September and the first nine days of October. At the time, the corn crop and much of the small grain crops were in the fields. Parts of Pekin were flooded in the first phase, and virtually all of Beardstown and Naples were engulfed for several weeks; East Peoria, Liverpool, Havana, Bath, Browning, Frederick, Chambersburg, Meredosia, and Valley City were damaged, also. The succession of early September floodings extended to the Pekin and La Marsh, Big Swan, McGee Creek, Valley City, Hillview, Hartwell, and Mauvaise Terre, and Scott County drainage and levee districts (map 6.4). Although the destruction of levees, crops, and livestock at the Mauvaise Terre and Scott County districts, and the flooding of Naples, could be attributed to the surge of water from a failed reservoir near Jacksonville, the same effects resulted from unprecedented runoff at the Big Swan, Hillview, and Hartwell districts. It may have been the case at nearly every district. In essence, the creeks broke the levees through which the rising river backed subsequently. The flooding of the McGee Creek district resulted from breachings in adjacent levee systems, as was the case in October for the Kerton Valley district, flooded from the West Matanzas district. The October inundations were, in order, at the Lost Creek, East Liverpool, Fairbanks (Keach), Banner Special, Kelly Lake, Lacey, Langellier, West Matanzas and Kerton Valley, Meredosia Lake, and Chautauqua districts. In all, 21 drainage and levee districts had levee breaks (table 6.5). As much as 15 feet of water lay on the fields. There was no loss of life because residents of the lowlands had ample time to retreat to high ground with their possessions. Some later returned to their fields in small boats to shuck out what corn could be found. Meanwhile, a plague of sightseers in an estimated 4,500 motorcars saturated the bluff grade road between Rushville and Frederick on one day.[30] Convergences of the curious probably occurred elsewhere.

At the time of the 1926 flood, there were about 196,000 acres of improved land in the drainage and levee districts below Peoria—about 63 percent of the 312,000 acres of floodplain and lakes. About 103,500 acres were in the districts where levees broke. Districts with another 48,000 acres sustained damage to drainage works, crops, and improvements; over 6,000 acres of private reclamations were damaged. It is doubtful that any of the organized districts or private drainage enterprises escaped injury. Assuming that the damage at unreported entities was in proportion to the damages

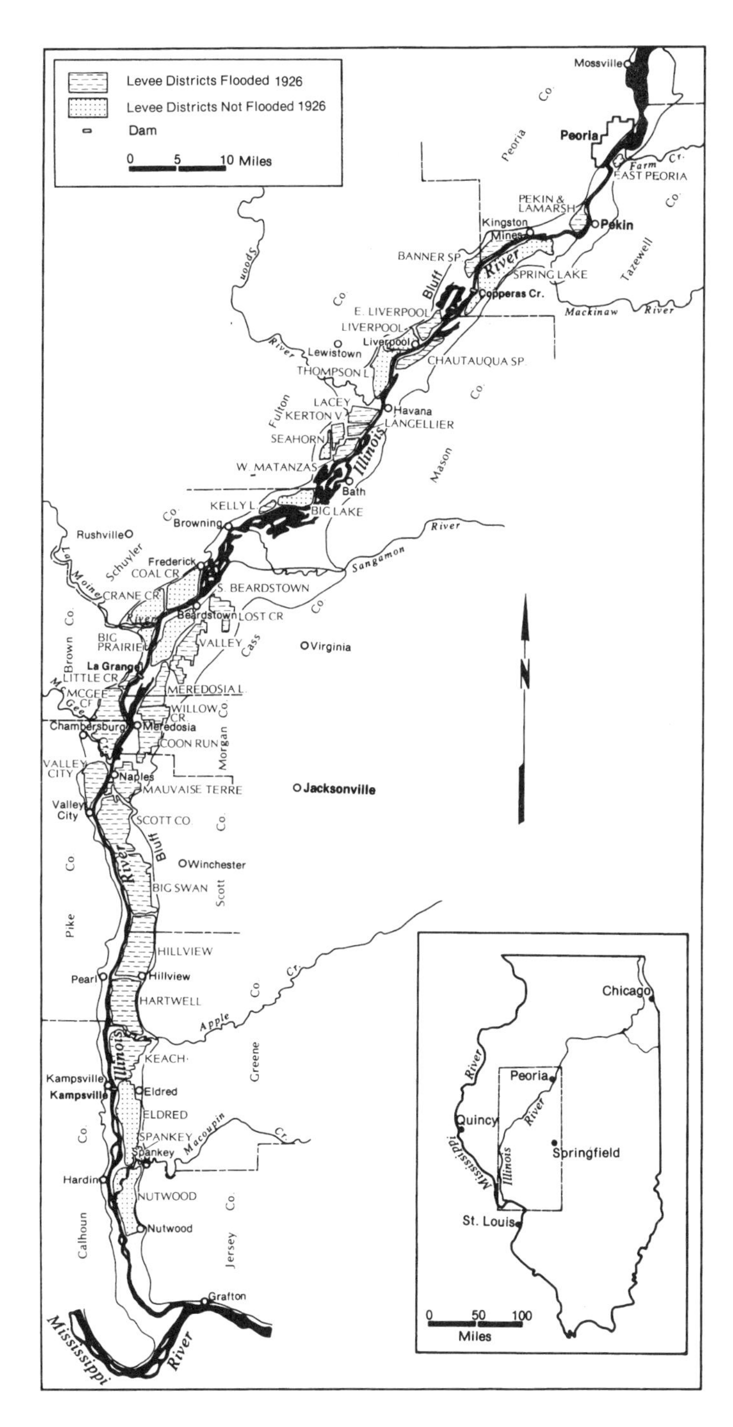


**Map 6.4.** Drainage and levee districts and the flood of 1926

**Table 6.5.** Estimated flood costs, drainage and levee districts, September–October 1926

| Drainage and levee districts | Flood fight cost ($) | Levee break | Damages ($) | | | Total cost ($) |
|---|---|---|---|---|---|---|
| | | | Levee | Crops | Other | |
| Banner Sp. | 5,000 | X | 74,000 | 99,000 | 20,000 | 198,000 |
| Big L. | 3,000 | | ... | ... | ... | 3,000 |
| Big Prairie | ... | | ... | ... | ... | ... |
| Big Swan[a] | 35,000 | X | 40,000 | 450,000 | 15,000 | 540,000 |
| Chautauqua | 2,500 | X | 125,000 | 75,000 | 50,000 | 252,500 |
| Coal Cr. | 13,000 | | 5,000 | 60,000 | ... | 78,000 |
| Coon Run | ... | X | 1,200 | 50,000 | 5,000 | 56,200 |
| Crane Cr. | 19,000 | | 16,000 | 60,000 | 6,000 | 101,000 |
| E. Liverpool[a] | 1,500 | X | 50,000 | 200,000 | 10,000 | 261,500 |
| E. Peoria | ... | | ... | ... | ... | ... |
| Eldred | ... | | ... | ... | ... | ... |
| Hartwell | 1,000 | X | ... | 160,000 | ... | 161,000 |
| Hillview | 60,000 | X | 40,000 | 210,000 | ... | 310,000 |
| Keach (F'banks) | 2,000 | X | 5,000 | 150,000 | 10,000 | 167,000 |
| Kelly L. | ... | X | ... | ... | ... | ... |
| Kerton V. | 500 | X | ... | 76,000 | 300 | 76,800 |
| Lacey | 6,000 | X | 130,000 | 60,000 | 60,000 | 256,000 |
| Langellier | 1,000 | X | 45,000 | 65,000 | 30,000 | 141,000 |
| Little Cr. | 1,000 | | 7,000 | 25,000 | 15,000 | 48,000 |
| Liverpool[a] | 17,163 | | 10,142 | 20,000 | 50,000 | 97,305 |

*(continued)* **Table 6.5.** Estimated flood costs, drainage and levee districts, September–October 1926

| Drainage and levee districts | Flood fight cost ($) | Levee break | Damages ($) | | | Total cost ($) |
|---|---|---|---|---|---|---|
| | | | Levee | Crops | Other | |
| Lost Cr. | ... | X | ... | ... | ... | ... |
| Mauvaise T. | 2,000 | X | 10,000 | 25,000 | 10,000 | 47,000 |
| McGee Cr. | 6,000 | X | 28,000 | 351,510 | 47,000 | 432,510 |
| Meredosia L. | 10,000 | X | 30,000 | 75,000 | 10,000 | 125,000 |
| Nutwood | ... | | ... | ... | ... | ... |
| Pekin & La Marsh[a] | 4,000 | X | 50,000 | 100,000 | 6,000 | 160,000 |
| Scott Co. | 6,000 | X | 25,000 | 225,000 | 100,000 | 356,000 |
| Seahorn | 6,000 | | 5,000 | 15,000 | 50,000 | 76,000 |
| S. Beardstown | 2,000 | | 1,000 | 1,000 | 2,000 | 6,000 |
| Spankey | ... | X | ... | ... | ... | ... |
| Spring L. | 12,000 | | 25,000 | ... | ... | 37,000 |
| Thompson L.[a] | ... | | ... | ... | ... | ... |
| Valley | ... | | ... | ... | ... | ... |
| Valley City | 2,500 | X | 1,000 | 1,000 | 2,000 | 6,500 |
| W. Matanzas | 1,500 | X | 50,000 | 100,000 | 30,000 | 181,500 |
| Willow Cr. | ... | ? | ... | ... | ... | ... |
| Private | 800 | | 12,000 | 30,000 | ... | 42,800 |
| Total | 220,463 | | 785,342 | 2,683,510 | 528,300 | 4,217,615 |

[a]Combines 1926 and 1927 data.

at the tracts where levees were not breached and for which there is information, the total exceeded $4.75 million, rather than the $4.38 million shown in table 6.5. Adding $740,000 damages at Beardstown and $1,760,000 for other urban and industrial damage along the river, it is reasonable to estimate that more than $7.25 million in losses and expenditures were incurred in the study area. The estimate by the Corps of Engineers, shown in table 6.3, put the total at over $8 million.[31]

The flooding of most districts and Beardstown began as levee overtoppings, suggesting that the engineering design, if not the height, of the structures was satisfactory. It is unclear, however, how many levee systems that failed had not yet been fully restored from the 1922 disaster because of financial difficulties in the districts. The overtoppings began across diversion and flank levees in 9 cases and across the river levees in 12. The archival record is fairly clear that bluff ditches and creek channels had "a perceptible filling [of sediment] of considerable magnitude" soon after entering the Illinois Valley. Such evidence of neglected maintenance is consistent with the suspicion of contemporaries that rodent runs caused weakening where 4 levees failed. Nevertheless, the professional view was that improper maintenance had not contributed in any great degree to levee failures.[32]

It was difficult for the districts to undertake repairs of levees and pump houses during the winter of 1926–27 because they were in poor financial condition and because the river remained bank-full, or nearly so, for much of December through March. The exceptionally rainy winter was extended by an April during which precipitation was about 90 percent above normal. The persistently high water, together with recurring strong winds, caused extensive wave damage to the saturated levee systems, notably in the spring of 1927. The culminating blow was the crest that moved through the lower Illinois River in late April with stages that were exceeded only in 1926 and 1922, as may be noted in table 6.1.[33]

The flood event of April 1927 inundated the city of East Peoria and three districts that had restored levee systems that were broken in the preceding fall—the Pekin and La Marsh, Kelly Lake, and Meredosia Lake districts. The East Peoria district was flooded for the first time by way of the flank levee on Farm Creek, and the private Crabtree district was flooded by the Spoon and Illinois rivers. The East Liverpool, Kerton Valley, and Little Creek districts suffered levee breachings during the repair efforts begun earlier in the winter. Yet another three districts managed to hold repaired levee systems in the spring. Otherwise, all of the districts that flooded in the fall remained awash—most of them until July 1927 or so. The extent of rural damage is shown in table 6.6.[34] Assuming that the districts not of record incurred damages and flood fight costs equivalent to the $10.06 average per acre costs for districts that did not have levee breaks, about $305,210 should be added to the total attributed to the flood of 1927. The $2.55 million result is reasonably close to the estimate in table

6.3. When urban, industrial, and other costs are added, a total of $4.2 million in damages may be tallied against the flood.

The exceptionally wet winter and spring of 1927 were experienced widely over the subcontinental span of the Mississippi River's watershed. By late April, some 26,000 square miles of floodplain were inundated downstream of Cairo, resulting in the displacement of 700,000 people, almost half of whom had to be rescued. Direct and indirect losses exceeded $436 million—over one hundred times the loss in the Illinois Valley. The experience, given public opinion and the sustained message of proponents of a larger role in flood control matters for the United States government, resulted in passage of the historic Flood Control Act of 1928, which provided for a comprehensive remedial program in the lower Mississippi Valley, as it did for California's Sacramento Valley, the problems of which are related later.[35]

## State and Federal Flood Control Assistance

The reverses of 1926 and 1927 so deepened the financial distress of the districts and the landowners that the General Assembly enacted a $1.5 million flood relief bill in July 1927. This aid was to repair and restore levees at districts along the Mississippi and Ohio rivers, as well as along the Illinois River. The restoration of flooded farmland and buildings remained the responsibility of the owners.[36]

Projects designed to restore the levees to the height and cross section of 1926 largely were completed along the Illinois River by the end of 1929. In the process, over 1.05 million cubic yards of spoil were emplaced, 82 percent of it by dredges, 13 percent by draglines, and the remainder by men with teams. Meanwhile, in mid-1928, the state completed the concrete seawall and enlarged levee system at Beardstown, for which $350,000 were reapportioned in 1927 when the city and its two railroads agreed to pay $51,000 for related urban drainage facilities and raised embankments.[37]

In addition to funding the protection for Beardstown and the levee restoration work for the districts in central and southern Illinois, the General Assembly appropriated $50,000 in 1927 for an engineering study of the flood events and a flood prevention plan. The plan was to be formulated in cooperation with agencies of adjacent states and the United States. The resulting *Flood Control Report* by Jacob A. Harman et al. was endorsed and published in 1930 by the Illinois Division of Waterways. Among its recommendations were the setting back of levees, the selection of reclaimed tracts for flood-crest storage and for game and fish preserves, and the designation of an agency, such as the Division of Waterways, to control private construction in floodways, as well as to represent the state in collaborative work with federal agencies and the districts. Meanwhile, Congress passed the Rivers and Harbors Act of 1927 and the Flood Control Act of 1928, which together with an act approved in 1922 authorized the Secretary of

**Table 6.6.** Estimated flood costs, drainage and levee districts, April 1927

| Drainage and levee districts | Flood fight cost ($) | Levee open/break (O/X) | Damages ($) | | | Total cost ($) |
|---|---|---|---|---|---|---|
| | | | Levee | Crops | Other | |
| Banner Sp. | 2,000 | 0 | 30,000 | 109,000 | 12,200 | 153,200 |
| Big L. | 2,500 | | ... | ... | ... | 2,500 |
| Big Prairie | ... | | ... | ... | ... | ... |
| Big Swan[a] | ... | 0 | ... | ... | ... | ... |
| Chautauqua | ... | 0 | 20,000 | ... | ... | 20,000 |
| Coal Cr. | 12,000 | | 5,000 | 160,000 | ... | 177,000 |
| Coon Run | ... | | ... | 25,000 | ... | 25,000 |
| Crane Cr. | 4,000 | | 12,000 | 60,000 | 6,000 | 82,000 |
| E. Liverpool[a] | ... | 0 | ... | ... | ... | ... |
| E. Peoria | ... | X | ... | ... | ... | ... |
| Eldred | ... | | ... | 40,000 | ... | 40,000 |
| Hartwell | ... | 0 | ... | 50,000 | ... | 50,000 |
| Hillview | 20,250 | 0 | 5,000 | 135,000 | ... | 160,250 |
| Keach (F'banks) | ... | | ... | ... | ... | ... |
| Kelly L. | ... | X | ... | ... | ... | ... |
| Kerton V. | 1,200 | 0 | 6,000 | 85,000 | 18,000 | 110,200 |
| Lacey | ... | 0 | ... | 89,000 | ... | 89,000 |
| Langellier | ... | 0 | 5,000 | 65,000 | ... | 70,000 |
| Little Cr. | ... | X | 1,000 | 30,000 | 5,000 | 36,000 |
| Liverpool[a] | ... | | ... | ... | ... | ... |

(*continued*) **Table 6.6.** Estimated flood costs, drainage and levee districts, April 1927

| Drainage and levee districts | Flood fight cost ($) | Levee open/break (O/X) | Damages ($) | | | Total cost ($) |
|---|---|---|---|---|---|---|
| | | | Levee | Crops | Other | |
| Lost Cr. | . . . | | . . . | . . . | . . . | . . . |
| Mauvaise T. | 2,500 | 0 | 9,000 | 15,000 | 15,000 | 41,500 |
| McGee Cr. | 800 | 0 | 27,000 | 400,000 | 100,000 | 527,800 |
| Meredosia L. | 8,000 | X | 10,000 | 120,000 | 15,000 | 153,000 |
| Nutwood | . . . | | . . . | . . . | . . . | . . . |
| Pekin & La Marsh[a] | . . . | X | . . . | . . . | . . . | . . . |
| Scott Co. | . . . | 0 | . . . | 225,000 | . . . | 225,000 |
| Seahorn | 3,000 | | 1,000 | 10,000 | 36,200 | 50,200 |
| S. Beardstown | 500 | | 1,000 | 2,000 | 2,000 | 5,500 |
| Spankey | . . . | 0 | . . . | . . . | . . . | . . . |
| Spring L. | 14,000 | | 25,000 | . . . | . . . | 39,000 |
| Thompson L.[a] | . . . | | . . . | . . . | . . . | . . . |
| Valley | . . . | | . . . | . . . | . . . | . . . |
| Valley City | 5,250 | | 15,000 | 50,000 | 10,000 | 80,250 |
| W. Matanzas | 500 | 0 | 10,000 | 100,000 | . . . | 110,500 |
| Willow Cr. | . . . | ? | . . . | . . . | . . . | . . . |
| Private | . . . | | . . . | . . . | . . . | . . . |
| Total | 76,500 | | 182,000 | 1,770,000 | 219,400 | 2,247,900 |

[a]Combined with 1926 data on table 6.5.

War to order flood control studies and work on tributaries of the Mississippi River, including the Illinois River. Already, the Mississippi River Commission had begun to consider the problems of those reaches of the tributaries that were affected by backwater from flood stages on the Mississippi River. The jurisdiction of the Corps of Engineers, the commission's successor (1928) in civil works planning and development, was extended to Beardstown and then to Havana. This inclusion of the lower Illinois river was solicited by the governor and the General Assembly and evidently was orchestrated by congressmen from Illinois. The achievement came about as part of a general response to very serious flood events throughout the basin of the Mississippi River and elsewhere. The turning point in moving Congress and the Corps of Engineers to accept the idea that control of floods required federal involvement occurred in the wake of the great flood events of 1912 and 1913, which coursed through the watershed of the Ohio River and across large areas of the Mississippi River floodplain. In 1917, Congress passed the first federal flood control act. The act extended the Mississippi River Commission's accustomed responsibility in navigation matters to flood control concerns on the Father of Waters. Also, the Flood Control Act of 1917 acknowledged federal responsibility to plan and develop a systemwide flood control program in the Sacramento Valley. Federal involvement with flood control in the Golden State was the outgrowth of experience gained by the Corps of Engineers and state authorities following the creation of the federal California Debris Commission (1893) to study and mitigate the enormously disruptive flow of flood-borne debris from hydraulic gold mines in the Sierra Nevada. The act of 1917 established that the Corps of Engineers' task of navigation improvement was but a facet of comprehensive watershed management; before long, the agency assessed streams in terms of flood control, drainage and irrigation, and hydroelectric power generation.[38]

A multifaceted survey of the Illinois River was solicited from the War Department by Congress in the Rivers and Harbors Act of January 21, 1927, and under provisions of the act of May 15, 1928. The opportunity for such a survey arose in the wake of the adoption of the Water Power Act of 1920, when Congress recognized that the development of hydroelectric power in the nation's watersheds had to be coordinated with navigation and flood control considerations. The Corps of Engineers' survey of the Illinois River was contemporaneous with the flood control investigations conducted by Jacob A. Harman and associates for the State of Illinois. However, only the Corps of Engineers assessed matters of navigation, power development, and irrigation. While the last two concerns were on the long-time agenda of Congressman Rainey, assessments of public interest in these areas were common to all federal river surveys. Power development had occurred on the river above Utica, and large farms in the lower valley were experimenting with irrigated rice production in the late 1920s. Nevertheless, the investigators and chain of command in the Corps of Engineers recommended against federal participation in power development; they dismissed irrigation as a nonexistent matter.[39]

The survey by the Corps of Engineers found that flood control work could be justified under existing laws downstream of Beardstown, and that work upstream was more a political than a technical problem. The proposed work, expected to reduce flood crests 1 or 2 feet, included raising most levees in situ and creating additional floodway by setting back almost 31 miles of levee at 11 drainage and levee districts. The adopted standard for enlarged and new levees included an 8-foot crown and 1:3 slope for both levee faces; levee crowns were 3 feet above the maximum flood level. The Corps provided yardage and cost estimates for necessary work above Beardstown, as well. About 45 million cubic yards were expected to be emplaced in levee systems then aggregating over 27 million cubic yards. The proposed flood control work did not impinge upon the 9-foot navigation project then in progress between Grafton and Utica, completion of which was recommended. Additionally, channel dredging was proposed for navigation pools in the Illinois State Waterway above Utica.[40]

Congress required that federal funds to raise, strengthen, and relocate levees be matched by local donation of rights-of-way, plus one-third of project costs and assumption of responsibility for maintenance of the structures. This concept of cost sharing between landowners, a state, and the United States was established in the Flood Control Act of 1917. While the landowners, as represented by the districts on the bottoms of the Illinois River, were unable to contribute new money for levee restoration, they had made substantial contributions theretofore. The General Assembly appropriated $1 million in matching funds in 1929. By mid-1932, over $918,100 were committed through the State Division of Waterways to contracts and purchases of rights of way to enlarge levees along the Illinois River. Most of the funds were used at districts downstream of Beardstown. About 11 percent of the state's monies were committed to work at districts located upstream of Beardstown, pending Congressional authorization (1936) of work above there.[41]

The better part of the cooperative program of levee enlargement, relocation, and construction was authorized by Congress in the flood control acts of 1928, 1936, and 1938. The work was undertaken in the 1930s and 1940s by the Corps of Engineers. In the Illinois Valley, as elsewhere, the work focused on building systems of massive levees between which to confine flood stages. Over $1.5 million in federal funds were expended at the 12 districts near and below Beardstown, where work was completed in the 1930s. Nearly $3.6 million were expended on projects at 16 districts between Beardstown and Peoria in the 1940s. Another $1.9 million in federal work was done at the Coal Creek district by 1950, and the defenses of the City of Beardstown received $2.3 million of federal assistance. This heavy commitment to the protection of Beardstown was complemented by setting back and enlarging major sectors of levee at the South Beardstown and Coal Creek districts. The need to extend the flood control work was reinforced by the 1943 flood, but only the work arising from flood events in the 1920s is within the purview of this account. Clearly, the landowners were in no position to match with new money the federal commitment of $7 million in the 1930s and 1940s to achieve the scale of flood defense prescribed to protect urban centers upstream to Peoria, as well as to protect the sparsely peopled districts. The required project levees had to have 8-foot crowns at least 4 feet above the high stage of the river in 1926 and slopes of 3:1. Meanwhile, starting in 1935, federal aid to the districts included brush and tree removal from levees and seasonal waterways and the cleaning of drainage ditches by crews from camps of the Civilian Conservation Corps located at Havana and Eldred. Whenever possible, the districts furnished draglines and construction materials.[42]

## Market Disaster and Early Aftermath

The pursuit of drainage and farming on the floodplain of the lower Illinois Valley was related primarily to high prices for corn, which bottomlands yielded heavily. Beyond use on the farms, the market largely was comprised of livestock feeders in the valley and adjacent upland areas. Wheat and other crops were profitable, as well. The returns to farmers and landowners were so attractive, even with paying for drainage works and their maintenance and operation, that the land sold readily for $75 to $175, and for as much as $200 to $300, per acre in the period 1910 to 1920. In general, drainage bonds sold well and promptly, which was the case for equivalent issues bearing 6 percent interest that were launched in Missouri and elsewhere. Also, farmers seem to have had little difficulty borrowing from banks and insurance companies to make improvements and to underwrite crops.[43] The use of credit expanded sharply in the buoyant period of rapid price increases for farm commodities that marked World War I. Actually, the sensational price rise for corn that occurred during the war years was preceded by a steady increase since 1898, as shown in table 1.2.[44]

Farmers in Illinois were staggered by the decline in corn and other commodity prices that occurred in the early 1920s. Corn, which sold for $1.49 per bushel in 1919 and $1.35 in 1920, plummeted to $.49 per bushel in 1921. As little as $.23 per bushel was paid at country elevators in the study area. Just as corn prices rose at a faster rate than commodity prices generally, the decline in 1920 and 1921 was sharper than for prices generally.[45] However, there were no corresponding reductions in the interest charges on money borrowed by the districts or by the farmers. Moreover, property taxes and operating costs for farms and drainage districts remained at a high level. Relative to farms on the prairie uplands, farms on the bottoms had heavy taxes.

The assessments on drained bottomland seemed especially onerous between 1922 and 1930 because the general flood events of 1922, 1926, and 1927 were interspersed with a series of rainy years during which many districts reduced costly pumping activity. The spokesman for the drainage and levee districts who claimed before a committee of Congress that he had produced only one good crop between 1920 and 1930 probably was not alone among farmers in the bottoms. For that matter, the farmers of floodplain lands elsewhere in the Midwest, such as the developers of farmland out of cutover timberlands on the floodplain of the Mississippi River in Missouri and southward, were having the same problem.[46] Understandably, the local counties of Cass, Greene, Schuyler, and Scott led the state in acreage of failed crops and idled land reported in the census of 1920; they reported from 11.9 to 15.3 percent of all farm acreage out of production. By early 1930, a quarter of all landowners in the drainage and levee districts were delinquent in paying their assessments, and about a quarter of these had faced foreclosure. The Federal Land Bank of St. Louis, like local banks, had stopped lending money on farms in the districts. Local banks were said to be pleading for debt repayment in 1930, and insurance companies and large banks were foreclosing. Among the larger institutions that began to take over farm operations were the Continental Illinois Bank and Trust, the Federal Land Bank of St. Louis, and the Aetna and Springfield life insurance companies, which added to the anxieties of solvent local farmers and investors in land. They perceived that a competitive advantage would be gained by the better capitalized enterprises, and that production by corporate farms would hurt prices in a market already grown soft.[47]

Insolvency among the drainage and levee districts of the lower Illinois Valley was no worse than elsewhere in the state and in the nation where large-scale reclamation had been undertaken; the problem affected irrigation districts as well. State and national drainage associations, elected representatives, and financial institutions tried to formulate programs to refinance bonded indebtedness. Progress in the matter culminated in Washington in 1933 and 1934, when the Reconstruction Finance Corporation received appropriations from Congress to begin the recovery process for qualifying districts.[48] By then, landowners in 15 districts on the Illinois River again were meeting their assessment charges for debt retirement, but at 16 districts delinquencies were still a problem. Most of the financially distressed entities lay adjacent Beardstown and for 20 miles downstream. Shortfalls aggregated between 5 and 49 percent of all assessments due in 13 districts. While 5 of the entities reported flood event damages and costs totaling $400,000 to $1,331,000 in 1922, 1926, and 1927 (tables 6.4, 6.5, 6.6), most of the lower valley's financially troubled districts reported modest or no flood-related losses in the decade. Perhaps the financial deficiencies were related more to a loss of confidence in the future of farming where drainage costs were high and where freedom from ponding and flooding so uncertain.[49]

In view of the financial circumstances of landowners and of some districts in Illinois by the late 1920s, and the reported low morale of many individuals, there was consideration in some quarters of purchase by the State of Illinois of floodplain areas so that the tracts could be returned to hunting and fishing preserve functions. As such, the wetlands areas would increase flood-water storage and mitigate crest levels.[50] However, as chronicled in chapter 8, nothing happened for years. Wildlife conservationists who questioned the wisdom of committing state and federal funds to levee restoration and enlargement at the drainage and levee districts[51] seem to have been less persuasive with members of the General Assembly and Congress than were the advocates of public works projects to preserve the farming that private capital developed on the bottoms.

As it turned out, the reversion of leveed and ditched bottomland to its natural functions proceeded more rapidly through private arrangements than through public agencies. Areas of reclaimed floodplain that reverted were located in the lower reaches of the Sangamon and Spoon rivers, where the incidence of high water was related to the regimes of the Illinois River and the tributaries. Illustrative were the Lynchburg and Sangamon Bottom Drainage and Levee District, which occupied some 2,000 acres just above the stream's confluence with the Illinois River, and the Crabtree "district," a privately reclaimed 1,200-acre piece of bottoms located between the converging Spoon and Illinois rivers. The Lynchburg and Sangamon Bottom district, organized in 1901, adopted a pumping capability in 1911 because of the recurring crop losses attributed to the discharge of the Sanitary District of Chicago. Levee breaks in 1913, 1922, and 1926 were followed by the dissolution of the district in 1928 and its transformation into a private hunting preserve. An investment of about $143,000 in drainage works was given up as the heavily alluviating and shifting Sangamon River recaptured the leveed land. The rising channel floor, to judge from a level profile surveyed in 1939, was 10 feet higher than the levee crown known in 1910 at mile six above the junction with the Illinois River; at mile fourteen, the channel floor was 6 feet higher than the levee completed in 1908; and there was appreciable alluviation for miles upstream. For similar reasons, the Crabtree area became a hunting preserve in the wake of the flood of 1922. The investor in drainage was unable to pay $36,500 in repairs to restore a $91,000 investment in drainage works made after 1914.[52] Other privately held, small, ditched and leveed floodplain segments probably became private hunting and fishing preserves, which recreational function has predominated wherever the floodplain was not cleared and planted.

Between the collapse of commodity prices and the stunning experience with the flood of 1922, new capital could not be raised to reclaim segments of the residual bottoms. The devastating floods of 1926 and 1927 reinforced investor disinterest in new reclamations. The approximately 102,000 acres of floodplain that were unleveed in the 1920s were appreciating in value, however, for use as private hunting and fishing grounds. The lands were worth at least $15 to $35 per acre in the 1930s, by which time they were held largely by individuals or clubs. Between the acquisition of land by these private interests, the preservation of private landownership in the drainage and levee districts, and the priorities of those who could engage the public sector, the opportunity to rededicate former public lands to public use was missed. Only 6,805 acres of bottomland were held in federal or state hands in 1939.[53]

7

# Agricultural Activity and Settlement in the 1920s

In the course of the past century and a half, there were three phases to the modification of woodland and prairie and of the relationships between water and land in the lower Illinois Valley. Pioneer Anglo-American settlers began to replace the pristine cover with cropland and pasture in the late 1820s and the 1830s, especially near the bluffs. Within several decades, the evolving commerce between St. Louis, Peoria, the Illinois and Michigan Canal, and Chicago required that the shoaled river be modified with dams and locks. The resulting navigation pools impinged upon adjacent floodplain and cover, while improving market access for the wheat and cordwood harvests of the vicinity. The encroachment by water grew after 1900 as a result of diversions from Lake Michigan and water released by increasingly effective land drainage in the uplands and bottoms of the Illinois River watershed. This phase of water encroachment occurred in the wake of a decade or more of expanding cultivation on the higher floodplain areas that remained free of overflow water at planting time.

Already by 1900, the third and most far-reaching phase in the modification of land and water relations and in changes to natural vegetation had begun on large tracts of perennially and seasonally wet floodplain. The physical appearance and functions of the floodplain were greatly altered by three steps of the land draining process. Watercourses from the eroding bluffs and uplands were diverted and rectified; large tracts were enclosed with dredged earthen ramparts and external moats; and elaborate ditch systems were excavated across the former wetlands in order to guide ground water and runoff to pump houses at riverside. Within the drainage and levee districts, the reduction and elimination of the natural cover and of myriad ponds, lakes, and sloughs occurred. Meanwhile, land developers and farmers extended the orderly geometry of fields wherein swards of winter wheat and mantles of summer's corn and lesser crops soon cloaked recently bared soil. The stamp of settlement on the landscape was marked further by farm and township roads and farmsteads. Stages in the processes of replacing the natural cover with crops and with the infrastructure of reclaimed land and of raising a navigation pool are discernible by comparing maps

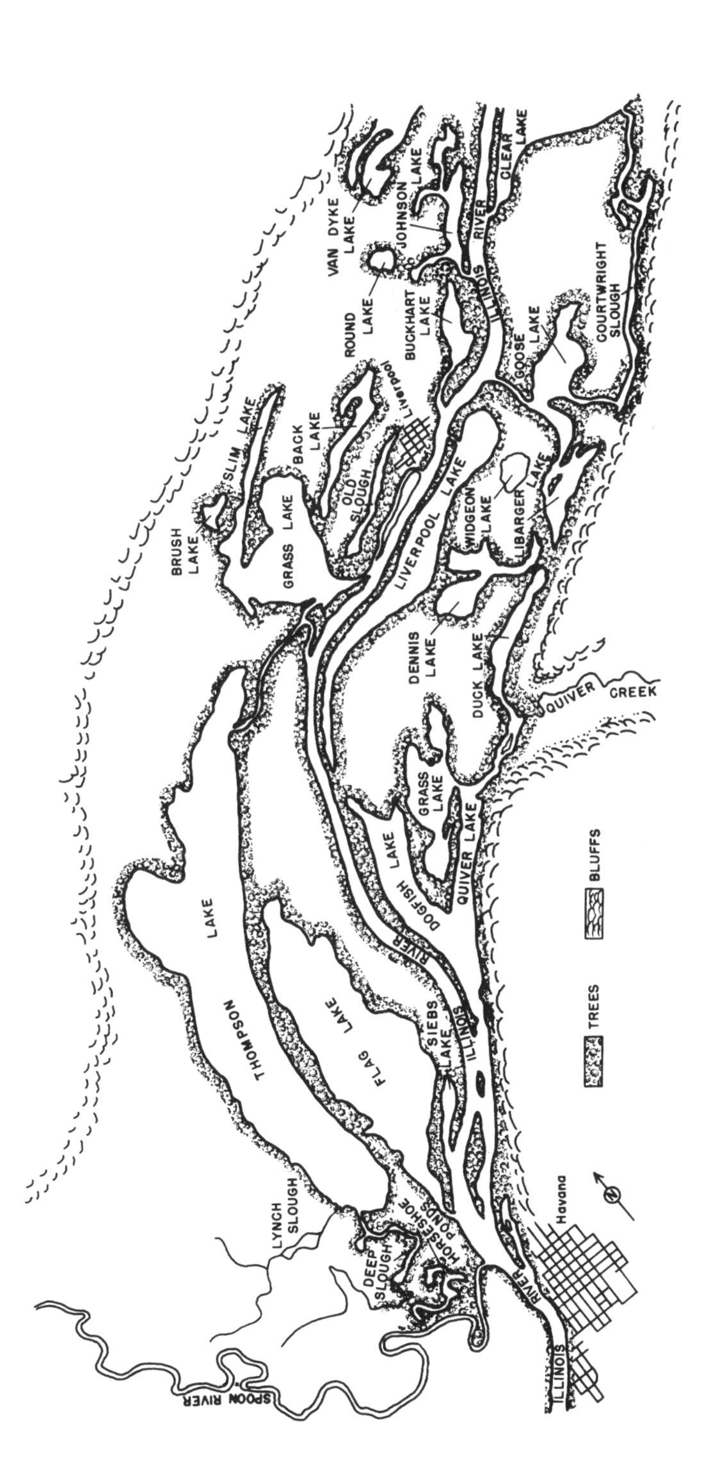


**Map 7.1.** Floodplain lakes between mile 119 and mile 131, Illinois River, 1902–1904. After Woermann, "Map of the Illinois and Des Plaines Rivers"; courtesy of Illinois Natural History Survey.

7.1 and 7.2, which show the floodplain area just upstream of Havana in 1902–4 and 70 or 80 years later.[1]

## The Measure of the Bottomlands

The venture in shaping drainage and levee districts so that arable land would be secure from overflow and elevated levels of ground water was the final phase to the spread of Corn Belt agriculture in the counties that flanked the Illinois River. The bottoms were the last large reserve of virgin land in the heart of the state. Understandably, the integration of the reclaimed bottoms into the regional agricultural system was achieved primarily by residents of the valley and adjacent uplands, who were landowners, tenants, and investors. Whereas farm development on the uplands did not require drainage, or was but partially dependent on on-farm tiling and on-farm and drainage district ditches, the cropland on the bottoms was dependent upon elaborate drainage works (including pumping stations) for productivity.

The formation of drainage and levee districts, as the *Jerseyville Republican* put it regarding the Nutwood district, was "by far the largest financial proposition ever organized" in the county, which was the case in most counties of the study area.[2] The *Rushville Times* applauded creation of a "Little Holland" in the Coal Creek district because of the revenue "of no small amount" that would accrue to the county.[3] Others saw the benefits in terms of enhanced land values and the growth of business for elevators and retail stores in the towns and villages of the valley and in county seats. Nevertheless, the degree to which corporate-type ranches and tenantry were part of the process was lamented by individuals who prized the yeoman tradition associated with family owned and operated farms. Perhaps the observers recognized the development of the drainage and levee districts as a retardant to emigration from the rural counties, which had been going on at least since 1900. One editor commented on the implications of land drainage for public health. However, with the exception of the developers of the Big Prairie district, between Beardstown and La Grange, the control of malaria or the improvement of public health and sanitation were not stated purposes in the petitions to reclaim land along the lower Illinois River.[4] Malaria was present but ceased to be a concern at large some decades earlier.[5]

By the mid-1920s, the reclamation effort created drainage and levee districts in the lower Illinois Valley with over 183,000 acres of assessable land. A good 90 percent of the leveed and ditched land was cultivated, which was exceptional for large areas of drainage and irrigation entities anywhere in the United States. The land was protected from overflow by over 330 miles of levees and was protected from ponding by pumps, to which unwanted water was delivered by more than 370 miles of open ditches and over 260 miles of tile drains, of which close to 40 percent were large-diameter lines

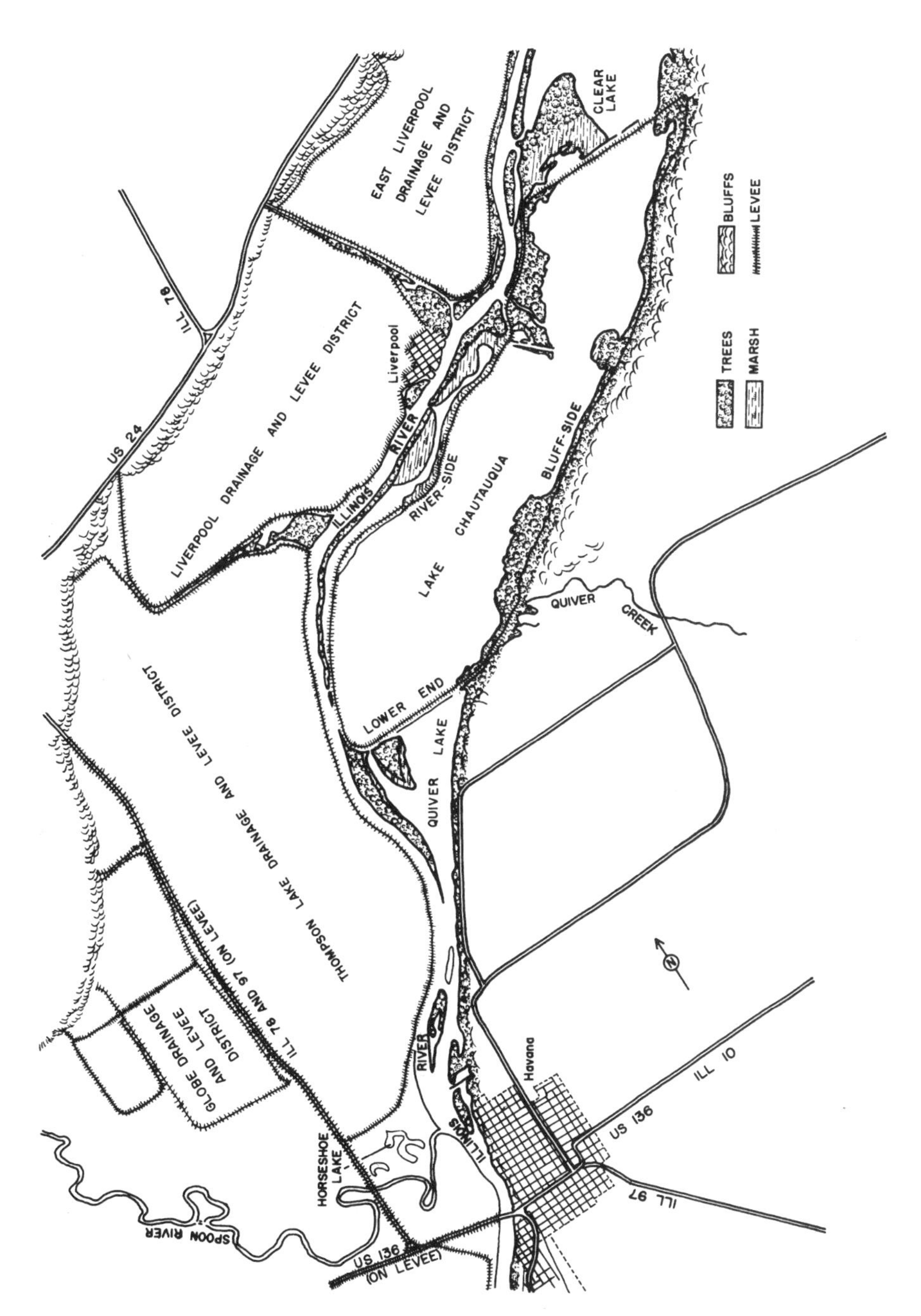


**Map 7.2.** Salient contemporary features between mile 119 and mile 131, Illinois River floodplain. Courtesy of Illinois Natural History Survey.

emplaced for the districts. Assuming that production on such land was worth $35 per acre (1928 estimate),[6] the reclaimed areas of floodplain yielded well over $6 million per annum in crops long after the palmy days.

Perhaps the significance of the reclaimed lands is measured best by noting that they aggregated about 7.9 congressional townships of assessable farm land. While not an appreciable area compared to the total of productive land in the state, these reclaimed lands were sizable additions to the resource base of the counties between Peoria and Grafton. Greene, Scott, and Fulton counties each gained between 1.0 and 1.5 townships in assessable land. The reclaimed bottoms in Scott County were equivalent to 18 percent of its total area (251 sq. mi.). Such assessable lands aggregated about 10 percent of the entire area of Greene County (545 sq. mi.); and in 7 other counties that shared the floodplain (Brown, Cass, Fulton, Jersey, Morgan, Schuyler, and Tazewell), the assessable land in the drainage and levee districts represented between 3 and 5 percent of the total land areas. In all of the counties that shared the Illinois Valley, the districts and independent reclamations constituted appreciable sources of corn, wheat, and oats, employment and trade, and revenue. In the span of three decades, the keepers of assessor's maps and of the maps in county atlases had to replace the symbols showing natural features of the landscape with symbols for systems of levees and ditches, section line roads, farmsteads, and new or spreading villages and towns.

## The Agricultural Scene

By and large, 50 to 60 percent of the area of newly reclaimed districts was planted to corn and wheat, which yielded handsomely. Exceptionally, 150 bushels of corn and 50 bushels of wheat were harvested per acre from the floors of drained lakes. However, 75 bushels of corn were closer to the average on virgin land, the yields being enhanced by the unusually fertile floors of lakes and ponds. Following about five to ten years of cultivation, yields averaged 50 to 60 bushels, which was the general yield estimated for all reclaimed land in 1914 and in the late 1920s. The average districtwide yields of corn compared favorably with a statewide average of 36 bushels to the acre. At the time, districtwide yields averaged 25 to 30 bushels of wheat, while the average wheat yield in Illinois was 17.3 bushels. Oats generally averaged between 35 and 50 bushels, somewhat greater than the average of 30 bushels per acre for the state. Timothy and clover yielded 2 to 3 tons of hay per acre, up to twice the statewide average.[7]

The fine yields on recently drained soils implies that systems of tile underdrainage went into place on the farms about as soon as the district pump houses began to draw down water levels. However, the timing of laying and perfecting the field tile systems is not documented.

Instructive data on the agricultural scene in 1920 are preserved for Bluffdale Township, in northwestern Greene County, where the Columbiana Ranch and over 70 farms occupied approximately 22,000 acres. Corn was planted on 31 percent of the land; lesser proportions of land were planted to wheat (23%), pasture (19%), and timothy and clover (5%). Woodland yet occupied 12 percent of the township. Silos, which became fairly common in the lower Illinois Valley after 1909, numbered 29 in the township by 1920. Two or 3 stood on larger properties that finished cattle or had dairy stock, while but 1 served smaller farms. Allocation of land to grain crops on the 2,700-acre Columbiana Ranch was representative—corn (32%) and wheat (26%). Woodland occupied 13 percent of the ranch. The acreage in timothy and clover (26%) and in pasture and oats (3%) reflected a livestock operation involving 500 hogs, 90 of them brood sows, and 40 mules and 20 horses. The work stock was soon supplanted by recently adopted tractors. At the time, family farms in the township averaged about 145 acres and yet depended upon a team of mules and 4 horses for motive power. On the average, such farms had a couple of milk cows, a handful of beef cattle, 4 brood sows and progeny, and some 60 poultry.[8]

These data for Bluffdale Township, when compared with less comprehensive data for upland townships within Greene County, show that corn and wheat acreages were of similar importance and that timothy and clover acreage was much more important (8–10%) on the small farms in the uplands. Dairy stock and brood sows, but not beef cattle, were relatively more important in the uplands, too. In Greene County overall, the average farm was 149.8 acres, and 55.4 percent of the farms were operated by owners.[9] The valley floor of the lower Illinois River was developing as part of the regional farming system.

There was a greater incidence of large livestock ranches in the drainage and levee districts than in the uplands. The large operations tended to be early adopters of steam- and gasoline-powered machinery; they made provision for tenants on a relatively grand scale, and they were among the experimenters with new crops. The ranches were run by the owners or managers. Notable among the well-capitalized early enterprises were ranches at what became the Hartwell and Hillview districts, the Christie and Lowe operation at Coal Creek, the Fairbanks Ranch, and large properties in the Nutwood district that were owned by men of means in Jersey County. A later, model operation was developed by "salt king" and Chicago financier Joy Morton at the Thompson Lake district after 1921.

The best documented of the early large enterprises was the 8,440-acre Fairbanks Ranch. Its assemblage of farm buildings included the headquarters residence, boarding houses, barns and silos, and machinery sheds. The ranch was leased (1901) and acquired (1903) by the brothers William D. and Loresten M. Fairbanks from the widow of J. R. Keach, a rancher whose success was accompanied by the extension of nuclear holdings inherited in 1877. The Fairbanks family, which was experienced in land drainage and banking on the wet prairie uplands of Piatt County, determined to buy the property following successive years of water damaged crops on some 2,000 acres of cultivated land, where some 600 hogs and 200 head of cattle were kept. Production intensified greatly in the course of the five years of reclamation that ended in 1909, when the pumping station went on stream. The corporate farmers adopted steam traction engines after 1902, in part to use them for ditch excavation and land breaking. Their first tractor entered service in 1909. The adoptions were among the first in the lower valley. Cropland expanded to 5,000 acres in 1911 and to over 7,000 acres in 1913; employees increased from 160 in 1911 to 300 in 1913–14, by which time frame houses costing between $1,000 and $3,000 each were built across the property to accommodate 50 tenant families. In the interest of tenant satisfaction, the proprietors had a school and a church built. The tenant farms and a pair of recently formed (1909) owner-operated farms aggregating 720 acres were worked with 300 mules and horses. Another 300 mules and horses worked the Fairbanks Ranch, where 1,000 beef cattle, 1,700 hogs, and 500 sheep were maintained on the valley floor, and 300 Angora goats were kept in the bluff area.[10]

To judge from arrivals, field trials, and accidents reported in the newspapers of provincial towns, the steam-powered traction machines were not numerous. They were owned by two or three of the ranches and were purchased on a cooperative basis by several producers of wheat. Also, the behemoths, which took their toll of county bridges, were owned by contractors who harvested, threshed, and separated wheat, hauled grain, logs, or other heavy freight, and the machines could provide power for a portable sawmill. The adoptions, which occurred relatively late in the era of steam traction engines, probably were made possible by the large scale of farming on very productive lands that were flat and firm. As was the case nationally, the steam traction machines soon were supplanted by tractors on the large properties, local adoptions of the machine with an internal combustion engine beginning about 1908 to 1914. While as many as seven Fordson and three other tractors harvested the 1919 wheat crop in the Hillview district, tractors were enough of a novelty in 1919 that a crowd of 1,500 was drawn to a competitive field trial in central Mason County.[11] Diverse assemblages of working machinery attracted audiences from ten or a dozen miles around to the Coal Creek district in the 1915 harvest season and probably earlier. A Reeve's gasoline-powered threshing machine was in service there in 1910. Typically, each threshing rig required four pitchers in the field, ten wagons on the move, and a half-dozen stackers and sackers to keep the machine running near capacity. When steam engines were the power source, they had to be served by a water wagon and a water boy as well. The animated scene centering on threshing machines at Coal Creek in 1915 was rounded out with a 60-horsepower Holt crawler tractor drawing four

McCormick binders and a 22-horsepower Hart-Parr cleaning up with a half-dozen 19-inch plows. Thus, the era of the internal combustion engine had begun on the farms, as it had in the pump houses of the districts. The era was well under way by the late 1920s, when motor trucks were reported to be appreciably reducing the shipping of livestock, coal, and lumber by water carriers.[12] The competition from trucks must have been sensed by the railways, as well. Already in 1920, a few men made a living in the towns of the valley as drivers of automobiles and trucks, as mechanics and tire vulcanizers, and as automobile salesmen.

The process of settlement by tenants and individual landowners that accompanied drainage of the floodplain is suggested by the chronology at the Fairbanks Ranch and at the Nutwood and Coal Creek districts. At the Nutwood district, preparation for tenant occupancy followed the closing of the rough levee system (July 1909) and the placing of the pump house on stream (December 1909), which was well over two years after the first assessment (May 1907) for reclamation. Land clearing went on apace through the winter of 1909–10, at which time 10 pairs of tenant houses and barns were raised. At least another 10 of the paired structures, each nuclear group costing $3,500, were built during the summer following.[13] At the older Coal Creek district, the original levee and canal systems were in place by the end of 1897 and were greatly improved in the winter of 1899–1900. Already, the farmers in adjacent bluff and upland areas of Schuyler County were commuting to the Coal Creek bottoms in order to plant, cultivate, and harvest corn for use on the home properties, when it was not sold. In 1898–1900, the land developers built houses, barns, sheds, and cribs to accommodate settlers. A primary school was in place in 1910. By 1911, at least 25 tenant families were in residence; there were about 200 people in the tract in 1914. They were served by a good, all-weather, east-to-west road across the center of the district (now Illinois Highway 103), which gave access to Rushville, Beardstown, or Frederick (via Illinois 100). Substantial parts of the estimated corn crops of 100,000 bushels (1910) and 140,000 bushels (1916) and lesser volumes of wheat moved over this road, as did tile out of Beardstown and gravel or crushed road metal from the bluffs.[14] The directions of trade in such bulky goods were akin to what occurred elsewhere in the valley, where large properties fostered cash-grain production and where field tiling was done.[15]

There was some shifting about of tenant farmers, but the degree to which a sorting process drew better tenants to the best land in the districts is unknown. More certain was the timing of moves to new abodes by tenants and farm owners—about March 1. The process was marked by the movement of four-horse teams with wagonloads of household effects, tools, and implements—trailed by a cow or two. Sometimes the traffic reflected a change in ownership of a ranch, which occasioned the unloading of carloads of livestock, implements, supplies, and feed at railway sidings near the properties.[16]

## The Texture of Human Occupance

The patterns of population, employment, and household composition in 1920 reflected the culmination of the land drainage effort. They describe the situation on the eves of the crash in commodity prices and of a cycle of wet years and floods that ended the ebullient era of development. Data for 11 townships and precincts are drawn from the enumerators' sheets used in the 1920 census of population. The selected townships and precincts were scattered through the study area from just south of Pekin to opposite Hardin; most consisted of areas of the valley floor and the uplands. In the aggregate, there were over 7,650 rural residents (table 7.1),[17] of whom 55 percent were male, all were white, and nearly all were born in Illinois. A third of the cohort was gainfully employed, including women who were teachers, domestics, seamstresses, and laundresses. Wives of the heads of households and some widows represented about 22 percent of the rural population. The households averaged about 4.24 residents but ranged from 1 (usually a widower or unmarried male) to 11 members.

The heavy preponderance of whites and native-borns in the population of the townships adjacent the lower Illinois River differed little from the proportions of whites and native-borns in the counties of the area, which was somewhat more pronounced locally than in the state overall in 1920. The rural population in Illinois was 99.0 percent white and 91.3 percent native-born. The male-to-female ratio in the townships was higher than the statewide average of 108.9:100, perhaps a reflection of the circumstances of tenure and of the opportunities for employment in an area that was in-filling at a time when the various counties had been losing rural population for a couple of decades. Perhaps a reflection of the employment opportunities in the former wetlands, as against established upland agricultural areas, was the slightly larger household size (4.24 persons) than overall in the dozen counties (4.06) that touched the lower river.[18]

The men and youths of 15 years or more who were employed gainfully were engaged in agriculture, almost to the exclusion of other callings. Farm operators, including salaried farm managers, accounted for 48 percent of the employed, but the range for the townships was 38 to 58 percent, except in a Pike County township (28%) where employment in a large apple-packing house skewed the data. Most operators were general farmers, but a few reported specializing in grain, livestock, fruit, livestock trading, and truck or poultry farming. Another 39 percent of the gainfully employed (range of 25–50%) were farm and general laborers. These included male kin by blood or marriage of the heads of households and unrelated boarders, together with the heads of separate households. Domestics, most of whom were female housekeepers or maids, and school teachers (some of them males) were fairly common among rural residents. Other callings were represented here and there. Stationary and electrical engineers operated the pumping plants; traction engineers, like some mechanics, blacksmiths, carpenters (construction trades), and tile layers either were associated with very large farm operations or were in business for themselves. Here and there, too, were owners and clerks of country stores, ferrymen, and bridge tenders. Rural residents whose jobs took them elsewhere included a few coal miners, employees of governments or industries, and the odd dredgerman or marine engineer. Woodchoppers and fishermen tended to cluster near a sawmill or riverside town.

Most households in the rural townships that embraced the bottoms and adjacent uplands occupied rented abodes. Whereas 56 percent (range of 50–68%) of the heads of households in the rural townships were renters of houses (table 7.1), only 40 percent of heads of households in the villages and small towns of the area were renters (table 7.2). Most renters were farmers or farm laborers. The preponderance of house renting over ownership in the rural townships probably reflected the degree to which farm tenantry existed. Farm tenantry, estimated to have well exceeded 55 percent of all farm operations, may have topped 75 percent in some districts. The proportion of tenant operations among farms on the bottoms was closer to the nearly 60 percent incidence of farm tenantry in such east-central Illinois' counties as Champaign, Piatt, and McLean than it was to the overall incidence in counties adjacent the lower Illinois River—about 38 to 48 percent. In 1920, as today, the owner-operated farm predominated in this area of the state where uplands are broken by well-defined valleys and draws.[19] One suspects that the deterioration of cropland in these uplands, together with the promise of drained land on the floodplain, drew young farmers to the drainage and levee districts. Indicative of the age and sex distribution of the population in and near the valley are the data (table 7.3) for Bainbridge and Waterford townships, in Schuyler and Fulton counties, respectively. The data for the two townships may be used also for examining the patterns of residential ownership by heads of households in different age groups (table 7.4). The Crane Creek Drainage and Levee District (organized 1908) lay in the former area, while the latter included Thompson Lake, which was being reclaimed in 1920. In both townships, the male-to-female distribution approached the median for a dozen such tabulated entities. The age distribution of the populations was indicative of the situation in the lower valley.

The Hillview, Big Swan, and Scott County districts appear to have had higher than average proportions of owner-operators. Except for Christie and Lowe, who set aside 2,000 acres in the Coal Creek district in 1910 to be sold through the Chicago-based Farm Land Development Company, there were no commercialized attempts to develop colonies of truck farmers and small general farms. The concentration of truck farms that developed earlier on the 700 acres of cropland in the East Peoria district, like the early presence there of brick and tile yards and railroad yards, reflected an early phase in the development of suburban land where industrial plants and blue-collar housing

**Table 7.1.** Population and gainful employment of selected townships and precincts, 1920

| County:<br>Township: | Tazewell<br>Spring Lake | Fulton<br>Waterford | Schuyler<br>Hickory | Schuyler<br>Bainbridge | Scott<br>Naples |
|---|---|---|---|---|---|
| Population | 1071 | 353 | 571 | 883 | 126 |
| Households | | | | | |
| own house | 69 | 36 | 58 | 102 | 9 |
| rent | 162 | 45 | 66 | 113 | 20 |
| unknown | 7 | 5 | 1 | 12 | 2 |
| Gainful employment | 373 | 123 | 174 | 304 | 55 |
| Farming | | | | | |
| fruit | . . . | . . . | . . . | . . . | 1 |
| general | 189 | 67 | 65 | 175 | 28 |
| poultry | . . . | . . . | . . . | . . . | . . . |
| stock | . . . | . . . | . . . | . . . | . . . |
| truck | 4 | . . . | . . . | . . . | 8 |
| manager | 3 | 4 | 6 | . . . | 1 |
| Laborer | 151 | 34 | 68 | 111 | 14 |
| Drainage | | | | | |
| pump house | 2 | 3 | 5 | . . . | . . . |
| tiler | . . . | . . . | 1 | . . . | . . . |
| Extractive | | | | | |
| fishing | . . . | 2 | 7 | . . . | . . . |
| timber | 1 | . . . | 4 | 1 | . . . |
| trap/hunt | . . . | 1 | . . . | . . . | . . . |
| Service | | | | | |
| const. trades | 5 | 1 | 2 | . . . | . . . |
| domestics | 4 | . . . | 5 | 3 | 1 |
| railroad | . . . | . . . | . . . | 1 | 1 |
| retailing | . . . | 2 | 4 | . . . | . . . |
| seamstress | 1 | . . . | . . . | . . . | . . . |
| smith/auto. | . . . | . . . | . . . | . . . | . . . |
| teaching | 3 | . . . | 1 | 6 | . . . |
| traction eng. | . . . | . . . | . . . | . . . | . . . |
| packing house | . . . | . . . | . . . | . . . | . . . |

| Scott Oxville | Greene Bluffdale | Greene Patterson | Greene Walkerville | Pike Chambersburg[a] | Pike Flint |
|---|---|---|---|---|---|
| 466 | 989 | 1102 | 969 | 616 | 511 |
| | | | | | |
| 44 | 75 | 93 | 113 | 65 | 51 |
| 61 | 135 | 170 | 118 | 79 | 74 |
| 2 | ... | 17 | 7 | 5 | ... |
| 160 | 307 | 340 | 341 | 182 | 168 |
| | | | | | |
| 1 | ... | 1 | ... | ... | 2 |
| 73 | 139 | 151 | 140 | 90 | 40 |
| ... | ... | ... | 3 | ... | ... |
| 1 | ... | 2 | ... | ... | ... |
| ... | ... | 2 | 9 | ... | 9 |
| 6 | 5 | 1 | 2 | 2 | 2 |
| 57 | 153 | 156 | 146 | 46 | 47 |
| | | | | | |
| 2 | 3 | 1 | 1 | 1 | ... |
| 1 | ... | 2 | 2 | ... | ... |
| | | | | | |
| ... | 3 | ... | 1 | ... | 1 |
| ... | ... | ... | 16 | 8 | ... |
| ... | ... | ... | ... | ... | ... |
| | | | | | |
| ... | 6 | 3 | 1 | 4 | 3 |
| 5 | 1 | 2 | 8 | 3 | 3 |
| 1 | 1 | 5 | ... | ... | 14 |
| 3 | 2 | 1 | 1 | 18 | 9 |
| ... | ... | ... | ... | ... | ... |
| ... | 2 | 3 | 1 | 2 | 3 |
| 3 | 8 | 5 | 3 | 6 | 2 |
| ... | 2 | ... | 2 | ... | ... |
| ... | ... | ... | ... | ... | 27 |

[a]Includes village of Chambersburg.

**Table 7.2.** Population and gainful employment of selected communities, 1920

| | Bath | Browning | Eldred | Fieldon |
|---|---|---|---|---|
| Population | 408 | 455 | 298 | 248 |
| Households | | | | |
| own house | 68 | 86 | 50 | 31 |
| rented | 44 | 39 | 32 | 32 |
| unknown | ... | ... | ... | ... |
| Gainful employment | 157 | 160 | 115 | 80 |
| button factory | ... | ... | ... | ... |
| const. trades | 9 | 12 | 4 | 4 |
| domestic | 6 | ... | ... | 2 |
| farming | 4 | 3 | 7 | 7 |
| fishing | 35 | 46 | ... | ... |
| labor | 25 | 15 | 26 | 32 |
| laundry, sewing | 4 | 1 | 6 | 4 |
| livery-dray | 1 | 6 | 3 | 3 |
| quarry | ... | ... | ... | ... |
| railroad | 2 | 30 | 16 | ... |
| retail[a] | 35 | 21 | 28 | 12 |
| smith/auto. | 4 | 3 | 4 | 5 |
| teaching | 4 | 3 | 3 | 3 |
| timber related | 1 | ... | 2 | ... |

[a]Includes employed in banks, barber shops, lodging houses, restaurants, and stores.

would soon predominate. The 250 acres of cropland remaining in the late 1920s was gone by 1934.[20] This concentration of market gardening was unique in the East Peoria district; however, a good handful of truck gardeners worked in 1920 in the Spring Lake district (near Pekin) and near Liverpool and Naples. There was some concentration of melon production in the South Beardstown district, and by 1911, as much as a 1,000 acres were planted in potato and truck crops on the big ranches now and again. The quest by large operators for a specialty crop included rice, which appeared under irrigation in the Eldred and Thompson Lake districts between 1925 and 1928. As much as 1,440 acres were planted in the former district in 1928, but the crop usually did not exceed an experimental 200 or 300 acres.[21] While the productivity of the bottoms was

| Frederick[b] | Hillview | Liverpool[b] | Meredosia | Naples | Pearl |
|---|---|---|---|---|---|
| 519 | 577 | 1012 | 812 | 384 | 669 |
| 76 | 57 | 153 | 133 | 69 | 108 |
| 55 | 73 | 78 | 85 | 27 | 62 |
| 2 | 2 | 3 | 2 | 1 | ... |
| 182 | 199 | 327 | 243 | 146 | 212 |
| ... | ... | ... | 28 | 10 | 16 |
| 5 | 4 | ... | 11 | 6 | 9 |
| 2 | 4 | 3 | 7 | 3 | 1 |
| 53 | 22 | 148 | 6 | 21 | 19 |
| 3 | ... | 32 | 15 | 8 | 10 |
| 33 | 77 | 86 | 42 | 41 | 38 |
| 3 | 3 | ... | 2 | 7 | 6 |
| 2 | 4 | 3 | 6 | 3 | 2 |
| ... | ... | ... | ... | ... | 34 |
| 21 | 14 | 1 | 3 | 4 | 25 |
| 9 | 23 | 7 | 54 | 7 | 28 |
| 3 | 9 | 3 | 9 | 2 | 5 |
| 10[c] | 5 | 3 | 5 | 5 | 7 |
| 10[c] | ... | 8 | ... | ... | ... |

[b]Town and rural not distinguished.
[c]Some probably employed in Beardstown.

advantageous for conventional field crops of the region, the production of other field and row crops on a large scale was not sustained. The outcome was consistent with the general experience on drained lands in the Corn Belt.

Rural settlement in the lower valley was concentrated on high ground adjacent the weaving and undulating roads that skirted the foot of the bluffs or that followed Pleistocene sand ridges and terraces somewhat removed from the bluffs. The villages and small towns that evolved away from the river as agricultural service centers tended to occupy sites where the axial roads were intersected by roads off the uplands. As a rule, such lateral roads followed gradients notched by creeks across the zone of bluffs. Riverward of the coves and axial roads, where older farmsteads and ranch headquar-

**Table 7.3.** Age distribution of population in selected townships, 1920

| Bainbridge Township, Schuyler County | | | | | Waterford Township, Fulton County | | | | |
|---|---|---|---|---|---|---|---|---|---|
| Age | Total no. | % of pop. | Male | Female | Age | Total no. | % of pop. | Male | Female |
| 80–89 | 8 | 1 | 4 | 4 | | | | | |
| 70–79 | 19 | 2 | 11 | 8 | 70–79 | 7 | 2 | 3 | 4 |
| 60–69 | 57 | 7 | 34 | 23 | 60–69 | 21 | 6 | 14 | 7 |
| 50–59 | 73 | 9 | 39 | 34 | 50–59 | 21 | 6 | 15 | 6 |
| 40–49 | 76 | 9 | 42 | 34 | 40–49 | 37 | 10 | 23 | 14 |
| 30–39 | 134 | 16 | 70 | 64 | 30–39 | 49 | 14 | 22 | 27 |
| 20–29 | 135 | 16 | 72 | 63 | 20–29 | 54 | 15 | 33 | 21 |
| 10–19 | 160 | 19 | 99 | 61 | 10–19 | 83 | 24 | 46 | 37 |
| +0–9 | 185 | 22 | 97 | 88 | +0–9 | 81 | 23 | 40 | 41 |
| Total | 847 | | 468 | 379 | Total | 353 | | 196 | 157 |
| Percent | | 100 | 55 | 45 | Percent | | 100 | 56 | 44 |

**Table 7.4.** Residential tenure of heads of households, selected townships, 1920

| Bainbridge Township, Schuyler County | | | | | Waterford Township, Fulton County | | | | |
|---|---|---|---|---|---|---|---|---|---|
| Age | Total | Own | Rent | ? | Age | Total | Own | Rent | ? |
| 80–89 | 6 | 4 | 2 | . . . | | | | | |
| 70–79 | 9 | 7 | 2 | . . . | 70–79 | 3 | 3 | . . . | . . . |
| 60–69 | 29 | 21 | 7 | 1 | 60–69 | 14 | 9 | 3 | 2 |
| 50–59 | 40 | 24 | 16 | . . . | 50–59 | 15 | 8 | 7 | . . . |
| 40–49 | 39 | 19 | 17 | 3 | 40–49 | 24 | 8 | 13 | 3 |
| 30–39 | 59 | 19 | 35 | 5 | 30–39 | 22 | 7 | 15 | . . . |
| 19–29 | 42 | 6 | 33 | 3 | 19–29 | 11 | 1 | 10 | . . . |
| Total | 224 | 100 | 112 | 12 | Total | 89 | 36 | 48 | 5 |
| Percent | 100 | 45 | 50 | 5 | Percent | 100 | 40 | 54 | 6 |

ters were located, the tenant farmsteads were dispersed. Sometimes, these clusters of frame houses, barns, and ancillary structures stood on sand ridges or other slight elevations. In the nether reaches of the artificially drained tracts was the pumping station, at times with the residence of the stationary engineer nearby.

While the agricultural sector was served by the largest towns of the valley, the functions of these places were so diverse that anything beyond a general delineation is outside the purview of this study. Peoria, Pekin, Havana, Beardstown, Meredosia, and Hardin were akin from the outset in that they shared to varying degrees the catalytic advantage of elevated sites at riverside, where ferry and bridging points coincided with the landings favored by steamboat operators. All but Meredosia assumed early prominence as county seats, an asset lost by Beardstown in due time. These relatively few pioneer crossing places of the river and wetlands linked central Illinois with counties of the Military Tract and favored crossing points of the Mississippi River. To these sites of the Illinois Valley (save Hardin) were drawn the railways which bridged the river. The larger northerly towns attracted freight yards, car shops, and manufacturing enterprises of some consequence.

Perforce, the focus here is on the nature of the smaller communities, especially those known to have benefited most from agricultural development on reclaimed lands. Indicative of the growing economic activity was the penetration of electric transmission lines to the Pekin and La Marsh district in 1909 and to both sides of the lower valley between 1912 and 1915. Lines crossed the river at Pekin, Havana, Beardstown, and Valley City between 1912 and 1913. By 1915, electric power was accessible to all communities and to the drainage and levee districts in the valley, the county seats, and the coal mines of the region.[22]

Perhaps the greatest economic and physical change touched nascent crossroad villages that lay near the bluffs in southerly Greene and Jersey counties. Hillview, a "hustling village" of 465 residents in 1912, graced with a station, an elevator, and a lumber yard on the Chicago and Alton line, also had a four-room school, a church, a bank, three general stores, and a hardware and implements dealer. Eldred, about 12 miles to the south, was not far behind in providing services to the hinterland. A bank was added in 1912 to the island of buildings that included a church (1907), a handful of grocery and general stores, a hardware store, a lumber yard, and an elevator. Already in 1911, the upland town of Fieldon, 3 miles from the bottoms and 13 miles west of Jerseyville, had a bank. Electric service reached Hillview in 1912 and Eldred in 1913. Meanwhile, lots were laid out and incorporation proposed for Haypress, at the edge of the Fairbanks Ranch (Keach district). The village of Nutwood was platted in 1913, apparently in anticipation that the Chicago and Alton Railroad would extend a branch southward from Eldred. A line did reach 3 miles southwesterly from Eldred to Clarke (1913), where by 1916, 100 souls congregated in a camp of woodchoppers, teamsters, and sawmill hands to produce mine props and construction lumber—the by-products of land clearing on levee and ditch rights-of-way and on farms. In 1924, a year following destruction of the mill by fire, the railway reached East Hardin, where another mill and Hardin's shippers of Calhoun County apples were prospering. The railway captured wheat shipments that theretofore tended to move by barge from various landings along the lower river.[23]

The small towns and villages of the lower Illinois Valley were dependent primarily on agriculture, on harvesting and shipping the river's products, and on tourism. The patterns of gainful employment for eight of them are profiled (table 7.2) because of their roles as centers of service and residence for participants in the agricultural, fisheries, and other sectors of economic activity supported by the lands and waters of the valley. Profiles for Bath (10 mi. south of Havana) and Pearl are added for perspective, although the communities were peripheral to the development of agriculture in the drainage and levee districts of the valley. To judge from the census enumerator data for 1920, by which time patterns of local economic activity were evolved, the gainfully employed living in the communities provided retail goods and services as well as labor and supervision in the agricultural and fisheries sectors. A few farmers resided in the towns. Larger numbers of farm and general laborers lived in the communities, and sometimes fishermen, hands employed in button-blank factories, or quarry and railway employees were present. Every community had a handful of men who engaged in the building trades, a few teachers, domestics, laundresses and seamstresses, and providers of lodging, drayage, or livery service; and there were blacksmiths or automobile mechanics and tire vulcanizers. The grain buyer and miller was present generally, as were postal service employees. Power or telephone company employees and a physician or two might be residents. The few gainfully employed residents who worked elsewhere included coal miners, traveling salesmen, public servants, and factory laborers.

In the aggregate, the gainfully employed who resided in the communities represented about 34 percent of the entire urban population; the remainder included wives, children, and a few older folks. Among those who were heads of households, about 60 percent owned their residences. Depending upon the community, home owners and renters alike had households that averaged 3.5 to 4.4 persons. Thus, communities consisted of 60 or 80 to 220 dispersed frame residences and a few commercial buildings. Some of the latter structures were made of brick.

In sum, the general crop and livestock farming of the bottoms revolved about the production of corn and small grains, hogs and cattle, some poultry and poultry products, and a little milk and cream. The proportion of acreage in pasture and hay crops, like the acreage of corn and small grains, was consistent with a regional farming system in which a three-year, three-crop rotation prevailed. There were one- and two-bottom plow tractors on about a quarter of the farms in the region by 1928, when their usefulness for belt work rivaled the small (up to 5 h.p.), stationary gasoline engines then in use.[24] While most of the corn was harvested for grain, some went into silos. A good deal of the corn was sold as a cash crop or used on the tenant's home farm in the adjacent uplands. Nevertheless, most of the corn probably was consumed on the farm. Among the small grains, wheat was much more important than oats in acreage and in cash sales.

In the 1920s, the rewards of wresting the floodplain from the Illinois River were eroded severely; reclamation came to a standstill. The nation's nurturing economic climate disappeared, while the river, its affluents, and a spate of wet years challenged the wisdom of compartmentalizing, ditching, and artificially draining the floodplain. The commodity crash of 1921, the general flood events of 1922, 1926, and 1927, and the intervening period of wet years devoured the capital and optimism of tenants, landowners, and investors when they did not destroy the very foundations of security against water on the floodplain. The expansive era of land drainage and settlement that transformed this bypassed area of the frontier was over. The usefulness of the steam-driven excavator and pump, like that of horse-drawn implements and vehicles, had peaked. The risk to living and farming on the bottoms, together with advances in motor vehicle usefulness and farm mechanization and electrification, fostered the consolidation of farm operations. Also begun was the era in which agriculture on privately drained floodplain land, as it did elsewhere in the nation, became a beneficiary of direct federal assistance and, to a lesser degree in Illinois, of direct assistance from the state.

# 8

# Harvesting Nature's Endowment

THE floor and bluffs of the lower Illinois Valley and the flanks of innumerable valleys and draws that carried streams from the uplands into the Illinois River supported woodland and grassland ecosystems ranging from the aquatic and swampy to the well drained (map 8.1). These ecosystems provided timber, fur-bearing and game mammals, and wildfowl and fish as commercial, subsistence, and recreational benefits to resident and transient people before and during the era of large-scale land drainage. The communal harvests of nature's bounty contributed significantly to the commerce of towns like Pekin, Havana, Beardstown, and Meredosia, where wagon and railway bridges or ferries crossed the river, and to such smaller communities as Liverpool, Bath, and Browning.

Harvesting nature's endowment in the wetlands, in floodplain lakes and river, and on the drier valley floor and bluffs engaged men and youths for the most part, but a few women were employed in the small factories where button blanks were cut from mussel shells. Some of the extractive pursuits were seasonal but most went on year-round. A person's declared occupation did not preclude engaging in kindred activities. Farmers and their sons, for instance, might hunt, trap, and set nets to supplement their income, and they could cut timber. By the same token, farmers with teams could obtain employment by scraping and hauling dirt for drainage contractors after crops were laid by or the growing season was over. More to the point, self-employed fishermen doubled as market hunters, trappers, and guides for visiting sportsmen; and they could gather mussels. Most of their contemporaries in the day labor and agricultural labor force of small riverside towns could be expected to participate in the same activities, given the intermittent nature of their primary employment. In addition to the resident harvesters of nature's bounty, there were transient crews of full-time trappers and at least one large camp of woodcutters from elsewhere. It is likely, too, that many of the mussel gatherers were short-term residents.[1]

There was a seasonality to the different activities. Winter was prime time for timber cutting, trapping of fur-bearing animals, and commercial fishing. Nevertheless, the work of clearing rights-of-way for levees and clearing prospective fields might be done

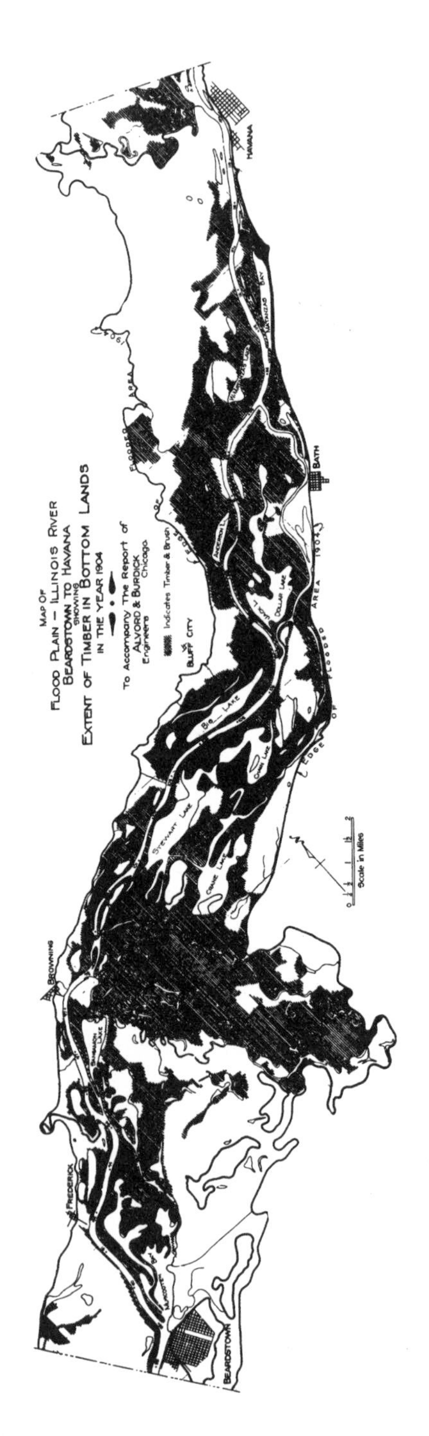


**Map 8.1.** A favored sector of the lower valley for hunting and fishing, 1904. Between Beardstown and Havana lay the broad, heavily timbered, and lake-studded bottoms that hunters and commercial fisherman favored. From Alvord and Burdick, *Illinois River,* fig. 30; courtesy of Map and Geography Library, University of Illinois.

at any time, which was also true of fishing. Ice, much of which was cut for market shippers of fish, was obtained in December and January in streams and ponds for storage, at least until ice-making plants appeared. Market and recreational duck hunting were most attuned to the seasonal passage of waterfowl along the broad Mississippi Flyway in the fall and spring. Catering to vacationing families with fishing poles or a Chautauqua in mind was a spring and summer activity, as was the valley's principal activity, farming.

To judge from an extensive reading in newspapers that chronicled human activity in the lower valley between 1900 and 1930, it was accepted that individuals who lived by the axe, trap, shotgun, seine, or sheller's gear, like the plowman, pursued what they perceived as their natural rights in harvesting the bounty of the floodplain and river. Use of the gifts of nature was unrestrained. A few reports of prodigious bags of waterfowl by market hunters and sportsmen, and of a trapper's wagonload of muskrat pelts arriving in town, were news. However, accounts deploring the unrestrained harvesting of fish and mussels and the conversions of major spawning and feeding grounds to leveed cropland were left to public servants and conservationists to make in agency reports and in the professional literature. Now and again, a newspaper reported on the scarcity of walnut trees or a hunter's recollections of richer rewards in yesteryear; and there were reports of wardens interrupting a sub-rosa trade in waterfowl or losing unattended patrol launches to fire. Encounters between commercial fishermen or hunters and the gamekeepers for private clubs appeared, as did instances of poachers being detained on newly leveed watery tracts where contract seining awaited more dewatering. Evident, as well, was some resentment towards the exclusiveness of the premises of rod and gun clubs whose stockholders were business and professional men in distant cities. Enough pieces are in the record to reconstruct the mosaic of how nature's gifts were harvested.

The rise and spread of water diverted from Lake Michigan and released by land drainage activity in the watershed induced local landowners to build levees in the lower valley and spurred nonresidents to secure residual wetlands for private rod and gun clubs. At the same time, the rising water profoundly altered the residual hunting and fishing grounds of the floodplain. The resentment and conflict involving landowners and the Sanitary District of Chicago had parallels among other economic sectors. People accustomed to essentially unregulated fishing and hunting in the valley before 1900 were forced to adjust to new realities—habitats for fish, waterfowl, and mammals were severely reduced by land drainage activity; residual habitats of the floodplain were submerged; and traditionally open areas were posted. Moreover, the Illinois River was increasingly burdened with urban and industrial waste and sediment.

This chapter recounts how nonagricultural pursuits in the lower valley evolved and operated prior to and during the period of accelerated environmental change that marked the 40 years before 1930. It recalls the livelihoods and land use activities of timber cutting, trapping, duck hunting, and commercial fishing and mussel gathering. It reviews the most relevant conflicts arising between competing users of the river and wetlands and the abuses that led to the entry of public agencies into fisheries and wildlife restoration and conservation activities. The excesses of the harvesters of the valley's resources were akin to the pursuit of self-interest by landowners of the drainage and levee districts and the undeviating pursuits of the Sanitary District of Chicago. Government intervention became more sustained and broader in scope as a sense of public interest sharpened after 1911, by which time the sentiment was widely shared across the nation. The conservation movement was in vogue.

## Timber Cutting

Timber cutting on the bottoms of the Illinois and Sangamon rivers and lesser streams was a commercial venture in its own right and an activity ancillary to land clearing, which the development of drainage districts accelerated. Timber cutting supported steam sawmills at Beardstown and Hardin, a sawmill camp at Clarke, which was located in the Eldred district, and at least one or two small portable sawmills that could operate anywhere. The products of the mills were for local use and shipment. Included were hardwood timbers and boards for construction and for tool handles, shooks and staves, and mine props. At least from the 1880s, the timber supply for the large A. E. Schmoldt mill at Beardstown was obtained on properties where the firm had exclusive rights. Its logging crews, barges, and towboat operated from between Pekin and Hardin on the Illinois River to about as far as Chandlerville on the Sangamon River, where the Chandlerville Lumber Company competed for cutting rights. It was Schmoldt's practice to remove all timber that was more than five or six inches in diameter, including limbs. What did not serve for lumber became cordwood. There are records of cutting in the Liverpool vicinity (1907), the Sangamon Bottoms (1914), and at Meredosia Bay Island in 1915, when the mill shut down. Because the work force, including logging crews, numbered about 90, the closing must have been quite a blow to Beardstown's economy.[2]

Although Schmoldt dominated the business that depended on purchased rights to timber, it also purchased logs cleared from the rights-of-way for levees and ditches and cut on prospective cropland, as was done at the McGee district (1907). The portable steam sawmills that operated on the Fairbanks Ranch (1902), near Bluffdale, and at the Columbiana Ranch, northwest of Eldred (1908), probably were representative of the arrangements made by owners of large properties to have the land cleared and the lumber readied for construction of tenant housing and other structures. Apparently, such a mill operated in Browning (1900) and at Valley City (1920) in connection with the development of the drainage and levee districts. They employed four to eight men. A barrel-stave mill with a daily capacity of 20,000 board feet and 8 employees was constructed at Hardin in 1901 by the Freeport Land and Timber Company, which owned rights to some 6,000 acres of timbered bottoms in southern Greene County. The venture appears to have been eclipsed in 1913 by the sawmill camp that began operating at Clarke, in the west-central part of the Eldred district. This enterprise, owned by a Peoria capitalist, soon employed 100 loggers, draymen, and mill hands. It was served by a railway spur. The community, which consisted of about 240 men, women, and children in 1920, dispersed in 1922. Although fire destroyed the mill, it is probable that not much timber was left by then.[3]

A good deal of timber appears to have been cut throughout the lower valley and its tributaries for local use and export, but much of the less valuable timber and trash were consumed in the great bonfires that followed all clearing. Indicative of the conditions early in this century was cordwood going begging at $2 per cord. Already in 1914 (but probably earlier), buyers were scouring the countryside to offer "tempting prices" for walnut trees yet standing about farmsteads and in residual groves near the bluffs and on the uplands. "In fact, almost every stick of timber has a market value, and the man who owns a grove or wood lot must be very much attached to it in a sentimental way if he can resist the temptation to clear it off, sell the timber and plant the ground in corn."[4] The work of the "wood butchers," decried by the editor of the *Hillsboro News* and echoed in the *Carrollton Patriot,* was not criticized elsewhere in the local press. However, another sign of the times in early 1917 was the serving of public notice by McGee Creek Drainage and Levee District that it would prosecute those who removed or molested timber standing in the strip of woods that buffered the levee against waves from the Illinois River.[5]

The local lumber industry could not be sustained on timber obtained through land clearing much beyond 1920. By then, major areas of residual timber near and upstream of the Sangamon River were prized by sportsmen's clubs. However, by about 1916 to 1918, mature bottomland timber was all but destroyed by high-standing water attributed to the diversion from Lake Michigan. Clearly, the large mill at Beardstown was vulnerable; it needed the sustained yield of 75,000 acres of timber to operate at capacity. At best, sawyers operating after 1925 relied upon power derived from tractors, as recorded for the Meredosia vicinity.[6]

## Trapping

Allowing for the likelihood that contemporary newspaper accounts about commercial trapping reveal the exceptional rather than the routine, the activity apparently was of some consequence, nevertheless. Trapping was a source of supplemental income for farmers and fishermen and of spending money for their children. Rarely did a resident claim trapping as the principal source of employment. However, an itinerant crew of trappers seems to have worked the valley now and again. The pelts moved by the thousands into Meredosia and Havana each winter and early spring. In 1905, for instance, Meredosia's major dealer was able to ship orders to Europe aggregating 11,000 muskrat, 4,574 skunk, 955 mink, 500 opossum, over 200 raccoon skins, and a few fox skins. While there were a number of lean years prior to 1915, Havana's dealer shipped as many as 2,000 muskrat pelts in an order that December, and a year later it was worthy of note that a single wagonload of over 700 muskrat pelts trundled into town, representing some weeks of trapping downriver. Meredosia's buyer shipped 34,000 muskrat pelts to London in December 1916, and he placed 100,000 pieces on the block in St. Louis the following winter. As recently as 1923, the loading of over $92,000 worth of pelts into a car at Meredosia was news.[7]

In spite of what seemed to be recurring harvests into the 1920s, a resident of Browning, in northeastern Schuyler County, wrote in 1899: "It is both noticeable and suggestive that the Sangamon Bottoms have pretty much quit reverberating with the sound of booming shot-gun and melodious bellowing of the coon hound as of yore. The shot-gun and steel-trap have not been beaten into sheep shears, tho [*sic*], nor do the hound and lamb lie down together; the depletion of game and fur animals simply places the gun and coon dog in the catalogue of old junk."[8] While it is risky to make much of such sparse data, the contrast between the diversity represented in shipments made in 1905 and the references to muskrats alone in 1915 and 1916 could well have reflected habitat change. There was increasing settlement and cleared land and the spread of exclusive clubs, on the one hand, and the expansion and deepening of aquatic habitat on the other. In any case, the lament out of Browning in 1899 applied to the valley thereafter.

## Waterfowl Hunting

The perspective from Browning in 1899 on the state of trapping and hunting followed by two decades the conferral of national visibility to similar bottomland just north of Havana as "the most noted sporting grounds in Central Illinois, if not the whole state," where the game was "of great variety and abundance." The quarry included numerous species of ducks and geese and a remarkable diversity of other wildfowl and fish, together with small fur-bearing mammals. The noted grounds lay on either side of the river for about eight miles but back from "its banks, which are open, hard, dry, and fine for camping, with plenty of wood, [and] . . . easily entered on almost every side" by "blind wagon roads." Here "camps are generally pitched so as to command several" of "almost twenty lakes, sloughs and ponds, varying from

**Fig. 8.1.** Hunting camp on sand ridge below Havana, March 1897. Just beyond this early camp scene in a stubble field may be seen a boat and high water. Courtesy of G. and V. Karl Collection, Havana.

three miles and a half, down to the ordinary pond" (fig. 8.1). Among these, "most prominent are Thomson's [*sic*], Johnson's, Slim and Duck Island, in Fulton County; Flag, Spring, and Mud and Clear in Mason County. . . . The lakes are generally shallow, and some may be waded; some are open, but most are broadly belted with wild rice, flags, grass, etc." Access was gained through Havana, or "steamboats often land parties right on the ground, hence Peoria and Pekin, from the north, are good initial points, where perfect outfit for camp may be purchased."[9]

Short periods of intense activity involved resident market hunters and recreational hunters when great flocks of mallards and other migrating waterfowl moved seasonally along the flyway that followed the Illinois Valley. The hunters freely tramped and fired, there being no constraining game laws and little posted land until after 1900. Already, breech-loading, double, chokebore shotguns were common, enabling hunters to tally large kills in a day. Recreational hunters came from as far as Chicago, Indianapolis, Columbus, and St. Louis. They included stockholders in the hunting and fishing clubs that became numerous after 1900 and maintained lodges and cabin boats on the leased or owned wetlands tracts that served the membership. Other sportsmen arranged for lodging, meals, and libations in rented cabin boats, lodges, and cabins and in the private homes, lodging houses, and restaurants of the riverside towns, which were "literally swarming with hunters" at the height of the seasons. In 1913, cabin boats could be rented for $1.50 per day in Bath, and one might board in Havana for $10 per week. Guides, live and carved decoys, and other gear could be rented. The accommodations performed double duty, serving summer tourists as well. There was also some entertaining of the business associates of absentee owners of large properties, to the discomfort of resident guides whose misfortune it was to be within patterns of shot unleashed by indiscriminate visitors.[10]

Before the 1920s, when the quality of roads and motor vehicles facilitated the weekend hunting trip, nimrods arrived in the area by rail, steamer, launch, or cabin boat. The same carriers served such summer tourists as Sunday fishermen or excursionists and families sojourning for a week or two. People from St. Louis and Alton came by steamer to Hardin and Kampsville or rode launches and their own motorboats to favored sloughs in the lower river. Residents of Peoria and Pekin ventured as far as Bath (10 miles south of Havana) in small craft or cabin boats, but Spring Lake (just south of Pekin) was more accessible for hunting and for black bass fishing. Otherwise, the duck hunters and the Sunday and summer visitors seemed to be drawn by rail and road to Havana, Beardstown, and Meredosia from towns nearby. In this respect, it is instructive that the residents of Naples voted in 1901 to have the town become "wet" so that they could vie with Meredosia for Sunday's visitors from Jacksonville. While the promise of splendid sport and a supply of good rental boats and fishing gear were important, "thirst parlors" were considered to be a town's prime asset.[11]

Reports of prodigious bags, or "slaughter" as it was put by some contemporaries, were fairly common until about 1905 along the river, as they were elsewhere in the Midwest. For example, trios of professional hunters were reported to have killed 2,200 ducks near Browning during three weeks in the spring of 1900 and 3,008 ducks in eight days below Bath in the fall of 1901. In November 1901, about 1,000 Mallards were obtained at Crane, Chain, and Stewart lakes by less than a half-dozen hunters from Decatur, one of whom shot 192 ducks in one day. Similar results were reported for a party that returned to Pekin in March 1902 with 25 dozen ducks, and a second trio shot 325 birds one day at Spring Lake. Whereas up to 100 ducks per hunter were shot at Duck Island in 1894, this favored location about 18 miles downstream of Pekin was reporting 5 ducks per day as fair shooting in 1905. Newsworthy accounts of hunting at Spring Lake included a wagonload of all kinds of ducks obtained by a pair of men in November 1903 but only 60 ducks for a party of seven in the following year.[12] Thereafter, newspaper accounts of hunting tallies ceased. More stringent regulations appear to have constrained commercial and recreational hunters—or made them more circumspect. In any case, a general depletion of waterfowl numbers, together with a loss of suitable feeding habitat in the valley, had occurred. A Peorian observed in 1906 that before the Sanitary District of Chicago began to divert water from Lake Michigan into the Illinois River system, "all of the favorite shooting grounds within fifty or sixty miles of [Peoria] in either direction afforded cover as secure as shooting from behind a barn. . . . Now these same marshes are open lakes of water, without a stalk of wild rice or a bunch of smart weed or buck brush to afford the hunter a natural blind."[13] It was conjectured variously that pollutants from Chicago's sewage were at fault; or that a wet cycle of precipitation had begun; or that the cover on more deeply flooded bottoms was plucked from the ground in the winter by the grasp of thick ice that rose and shifted with water level changes, readying the cover to be swept away when the ice went out on heavy flows of spring.[14]

While the year-round flooding submerged established lacustrine margin habitat, the spread of water over new areas fostered the development of rod and gun clubs where the feeding of migratory waterfowl became renowned. Meanwhile, large-scale land drainage projects were destroying a lot of wetlands. The hunters, too, continued to exact their toll, to judge from rare exposés of excesses. Illustrative was the confiscation in November 1913 and January 1914, respectively, of 500 and 300 ducks that hunters near Bath had packed into trunks and suitcases for delivery by rail to Chicago, a practice prohibited in 1904. The illicit traffic, handled more discretely by automobile, continued into the 1930s.[15]

## Commercial Fishing

At least from the 1860s, commercial fishing was the major employer and major source of income among the activities dependent upon nature's bounty in the valley. According to state fisheries officials, the relentless pursuit of the enterprise on both the Illinois and Mississippi rivers through the 1870s caused a decline in fish numbers and size "each year more marked." Buffalo (*Ictiobus* spp.), which had been so plentiful and inexpensive that it comprised "a large part of the food of the poorer classes," became "almost a luxury" for such consumers by mid-1880 because of advancing prices.[16] The annual harvest was less than a tenth of the 10 million pounds of buffalo reportedly taken in earlier years, when much of the catch was shipped in sugar hogsheads on riverboats to St. Louis. Sometimes, the volume taken in the spawning season exceeded shipping space and market capacity, which was costly for the shippers.[17] The waste was enormous, and catches of diminished size and annual volume were alarming.

Clearly, the Board of Fish Commissioners was created in July 1879 because the General Assembly feared that productivity and the fisheries sector were in jeopardy. The agency's principal tasks were to reduce excesses in the industry and to salvage and redistribute quantities of fish otherwise destined to perish each year as ponds and lakes on the floodplains dried into mud flats (fig. 8.2). The difficulties with the commercial fishing sector included a general indifference to closed seasons and a flouting

**Fig. 8.2.** Stranded fish, floor of Phelps Lake, 1894. Courtesy of G. and V. Karl Collection, Havana.

of regulations against the use of small-mesh nets. Not only were most of the fish taken with trammel nets and seines in the "rolling season" of the buffalo, but a considerable part of the eggs were destroyed. Alternatively, substantially all mature fish were intercepted in nets as they returned to the river through sloughs from spawning. Sometimes, word of the market collapse resulted in the discarding of catches in the river, creating a stench "for a great distance about."[18] "This is not the case with one lake alone, but in every lake along the rivers is the process repeated. For eight months in the year our rivers, lakes and streams are subject to this piracy. With so many loopholes through which to escape, the fishermen as a class are bold and defiant, regarding any attempt to enforce the laws as an infringement on their rights. So much so is this the case, that out of the great number of complaints filed with the Commission, of parties violating the law, not one percent of the parties informing consent to appear as witnesses, or allow their names to be used in any way in the matter."[19]

The Board of Fish Commissioners began in 1880 to rescue and redistribute fish of all species, notably working the floodplain water bodies of the Mississippi and Illinois rivers. Coarser species were returned to the rivers, but young bass, perch, and pike were placed in holding ponds and live boxes for shipment elsewhere by wagon, boat, or rail. The process continued intermittently. Exceptionally, as happened in 1889 and

1890, a seasonally dry floodplain locale near Meredosia yielded over 30 carloads of cans of game fish for shipment elsewhere. The fry of coarser species were returned to the river. Somewhat later, Meredosia was selected by the U.S. Bureau of Fisheries for one of a half-dozen fish-rescue stations operated in the basin of the Mississippi River in the early 1900s.[20]

In spite of the inroads made by fishermen and the vicissitudes of nature, the Illinois River sustained the second most important river fishery in the nation at the turn of the twentieth century, after the salmon fishery of the Columbia River. The catch, worth $327,500 in 1899, represented a quarter of the value of all fish marketed from the basin of the Mississippi River. In 1907, when the harvest from the Illinois River represented over 10 percent of the riverine catch in the United States, it was valued at $750,000. The valley's wetlands yield was estimated at $10 per acre. Five years later, over $1 million in coarse fish went to market by rail. Employment in river fisheries increased from perhaps 1,000 in 1899 to between 1,700 and 2,500 fishermen and shoresmen in the next decade. They included the self-employed and the employees of wholesale dealers who operated between Peoria and Hardin. Although commercial fishing was apt to be considered the lifeblood of the riverside communities, it is clear in the enumerators' census data for 1900, 1910, and 1920 that the towns supported service and bedroom roles for farming, as well. By the same token, farmers and farm hands contributed to recurring gluts of fish and a lot of waste.[21]

The congenial spawning and feeding conditions over the floodplain and the salvage and stocking efforts by public agencies helped to reinvigorate the industry between the 1880s and 1908. More important, however, was the rapid multiplication of the introduced German carp (*Cyprinus carpio*). The species, first shipped in 1879 by the United States Fish Commission to private parties in Illinois for pond stocking, was planted in public waters in 1884. Enthusiasm for carp culture was widespread, with hundreds of private ponds in rural and urban settings being stocked between 1880 and 1884. The view of the State Fish Commission that carp was the "coming food-fish" for Illinois was conditioned, no doubt, by the severe decline in buffalo stocks. Although the fisherman's and consumer's reservations about the merits of carp as food were acknowledged by the state commission in 1884, it was a moot point. The hardy, prolific, and rapidly growing species adapted very well to streams and lakes. Already in 1884, commercial fisherman at Pearl, Meredosia, and Pekin were taking large carp. Meanwhile, the commission undertook to educate commercial fishermen and consumers on the merits of the fish, in which task the *Prairie Farmer* was supportive. Consumers of modest means had little choice.[22]

The estimated marketable catch from the Illinois River rose from about 1 million pounds in 1889, through an annual average of 8.9 million pounds between 1894 and 1899, to an average of about 15 million pounds between 1900 and 1908. The latter average was skewed somewhat by a remarkable catch of nearly 24 million pounds in 1908, when exceptionally favorable conditions prevailed for the propagation and taking of fish. It is probable, too, that the record catch reflected a growth in the labor force in commercial fishing. A general decline followed, the catch being estimated at 8 million pounds in 1915 and something over 4 million pounds in 1921. A resurgence to 10.6 million pounds of marketed fish occurred in the flood year of 1922. Estimates made in 1930 were of the same order, perhaps because of the effect of a cycle of wet years and the river's recapture of large areas of the floodplain in 1926 and 1927 and the slow recoveries at drainage districts. Nevertheless, the area's one-term member of Congress, Guy L. Shaw, testified in 1924 that consumers' distaste for fish from the river required that vendors at Beardstown sell fish taken from the Mississippi River.[23]

The principal commercial species of the river and its floodplain lakes and sloughs were the native buffalo and the introduced carp, but buffalo declined from 55 percent of the catch in 1894 to 7 percent in 1908 and virtually was gone by 1913. The resulting pressure on game species such as the black, or largemouth, bass, crappie, and perch caused the General Assembly to enact laws prohibiting the use of any device other than hook and line to take game fish. Meanwhile, carp soared from 9.6 to 56.5 percent of the commercial catch between 1894 and 1897 and to over 64 percent by 1908. While carp represented 58 percent of the harvest in 1922, and buffalo 21 percent, the proportion of carp in the commercial catch of the late 1920s may have approached 90 percent. By then, buffalo, catfish (*Ictalurus* spp.), and bullheads (*Ameiurus* spp.) comprised most of the remainder. In the interim, the very desirable black bass (*Micropterus salmoides*) was being caught in declining numbers. The species represented about 20 percent of the commercial catch in Peoria Lake in 1898 but was uncommon in the lakes and river to the south of Pekin. It was virtually gone from Peoria Lake in 1904, when commercial fishermen took two black bass among thousands of pounds of haul. Like other game fish that were supposed to be taken by hook alone, the number of black bass appears to have declined because of intensified fishing, degradation of water quality, and destruction of spawning and feeding areas. Nevertheless, enough of the prized species survived near and downstream of Liverpool to support a vigorous trade in illicit consignment shipping by rail and boat to St. Louis and elsewhere in 1912 through 1914. The trade moved by automobile in the 1920s and 1930s.[24]

Most commercial fishing occurred between Meredosia and upper Peoria Lake, especially between Browning and Mossville (map 8.2), where 62 to 84 percent of the annual catch was landed between 1896 and 1908. This sector held such prime spawning and feeding areas as Quiver, Chautauqua, Thompson, Clear, and Spring lakes. Such fishing grounds in the lower valley were leased by wholesale dealers in fish, whose steam tugs or gasoline launches, rowboats, cabin boats, and nets were manned

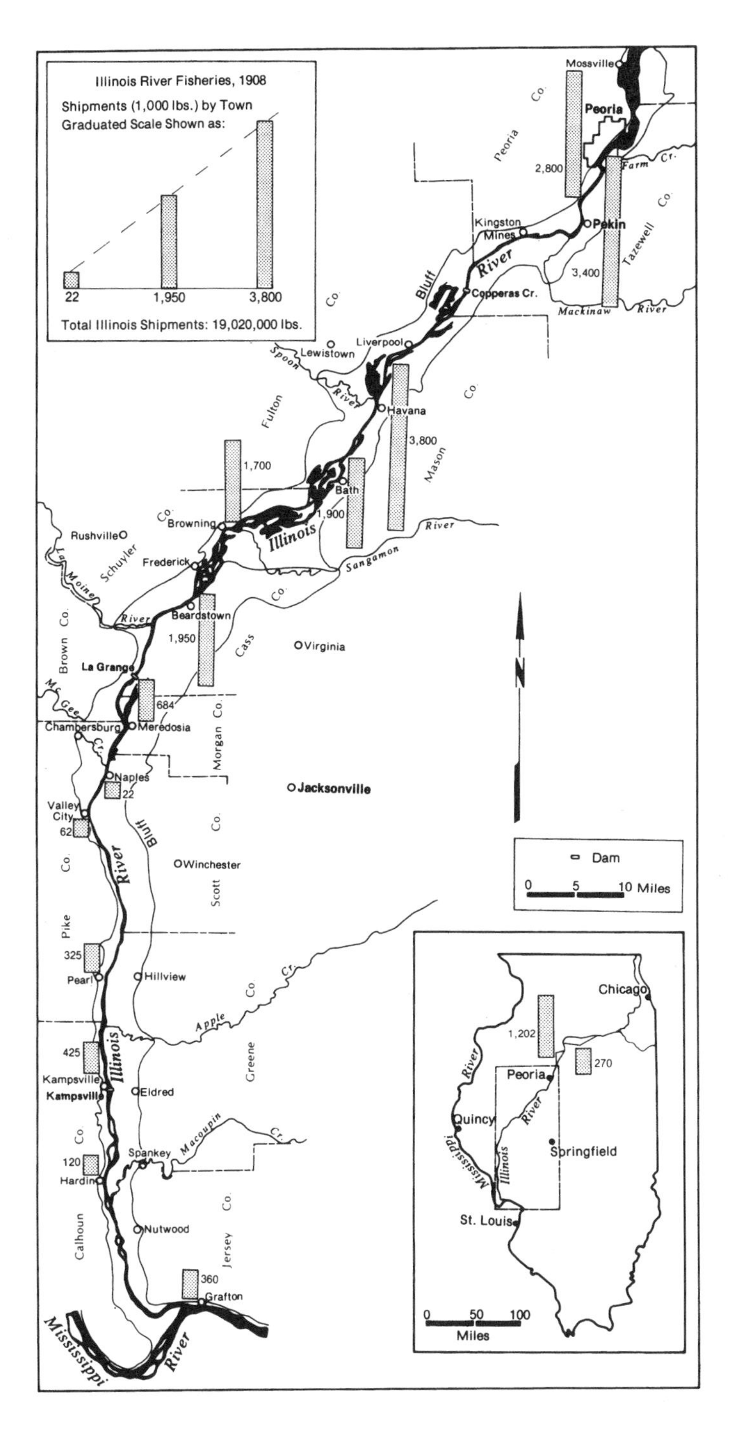


**Map 8.2.** Illinois River fisheries, 1908

by resident employees. The dealers also purchased fish from self-employed fishermen. The dealers operated with large "market boats" (floating packing sheds) and icehouses in a number of communities. The businesses supported clusters of fishermen numbering almost 20 in Hardin and 50 in Browning in 1899, 34 in Meredosia and 89 in Beardstown in 1900, and well over 100 within a 10-mile radius of Havana in 1900. Actually, Havana's resident commercial fishermen numbered 64 in 1900 and 71 in 1910. At the time, Havana's wholesale houses (fig. 8.3a) shipped more fish than any other town on the river. The trade filled at least 55 carloads (averaging 26,000 lbs. each) in 1899 and 100 carloads annually in the decade following. By then, live fish were shipped to the East in express cars fitted with tanks of aerated water. The trade, worth $100,000 in 1907, provided about half of the town's income. Havana's position was sustained for some time, as many as 250 to 350 people being engaged thereabouts by the industry in 1915.[25] The town's share of traffic from the Illinois River was about 19 percent, as may be noted in map 8.2. Important, too, were Pekin (14%), Peoria (13%), and Beardstown (13%). It is likely that between 1,250 to 1,750 people were gainfully employed in the industry along the river then.

The commercial fishing season began in July and peaked during the fall through spring. Prices were highest in winter, but the catch was largest in the spring. Most of the catch was taken in manually drawn haul seines, which ranged from about 10 to 20 feet wide and 200 to 1,000 feet long (fig. 8.3b). However, prior to 1917, seines 3,000 to 4,500 feet long were in use in Meredosia Bay, near Havana, and in Peoria Lake; they were winched ashore with power from gasoline engines. Fyke nets that were about 5 to 6 feet wide and 15 feet long were used in the confines of the sloughs. As a rule, the catch was sorted by species, some were dressed, and all were packed with broken ice in barrels or boxes of 250 pounds (100 lbs. of ice) for shipment in iced cars to consignees in such cities as New York, Boston, Philadelphia, Atlanta, Memphis, and Chicago. Also, substantial quantities of iced fish moved to towns of central Illinois with peddlers or in express consignments to dealers. Sometimes, the packing was done in market boats on the fishing grounds, the catch being removed from floating live-boxes into which they were thrown from the tightening circle of the drawn net, as is illustrated in figure 8.3b. More commonly, the packing was done on market boats moored in front of riverside towns and villages. Most of the ice was cut in nearby ponds or in tributaries to the Illinois River. Since an average of 11 tons of ice were used per carload, the cutting and storage of ice or its manufacture were important local activities.[26]

All market boats had live-boxes floating nearby. Their contents were replenished from large cribs or modified ponds located at the leased or owned fishing grounds and elsewhere. The practice of holding and feeding fish probably began as a response to early gluts of buffalo. However, holding ponds are first documented for Browning and Clear Lake in 1898 and near Meredosia in 1901 and Havana in 1904. A 25-acre

**Fig. 8.3a.** Wholesale fish holding and packing facilities, Havana, 1898. Market boats, cribs, and cabin boats line the waterfront. Courtesy of G. and V. Karl Collection, Havana.

**Fig. 8.3b.** Commercial fishing crew, Clear Lake, date unknown. The encircled catch is transferred to floating live boxes, such as the object to the right that is low in the water. Courtesy of G. and V. Karl Collection, Havana.

spring-fed pond was modified to hold fish near Pearl in 1905. In that year, too, the enthusiasm for rearing trout to marketable size resulted in ponds being shaped on the bottoms near East Peoria. However, the investment was lost because the trout tasted like the ground beef liver upon which they were reared. Otherwise, the investment in such holding facilities and their contents were subject to vagaries of the river and weather. Although investment in such ponds is documented as late as 1917, when one was dug by dipper dredge opposite Beardstown, it was more common to adapt borrow pits for short-term holding of carp and buffalo catches—which is documented as late as 1931 for Meredosia.[27]

A variant procedure of rearing fish for market was adopted at the Coal Creek drainage district shortly after the original creek channel was sealed behind the levee system. In 1900, the district's manager had five miles of ditches and canals swept of fish, mostly carp, which were placed in a 200-acre pond located near the bridge to Beardstown. The carp were fed stable sweepings and much the same feed suited then to finishing hogs. They were seined in 1901 and 1902 and sorted and packed in ice for express shipment to consignees in New York. The enterprise lasted at least until 1906 and was reported in 1905 to have shipped an average of $5,000 of fish per year. Apparently, the operating costs were too high to attract others into fish farming. General farmers were not responsive to the suggestion made by Alvord and Burdick (1915) that fish be reared as a sideline, like keeping poultry.[28]

Elsewhere along the river, the commissioners of newly leveed drainage districts sold rights to seine the leveed lakes and sloughs as pumps drew water levels down. The harvests were large, to judge from the experience at the Swan Lake district, near Hillview, in 1907. A first shipment of 250,000 pounds of fish warranted an estimate that a total 400,000 pounds of fish would go to consignees in New York, Philadelphia, and Louisville. Similar agreements to take all marketable fish were concluded between wholesale houses and commissioners of the East Peoria (1911), Scott County (1913), and Thompson Lake (1922) districts. It remained common practice, too, for the state's Fish Commission and successor agencies to transfer coarse fish to the river and to plant the game fish fingerlings elsewhere in Illinois. Probably the largest of such tasks for the Division of Game and Fish was the transfer in 1918 of 85 tons of fish from the South Beardstown Drainage and Levee District to the river.[29]

The survival rate is unknown for fish rescued by seine from the shallow lakes and transferred elsewhere in the state or to the more benign conditions of the river. There are descriptions of considerable losses for fish held in live-boxes, however. It was common for the water in live-boxes to be covered with dead fish in the morning. The losses were attributed to injuries received in netting and in the transfer to live-boxes, to crowding, and to warm water during the summer. As with the residue from dressing fish for market, the remains were thrown into the river when they were not fed to hogs or poultry. The coincidence of hot weather and market glut could be wasteful in the extreme; for example, a fisheries official recalled sailing along a continuous 50-mile trail of dead fish discarded at Beardstown.[30]

Thus, the commercial fishing industry along the Illinois River was built on the buffalo and sustained on the carp, which held its own against the excesses of commercial fishing, stream pollution, and the reduction in spawning and feeding grounds attributable to the land reclamation process that ended in the 1920s. Less resilient to heavy exploitation and habitat modification were the mussel beds, which supported an industry that ran its course quickly during the era of land drainage.

## Pollution and the Fisheries

The concern of residents in the valley of the Illinois River over the aesthetic, health, and economic effects of sewage and industrial waste from Chicago was touched on in chapter 5. The pollution problem bears amplification. It was the rule for towns and industries to discharge waste water into adjacent streams and rivers during the nineteenth and early twentieth centuries. At least by 1887, the Fox and Illinois rivers converging at Ottawa were undrinkable and malodorous at low water stages. The conditions were known at Peoria, too. The summer air at Beardstown in 1889 reeked of dead fish and decomposing garbage and vegetation that residents dumped on the town's watery margins. There was no landfill or organized waste collection system. Peoria's situation was more complex. Among its contributions of waste to the river was a very large volume of slops generated by a number of distilleries and breweries. The waste either was dumped directly into the river or fed to livestock in the town's yards (40,000 head of cattle in 1891) and then entered the river as animal waste. Pekin also contributed to the fish kills of summer that extended downstream to the lock and dam at Copperas Creek. The river and its banks were dark with decomposed organic waste, and accumulated material was dredged now and again at Peoria to facilitate steamboat and barge movement. Analyses of the river at and below Peoria in 1886 and 1899 revealed pollution resembling conditions in the stockyards branch of the Chicago River.[31]

The sewage and industrial waste discharged into the Illinois River system from Chicago after 1900 profoundly degraded the plant and animal life in the stream. Chicago's waste, which contributed 75 or 80 percent of the river's pollution load, overcame the buffering role of extensive Peoria Lake by 1912, destroying the commercial fishing industry by 1919. Another 17 percent of the stream's pollution load was contributed at Peoria and Pekin, degrading life downstream to Havana. Absorption of the decomposition products of sewage and sludge by fish resulted in a "gassy" or carbolic-acid-like taste to their flesh. At least as far downstream as Meredosia, commercial fishermen tried to rid their catches of the taint by holding the fish for weeks to months in ponds or borrow ditches that were fed by springs or tributaries. Nevertheless, as noted earlier, the industry's production in 1915 was but one-third of the 1908 level and in 1921 was but 15 percent.[32] The loss of breeding and feeding grounds to drained farmland was compounded by the degradation of the water.

The implications of stream pollution for public health and fisheries survival was brought to the attention of the governor and the General Assembly at least as early as 1900 by the Board of Fish Commissioners. However, the General Assembly took over a decade to authorize the Rivers and Harbors Commission to abate industrial pollution.

Compliance with enforcement efforts was drawn out because of labor and material shortages during World War I. In 1917, the successor agency, the Division of Waterways, began to press communities to install adequate sewage treatment facilities. In 1919, it sought state intercession with the Sanitary District of Chicago to require the treatment and removal of solids and offal from the flow diverted into the Illinois River system. Already, federal authorities urged the adoption of such measures to reduce the diversions from Lake Michigan. About four years later, the General Assembly required the state's towns and cities to construct effective sewage treatment plants. Helpful in marshaling public insistence that something be done was the newly formed Izaak Walton League of America, which in 1923 ranked the ending of stream and lake pollution, along with the curbing of the excesses of commercial and recreational fishing, among its top programmatic priorities. In late 1924, the Chicago-based Waltonians launched a campaign to save the Illinois River. By then, a scattering of large towns in the basin of the Sangamon River were building the state's first modern sewage treatment plants, aside from a facility in Chicago. Peoria followed suit, becoming in 1931 the first city to have a sewage treatment plant on the Illinois River. The impetus at Peoria was a campaign mounted in 1926 by the local chapter of Waltonians. By the early 1930s, numerous chapters elsewhere in Illinois and the nation pressed authorities and the public to support programs to keep local streams clean. However, economic realities delayed construction of sewage treatment plants in many municipalities.[33]

## Mussel Gathering

Mussels were gathered from beds in the Illinois River sporadically in the 1870s and 1890s to obtain shells from which mother-of-pearl buttons were made and some pearls and slugs were taken. The pursuit became intensive between 1904 and 1913. Heavy exploitation followed the arrival of hundreds of professional shellers from the Wabash River and other streams in the Ohio River system, where resource depletion was advanced. Other shellers penetrated the Illinois Valley as part of the activity begun on the Mississippi River at Muscatine, Iowa, in 1891, when the first pearl button factory began operating. On the new grounds, a man with a boat could make $2 to $7 per day. Shell production in the Illinois Valley reached on the order of 9,500 and 8,000 tons per annum in 1908 and 1913, respectively. It dropped to 5,700 tons in 1914 and dwindled to less than 300 tons in 1923. Although product value at the peak of activity probably did not exceed 25 percent of the value of commercial fishing, shelling employed a lot of people. The Cornucopia soon emptied, however; shelling was thoroughly destructive, and stream pollution and sedimentation were damaging. Such experience was representative for streams in the Midwest, although the chronology and relative roles of destructive factors may have varied. In any event, pollution is credited with having finished off the mussels as far downstream as Beardstown between 1913 and 1920. Less than half of the species once known to the river have reappeared.[34]

It is possible that the river's shells were the first of domestic origin used by pearl button manufacturers in the United States. The raw material was shipped from Peoria (1872), Beardstown (1876), and Meredosia (1892). Activity apparently was not sustained because the cost of imported pearl buttons remained lower until the enactment of the McKinley Tariff (1890) protected production on the Mississippi River. Once the activity reached the Illinois River, a bonanza occurred, the stream being described in 1912 as the most productive per mile in the country. The harvest involved a number of species, notably *Quadrula* and *Lampsilis,* which were of good quality for button making.[35]

Although the intensive phase was begun by shellers looking for new beds to exploit, local fishermen, laborers, and farmers with slack time were drawn into shelling. They employed as many as 2,600 launches and johnboats between Peoria Lake and Grafton at the peak of the boom. Perhaps 3,000 fishermen and shoresmen were involved. By 1912, however, only 867 craft were active. Many of these operated below Kampsville, where the late exploitation of the beds probably reflected their lower productivity relative to beds upstream. Almost 60 percent of the working craft were powered with gasoline engines in the 2 to 20 horsepower range, and it was the rule to have an adjustable underwater sail rigged aft to facilitate drifting over the beds. From such a boat, a man could harvest 400 to 700 pounds in a 10-hour day using the preferred apparatus, the crowfoot dredge or bar. Employed to good effect on Peoria Lake was a dip net with prongs that projected forward from its lower edge. Either gear was cheap to make, required modest skill to use, and was very effective.[36]

As a rule, shellers operated out of nondescript and malodorous camps where tents and shacks supplemented cabin boats and where crude cooking vats were set up for steaming open the mussels. While the shells were thrown into piles to await cleaning and sorting, the soft material was discarded following probing for pearls and slugs. Shells were sold to intermediaries or to factories where button blanks were cut. These small plants were located in river towns with railway facilities, which carried blanks to pearl button factories in the East. Local factories first arose in Peoria and Beardstown, but by 1912 there were 15 small and short-lived plants in those two towns and in Meredosia, Naples, Pearl, and Grafton. Beardstown and Pearl each had 5 plants where button blanks were cut, and Meredosia had 2. Most factories began operating after 1910, only Beardstown having a work force (about 115) so employed at the time of the 1910 census. Beardstown was the trading center for pearls and slugs, with resident and visiting buyers vying for the rarities.[37]

The original prime areas of mussel production were in Peoria Lake and between

Bath and Meredosia, but the latter zone was depleted by 1912 and the lake soon afterward. Dwindling harvests caused professional shellers to move on, soon to be followed by the factories. Thus, the unregulated extractive activity ran its course in about 15 years, leaving a greatly impoverished fauna and habitat. Indicative of the concern about the condition of the beds along the river, there was talk in 1912 about allowing them to recuperate through a selective ban on shelling for periods of three to five years. By then, the General Assembly had adopted a closed season and other protective measures. Also, the U.S. Bureau of Fisheries was then well along with a program of artificial propagation that, had it been successful on the Mississippi River, might have been expanded to the Illinois River. However, prospects for the recovery of the mussel beds were slim, given the virtual elimination of the fauna by 1920, the disruption of the cycle by which young mussels were disseminated, the growing pollution problem, and the degree to which topsoil was accumulating in the lakes and navigation pools of the Illinois River.[38]

## Development of Rod and Gun Clubs

Private fishing clubs and various hunting and fishing clubs were established before 1900 along the bottoms of the Illinois River. Some of the fishing clubs may only have owned elaborate houseboats, but the hunting and fishing clubs owned or leased floodplain property. These clubs became quite numerous right after 1900, to the dismay of shooters who were unaffiliated. Hunting and fishing clubs were either private commercial enterprises (charging per diem fees) or exclusive clubs. They occupied land owned and leased by associations of stockholders whose avid interests in mallards or black bass and other quarry conferred on overflowed land with sloughs or streams a value exceeded only by drained farmland. Some of the club properties were valued in 1907 at $100 or so per acre. Among the well-known commercial enterprises were the Duck Island Hunting and Fishing Club (incorporated 1886), the Spring Lake Hunting and Fishing Club, located about 18 miles downstream of Pekin, and the Treadway Lake, or Phelps, resort, located about 5 miles northeast of Beardstown. Like most such enterprises, the Duck Island and Treadway Lake resorts were purchased by late 1914 to serve as exclusive clubs. Club stockholders were business and professional men who lived in county seats adjacent or in distant cities such as Springfield, Bloomington, Decatur, Rock Island, Moline, Chicago, Indianapolis, and St. Louis. Some clubs with a membership base elsewhere took the precaution of inviting local attorneys and businessmen to affiliate. As a rule, the enterprises and clubs built lodges on the preserve, kept cabin boats, duck blinds, and the like and had the premises maintained and patrolled by caretakers. The clubs tended to set aside areas as waterfowl refuge or rest areas.[39] The premises were enhanced for hunting by contractors who "shot" ditches and ponds with dynamite, as was done in 1916 by at least two clubs located near Beardstown. Sometimes, small ponds were maintained with water lifted from shallow wells by a windmill. At least one club located near Bath countered the effect of a dry year upon its ponds by operating a rented gasoline-engine, 8-inch pump and barge in an adjacent slough. These approaches to readying duck ponds were more viable than asking one's senator to intercede with the Secretary of War to allow the Sanitary District of Chicago to increase the flow of water from Lake Michigan, as was reportedly done in 1901, ostensibly to save pastures for livestock.[40]

Some gun clubs licensed commercial fishermen or recreational fishermen and hunters to use the premises. The Rushville Hunting Club, for instance, obtained a quarter of the value of the catch from commercial houses that fished the premises. The income, $500 in 1911, was some compensation for the estimated $1,800 worth of fish lost to poachers, among whom self-employed fishermen were common. The approach to handling recreational fishermen and hunters that was adopted by the club on the Spring Lake bottoms, by owners of the prospective South Beardstown Drainage and Levee District, and probably others was to charge a nominal fee.[41] Had the tracts in question not been so close to a relatively large town with rail service and amenities for the visitor, the procedure might not have been necessary. In less populous and less accessible areas of the floodplain, verbal understandings probably sufficed between landowners or their tenants and acquaintances of the vicinity who wished to hunt, fish, or trap.

Rod and gun clubs with premises on the bottoms predate 1900, but their numbers increased rapidly after 1901 and 1902. The earlier associations tended to be formed by business and professional men of local county seats, such as those organized at Pittsfield (in 1892) and at Jerseyville and Canton (in 1895), whereas the later rush occurred when the fellowship of sportsmen elsewhere in Illinois and Indiana realized that water from Lake Michigan was engulfing the bottoms. In effect, the general economic growth that forced Chicago to develop a sanitary and ship canal and that facilitated the organization of drainage and levee districts enabled relatively affluent people to invest in lodges, property, and leased acreage on the wetlands. The individual premises of early clubs occupied between 1,800 and about 10,000 acres between the present Pekin and La Marsh district and the Nutwood district. Notable among clubs that were antecedent to the drainage and levee districts organized from the premises were: the Rushville Hunting Club (organized 1902), succeeded by the Big Lake Drainage and Levee District in northeastern Schuyler County; the Thompson Lake Rod and Gun Club (incorporated 1901), located between Liverpool and Havana; and the Spring Lake Hunting and Fishing Club (incorporated 1902), which occupied about 10,000 acres in southwestern Tazewell County. While the stockholders in the Thompson Lake club were notables from Indianapolis and Terre Haute, members of the Spring

Lake club resided in Morton, Mackinaw, Armington, Manito, and Chillicothe. Among clubs in which Chicagoans predominated were the Knapp Island Gun Club (organized 1907), located adjacent the outlet of the Sangamon River, and the 25-member Grand Island Lodge, which acquired property near Bath in 1901 and 1902. Similar clubs functioned before 1900 on the floodplain of the Mississippi River with membership drawn from Missouri and Illinois.[42] For some stockholders in the clubs, the words of a Pittsfield man captured the essential idea of belonging: "Just a nice, quiet, peaceful place somewhere; not too close to the man of care and not too far from the millionaire."[43]

The clubs became very numerous in the lower valley of the Illinois River, occupying by 1918 nearly an unbroken stretch of overflowed bottoms and backwaters for 30 or 40 miles above and below Beardstown. There were over 125 licensed clubs in Mason and Cass counties alone by 1930. By then, the most desirable wetlands sold in farm-sized tracts for $150 to $300 per acre, sometimes $400. Well-located, 40-acre pieces suited to dryland baiting fetched $100 per acre. Similar bottoms near the river's mouth were worth $50 per acre.[44]

## Conflict over Wetlands Use and Access

Among commercial fishermen and hunters, many of whom are presumed to have been the same people, the transformation of favored wetlands haunts into fields of corn and wheat, and the segregation of remaining areas into hunting reserves for non-residents, was a matter of vexation and deep concern.[45] In words attributable to R. M. Chiperfield, a Canton attorney and unrelenting foe of the gun clubs: "Hunters and fishermen have been annoyed no little by meddling people in their frequent prosecutions where really there was no excuse for doing so. Frequent arrests have been made and fines inflicted that ought not to have been, being mostly technical rather than real violations. These suits and arrests are too frequently made through spite rather than in the name of justice."[46]

The sensibilities of fishermen and hunters were abraded especially when posted notices, cables or other barriers across sloughs, and uncharitable gamekeepers reminded them that they were poaching. They found some kindred spirits among the recreational hunting and fishing sectors, to judge from circumstances that arose at Spring Lake, to the southwest of Pekin, and at Thompson and Clear lakes, near Havana. In the spring of 1909, a "duck trust" consisting of the operators of the larger hunting clubs was attracting waterfowl onto their premises by organizing feeding programs, by limiting shooting to selected weekdays, and by dispatching hired men with guns to cannonade with abandon at public hunting grounds on Sundays. The feeding programs, at least, were nothing new. They continued to such an extent that Aldo

Leopold called Beardstown "the baiting capital of America" in his 1930 game survey of the north-central states. Leopold reported that clubs spread up to 430 bushels of corn per acre to draw mallards around strings and pens of live decoys located near blinds where 6 to 10 guns were accommodated. It was common for the guides to contribute to the success of clients with an automatic shotgun. Such sport cost $5 or $10 per day per person from 1914 to 1928, and as many as two or three parties of shooters were rotated through a blind per day. This commercialized shooting was a far cry from the days when hunters roamed at will, built blinds that were sacrosanct, staked tame drakes in the water to decoy mallards, or used stalking horses to get close to geese feeding in fields. Baiting was banned in 1934.[47]

The conflict over access to areas that hunters and fishermen had been accustomed to frequenting resulted in a handful of law suits, some of them celebrated and drawn out. These arose out of attempts by landowners to stop poaching on lakes and sloughs to which they claimed title, as was the case at Thompson, Clear, and Mud lakes; or they arose because the organization of a drainage and levee district was expected to extinguish a renowned hunting and fishing area, as was the case at Spring and Thompson lakes. Regardless of the immediate cause, the opponents of land drainage held that the waters in question were navigable, which conveyed to them the status of public property; property that the United States had transferred to the State of Illinois to hold as a public trust when statehood was granted in 1818. While there was no arguing with the provision of the Constitution, the navigability of the sloughs and lakes in question was claimed to be a function of water levels raised by flood stages or by insertion of locks and dams on the Illinois River and water from Lake Michigan.[48] Moreover, similar floodplain features were reclaimed for agriculture without contests of the title issue by either the public or the State of Illinois. Ultimately, the view that seemed to prevail was that the development of settlement and revenue producing lands was more important than preserving swamp and overflowed lands for hunting and fishing.[49]

Intruding into the contest over who owned the lands and waters in question were the matter of resident use versus outsider use of local resources and the sense that people who earned their livelihood from the river were arrayed against wealthy urban industrialists and professionals, among whom residents of Indianapolis and Chicago were particularly objectionable as owners of playgrounds on the Illinois River. The sentiments were captured in the *Mason County Democrat's* editorializing now and again. In 1907, for example, the editor lamented that Havana's hostelries and shops had lost much of their conventional sportsmen's trade because of the spread of exclusive domains where not even the prestigious brands of spirits available in town measured up to what was consumed in the private lodges and railway cars. In 1909, the editor fumed over an embarrassing experience of the Republican governor, Charles

S. Deneen, and Nat H. Cohen, Secretary of the Illinois Fish Commission, on the Thompson Lake premises. In spite of a letter of authorization from Indianapolis carried by Secretary Cohen, the fishing party was denied access by "the ill-bred vassal of a band of corrupt millionaire politicians," among whom Harry S. New, the recently named chairman of the Republican National Committee who ultimately became a U.S. senator (1917–23), was best known.[50]

The principal contests arising from poaching on gun club properties, and which were appealed to higher courts, arose over Clear and Mud lakes and the lands between, in northwestern Mason County, and at Thompson Lake, in Fulton County. The former suit, *Schulte vs. Warren,* reached the Supreme Court in 1905, where it was decided that the landowner's title to the edge of the lake, which had been meandered by federal surveyors, was not voided as a result of the submergence by waters raised by the works of man; and that although such inundation afforded the public the right of navigation, it did not convey to the public the right to fish and hunt over the flooded lands. In rendering its opinion, the court summarized the events at Clear Lake: "Some years ago a lock and dam was built at La Grange, below the lands, which raised the water of the lake about eighteen inches, and afterward the [Chicago] sanitary district canal was opened, raising the water three or four feet more, so that the natural stage of the water in the river is about five feet higher than in its natural condition. The lands have been overflowed and rendered practically worthless for any purpose except hunting and fishing."[51] Shortly thereafter, Schulte's 1,000 acres were purchased for $30,000 by a resident of Terre Haute for use of the Clear Lake Outing Club. Schulte, owner of a water taxi and towage service and a fish wholesaling firm, retained fishing rights. Defenders of the customary access to areas of fishing and hunting sought legislation affirming the state's title to all water surfaces between banks that had been meandered by federal land surveyors. But the intent to effectually abolish hunting and fishing clubs failed.[52] Although the editor of the *Mason County Democrat* lamented that nonresident ownership of "nearly all of the available hunting and fishing grounds" had "come about so suddenly that our people are temporarily off their feet,"[53] the people never quite recovered.

In one way or another, the controversy over access to Thompson Lake continued for over two decades following incorporation of the Thompson Lake Rod and Gun Club in 1901. The legal contest that pitted the club against fishermen from Havana and Liverpool culminated in the United States District Court, to which the successor to the club (dissolved in November 1908) resorted in 1909, believing he could not secure relief in local courts as a nonresident of Illinois. In the course of this action, the federal court issued an order against trespassing pending the outcome of proceedings. These ended after four days before a master in chancery, the plaintiff withdrawing his request for a permanent injunction against trespassing, which resulted in the dismissal of the suit. As a consequence, an antecedent decision made by Lewistown's circuit court in 1908 stood; the club could not claim title to the lake because it was shown to be navigable.[54]

Although the outcome of the Thompson Lake suit was a victory for the fishermen, the contest entered a new phase in late 1911. Proceedings to form a drainage and levee district began; the district was organized in 1915.[55] "Tiring of their toy, they now propose to drain the lake," noted the editor of the *Mason County Democrat.*[56] Soon, the purchaser of the club's land obtained a permanent injunction against trespassing from the circuit court of Fulton County. The suit was carried to the Supreme Court, which sustained the lower court's judgment on the successor's ownership of the land to be reclaimed. Again in February 1922, an injunction was obtained in federal court to prevent Havana and Liverpool fishermen from crossing the levee for any purpose, least of all for seine fishing, which privilege was let to two Peoria firms. Fishermen were informed conclusively that the land and water within the district were no longer in the public domain in April 1924, when trespassers from Liverpool were fined and incarcerated.[57]

The prospect of losing Thompson Lake to drained farmland so concerned anglers and the local commercial fisheries sector between 1912 and 1915 that a concerted effort was made to marshal support among sports fishermen and sports hunters to persuade the General Assembly to take action. Illustrative of the position was a district game warden's published appeal in the *Rushville Times:*[58] "If you are a sportsman, now is the time to get busy with the members of the legislature, or if you are a commercial fisherman, you will organize and do something at this coming session of the legislature, otherwise all sporting and commercial fish will be a thing of the past. Also remember, if you eat fish—and it is a cheap and wholesome food, you will be compelled to pay higher prices for inferior and imported fish when our native fish are gone." The writer acknowledged that pollution and certain practices of the seine fishermen contributed to the depletion of the fisheries but asserted that the principal reason was "the law of the state which makes it possible to destroy the breeding grounds. This law allows private citizens to form drainage districts of the overflowed lands which are the natural spawning grounds." The writer followed with the roll call of destroyed spawning areas:

> Spring Lake . . . was one of the best fishing grounds in the state twenty years ago; now it is a levee district and the fish have been destroyed. Dan Hole's Field, which is northwest of Havana about four miles, was considered by experts to be the best spawning ground. Big Lake, 18 miles south of Havana, has just been levied [*sic*]. West Matanzas, in Fulton County, has just been put into a drainage district. I understand from reliable sources that all overflowed land from Clear Lake to Quiver Lake, a distance of about 15 miles, on the east side of the river, is ready to

> be put in a levee district, and all the land from Liverpool south to Thompson Lake will soon be put in drainage districts. Thompson and Flag lakes are also to be put in drainage districts as soon as the suits pending are disposed of.

The writer recommended that if the legislature could not be persuaded to stop levee building along the Illinois River, it should appropriate funds to purchase 25,000 to 30,000 acres of the best remaining spawning grounds, where preservation measures, including controlled public access, would prevail. The project's operating costs were to be paid through fishing license fees. At the time, the Illinois Game and Fish Conservation Commission was authorized to set aside preserves in public waters, but not on private properties.

The desirability of preserving Thompson Lake and kindred spawning grounds was a matter of no small moment to Havana's economy. As the *Mason County Democrat* commented, public access was being denied at a time when Havana aspired to continued growth as the premier summer resort of Illinois, while keeping its role in commercial fishing. It was asserted that the superior quality of the area's clear water lakes for black bass and other game fish outweighed the fishing, scenery, and climate that were offered by Wisconsin and Michigan. As testament to that end, there was a trade in vacationers and Sunday visitors valued at $75,000 to $100,000 per annum. This equivalency to the value generated by commercial fishing was believed to prevail everywhere along the lower river by World War I. The summer tourism that blossomed about 1900 continued through the early 1920s, by which time the July 4 to Labor Day season was marked by the arrival in Havana of "almost a parade" of automobiles heaped with baggage, fishing and camping gear. This growth of tourism was paralleled by the development of clusters of cottages near the catalytic hotels, dance pavilions, and boat liveries of the beaches at Quiver, Matanzas, and Baldwin and at Chautauqua Park (figs. 8.4a and 8.4b). Similar clusters of cottages appeared at Liverpool, Bath, Kampsville, and other riverside villages. Perhaps contributing to the tourist activity in Havana and the lower valley by the 1920s was the extent to which summer cottage owners, excursion boat passengers, and recreational users of Peoria Lake above Chillicothe, and recreational users of the river between Peoria and Pekin, were discouraged by the river's foul smell and unpalatable fish. It may have been a matter of degree; malodorous water in 1923 discouraged bathing, boating, and recreational fishing as far downstream as Beardstown.[59]

No untoward events appear to have arisen between users of the waters open to the public in Thompson Lake and owners of adjoining land between 1909 and 1915. In the interim, a prolonged conflict over the prospective interruption of navigation on Spring Lake, in Tazewell County, was resolved. The commissioners of the Spring Lake district agreed with the Attorney General's office and the Rivers and Lakes

**Fig. 8.4a.** The Quiver Beach Hotel, Havana vicinity, 1920s. Courtesy of G. and V. Karl Collection, Havana.

**Fig. 8.4b.** The barge *Pearl* approaching Chautauqua Park landing with excursionists, Havana vicinity, 1920s. Courtesy of G. and V. Karl Collection, Havana.

Commission that the district should complete its reclamation project while preserving the navigability of the lake. The compromise was agreed to in late 1913, about 18 months before the petition to organize the Thompson Lake district was approved in June 1915. While it is likely that attorneys who counseled landowners in the prospective district at Thompson Lake were aware of the resolution of the Spring Lake contest, to what degree events in Tazewell County affected the decision to form a district in Fulton County is unknown. In any case, these two drainage districts evolved with a continuity of land ownership from a period when they were the projects of rod and gun clubs. By the same token, the organization of the Big Lake (1912), West Matanzas (1914), and Liverpool (1916) drainage and levee districts was achieved when a substantial part of the acreage was yet controlled by either rod and gun clubs (in the case of the Rushville Hunting Club and the West Matanzas Shooting Club) or by stockholders who succeeded to land parcels (as happened at Liverpool).[60]

In 1902, the Spring Lake Drainage Association was organized by stockholders for whom about 10,000 acres of land was accumulated earlier for the Spring Lake Hunting and Fishing Club. The club appears to have permitted nonmembers to hunt and fish the premises and use its elaborate 40-room lodge overlooking Spring Lake as a means of defraying expenses. Nevertheless, the promise of converting wetlands into cropland resulted in petitioning (1903) the circuit court in Pekin to form a drainage and levee district. The intent was to drain the upper two-thirds of the tract, while reserving the lowest area for hunting and fishing. The decision to preserve a sizable area for continued recreation and revenue generation may well have been made to pacify the opponents of reclamation, among them commercial fishermen of the area from Pekin to Manito who formed the Fishermen's Protective Union of Manito in November 1903.[61]

The hue and cry of commercial fishermen, echoed apparently by recreational fishermen and hunters and by farmers who shipped grain by barge through the lake and a canal cut (1879) northerly to the river, resulted in intervention by the Attorney General of Illinois to preserve the lake's navigability and public access. However, the proponents of land drainage provided credible arguments that the lake's navigability was a function of high river stages, the completion of the Copperas Creek lock and dam in 1877, and the diversion in 1900 of water from Lake Michigan by the Sanitary District of Chicago.[62] A standoff was in the making.

Levee construction and ditching for the district began in 1906, after two years of delay. The work proceeded with an understanding (October 1905) that there would be no blockage of access and transit by water on Spring Lake. Levee construction sealed the northerly canal to the river (fig. 8.5), and the pumping station near the lower end of Spring Lake was well along in August 1909 when a dredge was positioned to build a dam across the slough that was the lake's remaining access channel.

**Fig. 8.5.** View toward levee across state-built canal into Spring Lake, ca. 1909. The levee shows a recent layering of spoil atop structure where brush rises behind piling. The cabin boats are representative of craft occupied by fishermen. Illinois Submerged and Shore Lands Investigating Committee; courtesy of Illinois State Museum.

However, angry commercial fishermen moved four cabin boats into the slough and appealed to the governor for help. The threat of an armed showdown was allayed by the intercession of sheriff's deputies, and the contest shifted to the courtroom, where the district filed a bill in equity against the Attorney General (January 1910). This action to forestall the state's enjoinment of levee system closure made possible the completion of the dam in mid-1910. Ultimately, the Supreme Court decided (February 1912) that the district's obligation was to preserve a navigation channel, not the lake. By then, investors in the district faced heavy financial losses had the state prevented the damming of the slough. Persuaded that the district acted legally in its pursuit of draining land with a prospective value of $1.4 million, the Rivers and Lakes Commission recommended in March 1913 that the state's proceedings be dropped. Negotiations continued through December 1913, by which time the district proposed to save the lake by surrounding it with a levee and to preserve access by building a marine railway and related improvements to move barges and small craft across the barrier. The facility, ready for service in October 1916, was never used other than to demonstrate its capabilities. Meanwhile, Spring Lake was enclosed with a levee and spillway, and the surrounding lands were drained and cultivated. Some tenants occupied homes built with lumber salvaged in 1913 from the hunting lodge, which lost money at a time when land drainage was very rewarding.[63]

## Redefining Public Interest

The valley residents whose livelihoods and recreation were supported by nature's gifts treated such resources as common wealth, except where they were rooted in land to which title was held privately. The navigable river and adjacent lakes, like the air, wildlife, and waterfowl therein, were considered common property. The creation of navigation pools, the diversion of Lake Michigan water into the river system, and the drainage of land in the watershed deepened and extended the water across the floodplain permanently. The fishery was expected to benefit materially, but commercial fishermen chafed upon discovering that many fish were "out in the woods" rather than in unobstructed lakes. Also, while access to inundated areas improved, the freedom to fish or hunt over them was constrained by continuity of title. The accustomed freedom to fish was abridged especially by the spread of private rod and gun clubs.[64] The loss of large areas of floodplain to drainage and levee districts after 1904, together with the expectation that the process would absorb most of the remainder, put a premium on areas where habitat suited fish and waterfowl breeding and feeding. Meanwhile, the unscrupulous and indifferent despoiled the fisheries, nesting and migratory waterfowl, and mussel beds. Deleterious as well was the penetration of the area by urban and industrial waste—especially about the time of World War I. In essence, a growing urban and industrial society and the spread of intensive farming were transforming the last large area of wetlands remaining in the central part of the state.

Understandably, the economic development process was accommodated by the General Assembly. Among its decisions in the last quarter of the nineteenth century that were particularly relevant for the valley were those related to navigation improvement, land drainage, creation of the Sanitary District of Chicago, and commercial fishing. The resulting construction of civil and private works that modified the river and floodplain were discussed at length earlier. Here, the focus is on the renewable resources of the valley.

The Illinois Board of Fish Commissioners, it may be recalled, initiated a program to preserve commercial fisheries on the Illinois River. The agency rescued large numbers of stranded fry and fish, returning coarse species to the river and planting game species elsewhere. It cooperated in this work with the United States fisheries agency, as it did to introduce the German carp, which replaced the depleted native buffalo as the principal commercial fish. Acceptance of the carp was aided by the state agency's educational activities. On the other hand, there was little choice. The Board of Fish Commissioners maintained a corps of wardens, but the system of law enforcement was ineffectual.

The state's fisheries agency occupied an unenviable position. Many anglers believed that it served the market fishing sector rather too well; that is, even beyond the evidence that some of its personnel were crooked. For that matter, there was a suspicion that legislators in Springfield were being suborned by the market fishing sector. Moreover, when the desires of the commercial sector did not prevail as laws were reshaped in Springfield, effective enforcement was by no means assured. Wary and belligerent fishermen and dealers and unsympathetic agents of express companies and common carriers, together with local justices and juries, often made enforcement difficult, if not hazardous, for the minuscule staff of wardens. Sentiment favoring better law enforcement developed slowly, even reaching into the ranks of the Illinois Fishermen's Association, which represented the market fishing sector along the river from late in the nineteenth century. While sustaining the productivity of fisheries in the rivers and in Lake Michigan was desired by all, the matter of law enforcement seemed to eclipse consideration of the injurious effects of pollution upon fishing, to say nothing of its effects on summer bathing and the quality of ice cut from rivers for domestic use. Difficult to stop, as well, was the transformation of the breeding and feeding grounds for the fishes into leveed farmland.[65]

While concern for the conservation of fish and game led to the formation in 1874 of an Illinois State Sportsmen's Association in Chicago, it evolved over a quarter of a century into little more than a competitive trapshooting association. Also, beginning in 1886, there arose in the Chicago area the antecedents to the Illinois Audubon Society (1897); but these groups had limited resources, and their primary concerns were with song birds, millinery plumage, and the education of children. A broadly based and more assertive conservation sentiment did not materialize until after 1911, when associations were formed to represent sportsmen, rod and gun clubs, groups of commercial fishermen, and wholesalers. Some of the associations published journals, notably the *Illinois Fisherman* (1912), antecedent to the *Illinois Fisherman and Hunter* (1913), and the *Illinois Fish Conservation Society News-Letter* (1912), which subsequently included "Game" and "Warden's Journal" in the masthead. The journals provided information about the state of commercial and recreational fishing, laws and their enforcement, and matters deserving communication with members of the General Assembly. In 1914, following congressional enactment of the first international migratory game bill (1913), at least seven other leagues of sportsmen were organized with particular representation from counties of the lower Illinois River and of the Mississippi River. Some appear to have been in league with the game-market dealers in St. Louis, the rival to Chicago as a center of such trade. Their journal, the *Illinois Sportsman* (ca. 1915), appears to have replaced the *Illinois Fisherman and Hunter*. The interstate Mississippi Valley Fishermen and Market Dealers Association came into being (1913), and the Illinois Commercial Fishery Association appears to have succeeded the earlier organization of Illinois River fishermen and dealers. The mobilizations of disparate groups on behalf of conservation were a reaction to the perception of effective long-term lobbying in Springfield by elements of the commercial sector that resisted change, and there were conservationists in Peoria who revived the proposition that all rod and gun club properties revert to the state for use in wildlife propagation and protection. However, most groups formed subsequently appeared to be motivated to correct or derail enforcement of the perceived inequities in new federal or state legislation. The prevailing attitudes among commercial and recreational gleaners of nature's gifts were challenged as never before; laissez-faire was eroding. The perspective gaining currency could be summed up in the comments made (October 1911) during the discussion of a paper on the decline in Illinois River fisheries. The speaker believed that "a new and untried and perhaps an unwelcome" solution would be the state's acquisition of swamp and overflowed land, much as it might purchase "tracts of beautiful mountain or forest land" to be preserved for future generations.[66]

The mobilization of the conservationists was consistent with the philosophy of the national Progressive Movement, and it affected the Illinois Valley at a time when enthusiasts asserted that ennobling the character of American youths required the preservation of good, clean, outdoor recreation in the company of the exemplary democratic fraternity of anglers. Such themes persisted as activists later organized the Izaak Walton League of America (1922), which renewed the campaign for wildlife and natural area conservation in the Illinois Valley and nationally. But the groundwork was laid early in this century, when sportsmen were practically unanimous in believing that the use of seines and other nets should be abolished, which would have put an end to commercial fishing. Early conservationists wanted the state to enforce a ban on duck hunting in the spring and on the marketing of all game fish and birds, among other protective regulations. They supported resuscitating the program to rescue and redistribute fish, increasing the propagation from hatcheries and from managed remnant sloughs and lakes within levee systems, and encouraging a stronger state effort to designate remaining floodplain lakes as preserves from which commercial fishing would be excluded. Although the Illinois Fish Commission designated Matanzas Lake a preserve and began a campaign to destroy the voraciously predatory and unmarketable gar (*Lepisosteus* spp.), the agency was terminated in 1913 as part of a general administrative housecleaning by Governor Edward F. Dunne. Its successor, the Game and Fish Conservation Commission (1913), received new monies and mandates and a salaried professionalized staff to enforce more protective fish and game laws. The agency rehabilitated an abandoned fish hatchery at Havana (built in 1908), undertook the stocking of important fishing grounds with coarse and game species, rescued fry and fish from drainage and levee districts being dewatered after levee breaks, and launched a vigorous campaign to intercept illicit commerce in game fish and birds arising in towns along the river. The newsletters disappeared before long. Perhaps the concerns were eclipsed by the approach of war from Europe. In any case, except for the Izaak Walton group, the predominantly urban and suburban constituencies had insufficient visibility in the late 1920s to be identified in Aldo Leopold's survey of game (1931).[67]

The creation of the Game and Fish Conservation Commission seemed to reflect a growing appreciation in Springfield for the interests of recreational fishing and hunting; but it was another decade before associations of sportsmen such as the Izaak Walton League and the Federated Sportsmen of Illinois appeared to be satisfied with fish and game codes. The accustomed deference to the Illinois River commercial fisheries lobby in Springfield seemed to be waning.[68] However, a "return to nature" proposal that the state issue $20 million in bonds to purchase flooded bottoms (1927), including drainage and levee districts, was not embraced by the General Assembly. The vested interests in the drainage and levee districts of the lower valley, like the commitment in Chicago to operating a sewer outfall into the Illinois River system, were well established. These major institutional constraints to the productivity and enjoyment of nature's gifts by others had led the fish commission to urge in 1908 that "no single interest . . . destroy or overshadow any other" in using the Illinois River. "Measures for its utilization as a sewage outlet for great cities and as a commercial highway between the Mississippi and the Great Lakes, and for the reclamation of its enormously fertile bottom lands, should not be taken without due regard to its importance and promise as a perpetual source of cheap and healthful food to the people of the state and country."[69]

Contributing to an appreciation of the fisheries resource of the Illinois River and its preservation was the broadly conceived, seminal, and well-publicized research undertaken by the staff of the Illinois Biological Station, established near Havana in 1894 through the vision and industry of Stephen A. Forbes. Although supported initially by the University of Illinois and the Illinois State Laboratory of Natural History, the latter was given full administrative responsibility after 1897, by which time the research focus in aquatic biology was on the role of overflowed lands in fish spawning, the feeding of young fish, and the production of plankton. Forbes's interest in comparative research on the chemical and biological conditions of the river's water before and after the initiation of discharge into the Illinois River system by the Sanitary District of Chicago led to the funding of the State Water Survey (1895); and it was here that Arthur Palmer pioneered work in bacteriology related to environmental sanitation and water supply, in part supported by the Sanitary District of Chicago. Ultimately, the state agency's work led to the adoption of the kinds of sanitary reform measures noted earlier in the chapter.[70] Meanwhile, at least by 1910, Forbes pressed the state for protection of the Illinois River fisheries:

> [N]othing can be more dangerous to the continued productiveness of these waters than a shutting of the river into its main channel and the drainage of bottom-land

> lakes for agricultural purposes. It is fortunate for our fisheries when one of these lakes comes into the possession, or under the control, of a hunting or fishing club, for this insures its maintenance. The time has come . . . when the state should consider seriously the policy of preserving adequate breeding grounds and feeding grounds for our river fishes, even if it has to acquire and maintain them, since these waters are in imminent danger otherwise of being practically depopulated.[71]

Forbes's informed and persuasive positions helped to focus action by the state and the conservation-minded public.

The contests over preservation of navigability and public access to Spring Lake and Thompson Lake were contemporaneous with a sharpening awareness that the shores of Lake Michigan, the banks of the Chicago and Calumet rivers, and the floodway of the Mississippi River had been encroached upon for years by landowners' dumping of solid industrial waste and dirt. A convergence of concerns over damage to public interests caused the General Assembly to create the Submerged and Shore Lands Legislative Investigating Committee. The committee's report of 1911 documented the extent of the abuse. Its proposal for the creation of the Rivers and Lakes Commission to study, monitor, and preserve the state's interests in public waters was soon enacted. The committee's intent was that the fisheries and the remaining natural areas along the river be preserved as state parks. While the committee was supportive of agricultural development on lands that had been considered "waste" because of overflow, it urged the preservation of the capacity of the Illinois River to convey its natural flow, together with the volumes of water diverted from Lake Michigan by the Sanitary District of Chicago.[72]

Although the General Assembly addressed many issues in creating the Rivers and Lakes Commission in 1911, the agency was not given the modicum of police powers and resources needed to accomplish its missions, except for data gathering, until 1913. Among other tasks, the Rivers and Lakes Commission was to identify and stop the sources of encroachment and pollution upon public bodies of water (excepting where the Sanitary District of Chicago was involved); to provide information, advice, and assistance to prospective land reclaimers, whose construction plans were to be reviewed; to devise plans and recommendations for the legislature regarding the designation of natural areas for recreational and sporting use; and to foster the productivity of fisheries in public waters.[73]

The focus of executive agency attention on these pervasive problems along the Illinois River was sharpened by the general flood of 1913, described in chapter 6. Governor Dunne met on March 14 with the Game and Fish Conservation Commission, the Rivers and Lakes Commission, and representatives of the Natural History and Water surveys and relevant departments of the University of Illinois to review problems related to public waters, wetlands reclamation, fisheries preservation, and flood control. The most significant outcome was the decision to seek funding to enable the Rivers and Lakes Commission to initiate drainage basin surveys. The Illinois River study by John W. Alvord and Charles B. Burdick was remarkably comprehensive. It reviewed flood events, flood prospects, and flood control options, with particular attention to the implications for land drainage, the fisheries sector, and waterways.[74]

Contemporaneous with the Alvord and Burdick study, the Game and Fish Conservation Commission summarized the extent to which floodplain lakes were reduced. Of 46,005 acres of spawning and feeding grounds located between Grafton and Mossville in 1901, 18 percent was drained by 1907 and almost 25 percent by 1913. Yet another 4 percent was then being lost to reclamation. The agency projected loss of another 34 percent (15,540 acres) to land drainage projects. Until 1907, most of the losses (7,202 acres) occurred to the south of Meredosia; thereafter, the bottoms between Meredosia and Mossville bore the brunt of drainage work. Compounding this loss of lakes was the reduction of higher floodplain areas where plankton and fish flourished during periods of overflow.[75]

In May 1915, the Rivers and Lakes Commission published an advisory letter to Governor Dunne, describing Thompson Lake as "the greatest, if not the most valuable, body of water within the State of Illinois" and suggesting that its unexcelled qualities for the propagation and production of fish also made it a suitable area to provide recreation for the public and employment for commercial fishermen who worked between Havana and Peoria. The commission asserted that the state's title to the property was held in trust for all of the people and that the claim of private parties to ownership in fee simple only extended to the edge where land met water. At the time, the Rivers and Lakes Commission was taking inventory of the public waters, among which Thompson Lake (1,723 ac.), Spring Lake (1,300 ac.), and Meredosia Lake (1,182 ac.) represented 60 percent of the residual public waters adjacent the Illinois River.[76]

Major progress in shaping a framework for managing the recreational and commercial use of the wildlife of the floodplain occurred in 1917. Fish preserves for exclusive recreational use were designated. A new Game and Fish Code in 1917 provided for the first licensing of trappers and for more comprehensive licensing of commercial and recreational fishermen. The code incorporated such provisions of the Federal Migratory Bird Treaty Act (1917) as a prohibition on hunting migratory game birds in the spring. State acquisition and designation of waterfowl refuges was provided for. Indicative of the new perspective was the creation of the Department of Conservation in 1925. It was authorized to acquire submerged lands for the preservation and propagation of fish, mussels, game, and wildfowl and to monitor stream pollution. Already, jurisdiction over problems related to the waterways was assigned to the Division of

Waterways, created in 1917 to consolidate the tasks of the Rivers and Lakes, the Illinois Waterway, and the Illinois and Michigan Canal commissions. Thus, the state's responsibilities in navigation, drainage and levee district development, stream pollution and encroachment, and municipal sewage plant development were centralized for more effective management. In enacting the civil administrative code of 1917, the General Assembly signaled that the state would exercise responsibility over nature's bounty as never before.[77]

## Taking Stock

It is evident that the proponents of drainage district formation could not be prevented from transforming the floodplain into relatively flood-free agricultural land. By the same token, there seemed to be no recourse to the flooding of bottomland caused by dams and locks for navigation and by the influx of water attributed largely to the Sanitary District of Chicago. Also, there was remarkable public forbearance in the face of the deleterious effect of urban and industrial pollution on life-forms in the river and on public health and aesthetic sensibilities. For that matter, commercial fishermen, shellers, and hunters were allowed to exact a heavy tribute in nature's gifts for decades. Their pursuits, like trapping, were being constrained by an increasingly modified river and valley floor below Peoria. On the inside of levee systems, corn replaced timber and prairie, water and German carp, in the process reducing wetlands that harbored migratory waterfowl and local wildlife. Outside the levee systems, the river, lakes, and timbered wetlands were submerged, polluted, and alluviated by water levels raised year-round through artificial means. In reclaiming over 183,000 acres of floodplain, the land developers drained over 21,000 acres that had been water covered. Areas of perennially deep water grew in the late 1930s as a result of the construction of locks and dams at Alton and La Grange and above Pekin to replace the older structures. Bellrose et al. calculated that in 1969 there were 35,279 acres of perennially inundated floodplain below Peoria, double the unleveed water area of 1903. The spreading water so merged older water bodies that the number of lakes was reduced to one-third of what it was in 1903. At the same time that wetlands were supplanted by water surfaces, erosion in the bluff and upland areas produced heavy sedimentation in the lakes where fish yet spawned and fed, where migratory waterfowl fed, and to which muskrats and nesting birds were adapted. Already, the wealth of native aquatic species upon which waterfowl fed was depleted by the feeding carp.[78]

Although the preservation of fisheries, the protection of wildfowl and natural areas, and the unencumbered public access to the gifts of nature were matters before the public for decades, an appreciation of the common interest in sustaining the common wealth did not develop at large until about 1911. Diverse public-interest groups mobilized, echoing protests raised by naturalists and personnel in state agencies who warned that the heavy price exacted from the natural bounty by the pursuit of economic self-interest had to be reduced and the restoration of that bounty begun. Change seemed to be in the offing. The Alvord and Burdick study of 1915 suggested that the state invest in wetlands areas for parks and preserves. To that end, it recommended that acquisitions be worked out in collaboration with the Sanitary District of Chicago, then facing several million dollars in claims for damages attributed to overflow caused by the diversion of water from Lake Michigan. Also, Alvord and Burdick endorsed the idea of a rotational system in which leveed districts would receive some of the floods of spring to facilitate the reproduction and rearing of fish and to enrich the land with sediment, while reducing crests. This linking of fisheries preservation with flood control was conceived in a more manageable form in 1929, when Jacob A. Harman et al. recommended that some districts and all residual overflowed areas be designated permanently as flood-crest storage areas and as game and fish preserves. The recently flooded Chautauqua Drainage and Levee District was considered to be well suited to the functions. Otherwise, a restored, high, and strong levee system there assured dangerous conditions for Pekin and Peoria. The sporting sector proposed a state acquisition of bottomlands for preservation and public recreation through a bond issue, about the same time that Pickels wrote (1928) that landowners of some drainage and levee districts were discouraged enough to sell the tracts at reasonable prices. By 1933, the Illinois Division of Waterways considered the Big Prairie district ripe for acquisition by the Department of Conservation or by a hunting club.[79]

Nevertheless, the idea proposed decades ago that public agencies conserve substantial residual areas of floodplain for the common good is far from being realized. By the same token, there has been little acquisition of drainage and levee districts for the restoration of wildlife habitat and public recreational use. The concerted effort by sportsmen, conservationists, and certain public agencies to have the United States government acquire such lands in the 1930s lost headway when the Corps of Engineers determined that the Big Prairie Drainage and Levee District was alone among 34 tracts in offering the economic justification to warrant a return to flood-water storage functions and wildlife uses. In any case, it was the view of Corps of Engineers' investigators that illegal nocturnal seining by commercial fishermen was a larger factor in depleting fisheries than was land reclamation. The Corps of Engineers' cost effectiveness stance was no less firm than the landowning sector's opposition to land acquisition by the government. The landowners' view was supported by the railroads that served them and by the Federal Land Bank of St. Louis, which had assisted some districts to recover from the disastrous 1920s. Thus, an increment through land drainage of about 286 square miles of highly productive Corn Belt land, together with the interim resolution of Chicago's sanitation problem through the artificially swollen Illinois River, cost one-fifteenth to one-twentieth of the nation's wildlife resources, according to an estimate of the U.S. Division of Migratory Water Fowl. Contributing to the impoverishment of diverse ecosystems have been soil erosion and pollution caused by intensive land use on the prairie uplands, the excesses of hunters, fishermen, and mussel shellers, and the introduction of carp.[80]

The acquisition of bottomlands by public agencies in the valley between Peoria and the lower end of the Nutwood Drainage and Levee District, 146 miles distant, has resulted in a little over 8 percent of the study area being recommitted to natural functions. Land acquisitions began in 1936, when the abandoned (1926) Chautauqua Drainage and Levee District was purchased by the U.S. Biological Survey to become one of the early national migratory waterfowl refuges. The short-lived flurry of acquisitions by the state occurred between 1945 and 1950, during which time State Conservation Areas were created at Rice Lake (1945), Anderson Lake (1947), Sanganois (1948), and Spring Lake (1950). These important state projects to restore waterfowl areas lagged by 15 years similar initiatives in Indiana, Wisconsin, and Minnesota, where land drainage projects probably were less viable economically than they were on the bottoms of the Illinois River.[81]

The Illinois Department of Conservation yet hoped in 1950 to acquire over 48,960 acres of drainage and levee districts in the study area for conservation use and flood mitigation. The agency's approach in 1969 was less confrontational with the districts. It proposed no more than "an immediate program to preserve, protect, or restore the backwater lakes, sloughs, and bays along the river" and then to acquire unspecified large tracts of land along the river for "ultimate development into parks, refuges, and conservational areas," as well as "to afford the public access to the river"—all in the context of an Illinois River Corridor for recreation.[82]

Although the pristine floodplain in the lower Illinois Valley embraced about 425 square miles (table 1.1), the average spring overflows probably covered only half that area, and a quarter of the bottoms were in lakes, ponds, and sloughs anyway. To what extent the higher half of the floodplain was cleared for farming in 1890 is unknown, but much of it remained in timber and wet prairie in 1904. Most of the bottoms that flooded in the average spring bore natural cover then. With the compartmentalization of about two-thirds of the entire floodplain behind systems of levees by 1924, wetlands may have been reduced to about 20 percent of the entire floodplain. Water deepened by dams and locks and spread by diversions from Lake Michigan covered the rest. Thus, about two-thirds of the floodplain was committed primarily to agriculture, while a third of the residual floodplain remained in timber, prairie, shallow water, and lakes. The residual floodplain area, the river, and the meandered lakes to which the State of Illinois held title[83] remained the domain of the introduced carp, native fish species, wildfowl, game, and fur-bearing animals. Much of the residual wetlands habitat was held in private hunting and fishing grounds, however; and the privately owned lakes were apt to be leased as fishing grounds.

In the 1920s, agriculture remained the dominant land-use activity, employer, and supporter of commerce in the lower valley. Commercial fishing and trapping were on the wane. Timber cutting, mussel shelling, and commercial hunting had all but ceased. But demand for recreational fishing, hunting, and other water uses by the vacationing public were on the rise. It took the General Assembly another quarter century to establish conservation areas suited to the public's needs.

# 9

# Retrospect

THE valley of the lower Illinois River is much deeper and wider than the valleys cut by tributaries into the elevated glacial outwash plains and moraines on either side. The pronounced lowland corridor is floored by floodplain and outwash terraces and flanked for much of its length with bluffs that are scalloped by ravines and valleys, which head in the rolling upland plains. The continuity of bluffs is broken to the east of the valley's broad middle sector, where a rolling, sandy, and loessial mantle ascends to the upland plain. For the most part, the flanks and brows of the bluffs are wooded, but fields and grasslands interrupt the forest where gentler slopes occur. Similar open patches occur in the wooded draws and valleys that descend into the valley of the Illinois River.

The valley's prevailing backdrop of bluffs screens the horizon, drawing attention to the corridor, which is a landscape of fields, timbered areas, and water. With the exception of some strips of industrial land use that reach downstream of Peoria and Pekin and the modest urban spread of Havana, Beardstown, and smaller towns, the scene is rural. Only a few grain elevators, water tanks, boxy power-generating stations, and spans across the river break the continuity of cover and bluff lines against the sky. There is little residential development on the bluffs and in the draws away from Peoria and Pekin. Most people reside in the few, relatively compact riverside towns and villages, in dispersed strings of waterside cottages, and in a handful of small, platted communities on the high margins of drainage and levee districts adjacent the bluffs.

Settlers on the valley floor favored sites where high ground provided dry footing and some security from floods. Towns developed at riverside where dunes and terraces reached across the bottoms, providing landing and ferry points for routes between the uplands on either side. Farms and ranches were located near the bluffs, where woodland and prairie, springs and outcrops of stone, were near at hand. These ribbons of settlement fostered roadway development, with laterals ascending valleys and ravines to the uplands, leading to county seats and service centers such as Jerseyville and Carrollton, Rushville and Lewistown. The distances were shorter to navigable water, which gave access to other county seats and service centers at Hardin,

Beardstown, Havana, Pekin, and Peoria. Meredosia, too, was an early riverboat and ferry landing with a railhead linked to Jacksonville and Springfield.

By 1904, the ribbons of rural settlement near the bluffs had broadened with fields and improved pastures in a blocky and discontinuous fashion on the bottoms. However, recurring localized and general overflows delayed plantings and damaged or destroyed crops. The floodplain was yet occupied by numerous shallow lakes, ponds, and marshy and timbered wetlands.

The streams, the floodplain, the diverse natural cover, and the ecosystems of the valley were modified substantially by human activities in the quarter century following 1900. Land clearing and ditching began well before then, but the alteration of waterways and floodplain, of cover and ecosystems, accelerated thereafter due to the conjunction of economic opportunity and persistently high water levels. The spread of water beyond the margins of navigation pools created by dams and locks at Kampsville, La Grange, and Copperas Creek was marked after 1900. Although permanent flooding was attributed to large diversions into the river system from Lake Michigan, the flow from Chicago's Sanitary and Ship Canal was augmented by intensified land drainage activity throughout the uplands and bottoms of the watershed. Contributing to the surfeit of water was the enclosure of large areas of bottoms with levee systems and their dewatering through tile and ditch systems served by pumps belonging to numerous drainage and levee districts and to a few independent landowners. Meanwhile, the accelerated soil erosion in the loessial bluffs and uplands produced serious alluviation in streambeds, lakes, and ditches in the valley. Waste from cities near and far impacted water and life, and the excesses of commercial fishermen, shellers, and hunters took a heavy toll. In effect, the degradation of water bodies and wetlands resulted from the exuberant economic development in metropolitan Chicago, in the valley of the Illinois River, and in the fertile plains of the watershed. All sectors, resident and nonresident alike, used the resources of the valley with little concern beyond their immediate self-interest.

The artificially drained landscape on the bottoms of the lower Illinois River exceeded 183,000 assessable acres in 1924, about 65 percent of the floodplain. The artificially drained areas were unevenly distributed throughout the valley, as shown in maps 2.7 to 2.10. The leveed and ditched compartments provided new orders of local relief and water features and new patterns of surface drainage. Wherever possible, runoff from the bluffs was diverted by embankments into floodways flanked by the levees of drainage and levee districts. Commonly, too, beheaded sloughs and creek channels were integrated into the artificial drainage systems that conveyed unwanted water toward the pumps located just inside riverside levees. The continuity of drained surfaces was broken, as well, by grids of slightly elevated township and farm roads, some of them resting on sand ridges and artificial levees. By and large, the grids hewed to the rectangular survey lines that defined property and field, as they did throughout the Midwest. All such grids of roads accessed the integrating bluff roads, which bridged the floodways between drainage and levee districts. Most established farmsteads were along the bluff roads, but the residences, barns, and sheds used by tenants tended to be scattered throughout the districts. These drained compartments were more densely occupied in the 1920s than now. Then, as now, small relict groves interrupted the continuity of fields, while long galleries of timber beyond the levees, and the wooded bluffs, seemed to enclose each reclaimed tract.

Unreclaimed areas of floodplain and abandoned drainage districts aggregated about 100,000 acres in the mid-1920s, when the last drainage and levee district was completed. Possibly 20 percent of the floodplain was covered with water as a result of changes in the regime of the river and watershed made by well-intentioned institutions and individuals. The state and federal governments built locks and dams to facilitate waterborne commerce; the Sanitary District of Chicago addressed a burgeoning city's public health and sanitation needs with water diverted from Lake Michigan; and landowners and investors contributed to the expansion of Corn Belt farming, rural settlement, food production, and revenue through land drainage. Such enterprise on behalf of the betterment of life, region, and nation had unheeded environmental costs.

The permanent inundation of floodplain areas remaining outside the leveed and cultivated compartments destroyed timber and marsh vegetation and shifted the zones suited to biotic communities. The inundated areas soon became the domain of private hunting and fishing clubs, some of which permitted commercial fishermen and unaffiliated sportsmen to use the premises. Given the extent of land reclamation and of flooding, the intrusion of polluted water and sediment loads, and the tendencies of fishermen and hunters to maximize immediate returns, the valley's natural systems were stressed. Indicative of the situation after 1900 were plaints about the loss of cover suited to duck hunting and of natural haying and pasture lands; and fishermen and hunters were aware that the "catch" or "bag" of preferred species was dwindling. Indicative, too, were the efforts by state and federal agencies to save and propagate fish and mussel species, to seek more restrictive hunting and fishing laws, and to have preserves designated for fish and other wildlife. In the main, however, the public and the authorities accepted that the destruction and depletion of nature's gifts were compensated for by the more intensive use of waters and wetlands soils that resulted from metropolitan growth, flourishing agriculture, and the quest for economic well-being by dwellers of the counties that shared the valley. The outcome of contests in the courts and in the General Assembly between landowners and the Sanitary District of Chicago, on the one hand, and between landowners and fishermen, on the other, was a tenuous success for the Sanitary District of Chicago and an ultimate prevailing of agricultural land use on most of the bottoms. The commercial fisheries industry, the commercial and recreational hunting sectors, and the tourism industry made do with the residual and modified natural bounty.

The state and national legislatures, society's agents in such matters, facilitated the development of multiple and competing uses of river and floodplain and tolerated the destructiveness of unbridled extractive pursuits for a long time. Individuals and institutions were free to risk capital and enterprise to use nature's gifts in the quest for individual, sectoral, and regional well-being. The social implications of fostering mutually incompatible activities among different sectors were concerns of little moment generally, and there was even less appreciation of the disruptions to natural systems resulting from the activities.

The piecemeal development of floodplain segments into drainage and levee districts occurred at a time when it served the public interest to change the productivity of such areas through private endeavor. The initiators of large-scale projects tended to be owners of the larger and most flood-prone properties, parcels acquired at a fraction of the market value that drainage would ensure. Such enhanced value devolved from the land's exceptional productivity at a time when commodity prices were advancing smartly. The promise of the drained bottoms assured a good market for bonds issued to pay for constructing a district's defenses against overflow, runoff, and seepage. Speculation in wetlands development produced handsome returns, at least at first.

The drainage and levee districts, indeed all land-assessment entities designed to improve productivity through drainage, were organized under the auspices of the county courts. The legal framework that enabled numerous individuals to collaborate in developing such entities was in place well over a decade before physical and economic circumstances induced landowners to organize drainage and levee districts in the study area. Neither the processes of land acquisition nor of organizing and developing the districts seem to have been rife with fraud; but there were costly mistakes, and controversy arose. The mistakes reflected inexperience, incompetence, and corner cutting in applying a relatively new technology on a large scale to conditions of land and water that were not well understood. As noted in chapters 5 and 8, controversy flared over the damage to land resulting from persistently high water and over the destruction of habitats essential to sustaining the fisheries in the interest of developing agriculture. The land reclaimers were unable to persuade the Sanitary District of Chicago, the General Assembly, or the courts to abate the diversion of water from Lake Michigan or to provide restitution for damages in the manner established by the enabling legislation for the sanitary district. This experience of the landowners contributed to the animosities that characterize the downstate-Chicago relationship but was of little moment elsewhere in the state. While pollution had a negative impact on fisheries, recreational activity, and real estate and on the usefulness of naturally produced ice, it was not enough of a galvanizing issue to precipitate change. Concern over the effect of persistently high-standing water in altering habitats for wildlife and fish was all but eclipsed by the attention given to wetlands destruction and modification resulting from large-scale levee building, ditching, and land-clearing projects. Such losses, like the mounting risk to farmland security arising from the shrinkage of overflow areas and the constriction of floodways, probably were acceptable in view of the prospects for commercial agriculture on the floodplain.[1] Nevertheless, the environmental impacts drew attention at large to ideas of restoration and preservation of the natural bounty.

The rapidity and extent to which land drainage was accomplished on the bottoms was due to the adoption of power excavators and pumps. The floating dipper dredge moved most of the dirt that went into levees and came out of ditches. Teams, scrapers, and wagons were employed a good deal in ancillary tasks. However, by the time that most of the reclamation was achieved, floating dipper dredges, teams, and scrapers were being supplanted by tracked dryland excavators, notably draglines, powered with internal combustion engines. By the same token, the reclaimers of land witnessed a succession from steam to electric and diesel-powered pumping stations, as they did a gradual supplanting of teams and steam traction engines by tractors. Motor vehicles began to compete with rail and water carriers in hauling grain, livestock, coal, and contraband fish and game. These changes in technology did not begin in, nor were they unique to, the bottoms of the lower Illinois Valley, but the adoptions accelerated the process by which relatively undeveloped floodplain was transformed into an intensively farmed segment of the Corn Belt. The reclaimed land was as productive as were the upland wet prairies where tiling and ditching began a generation or two earlier. The delayed integration of the floodplain into the regional agricultural system, as was the case with floodplain drainage generally, came about as rising commodity prices placed a premium on making remarkably productive lands secure from seasonally or perennially high water. Landowners and investors alike expected that farmlands secured by the works of drainage and levee districts would yield worthwhile returns on capital for some time.

The land drainage venture on the floor of the Illinois Valley, while unique in many respects, was part of a process that brought over 65 million acres of wetlands into farm use in the United States between 1890 and 1930.[2] This era of land drainage, what McDowell termed "the National Drainage Movement,"[3] was roughly contemporaneous and akin in importance to the contribution of the irrigation movement in the subhumid to arid lands of the nation. Both approaches to land management compensated for environmental constraints on cultivation by investing in large-scale water management infrastructures. Whereas the national achievement in irrigation was aided by the direct involvement of the governments of the United States and western states, the accomplishments in the drainage of wetlands between 1890 and 1930 were attributable largely to associations of landowners who formed land assessment districts at the county level. Among public land states, only Minnesota and Florida had long-term direct involvement in the land drainage process by state agencies.[4]

It is curious that county and regional histories and bicentennial compendiums in Illinois virtually ignore the formation and development of the land drainage enterprises. Yet, except for industrialized Peoria and Tazewell counties, the individual drainage and levee district projects represented the largest single capital enterprises to develop in the counties that shared the bottoms of the lower Illinois River. Many of the landowners and investors in the drainage and levee districts were farmers, professionals, and businessmen of some local prominence. Along with nonresident landowners and buyers of bonds, they underwrote the creation of jobs, commerce, and new revenue for the counties. The introduction of corporate farms, the enlargement of farms where owners and tenants operated, and an increase in the number of absentee-owned tenant farms appears to have resulted. The farming system placed a number of residences, barns, and sheds on the reclaimed bottoms to accommodate the renters of family-sized farms. When these farmsteads were not destroyed or abandoned because of flood events in the 1920s, their viability waned as the Depression, farm mechanization, and electrification fostered farm consolidation.

A threat of damage to reclaimed districts by runoff and debris from the bluff areas and by freshets carried into the lower valley by tributaries of the Illinois River was ever present. Heavy precipitation could fall on limited areas at any time of the year. More predictable as to the season and the timing of peak flow once the event began were the general floods of the Illinois River. Among these, the record-breaking flood of 1904 occurred before much reclamation had been done. Before 1913, subsequent high stages of the river inundated no more than one or two drainage and levee districts per flood event, damaging tracts containing less than 6,000 acres of assessed land. In 1913, however, when about 45 percent of the floodplain below Peoria was enclosed by levees, and by which time there were notable constrictions in the floodway at Beardstown and below Meredosia, the levee systems were breached at 6 districts containing over 35,000 acres of assessed land. Again, 4 districts containing over 31,000 assessed acres had levee breachings in 1916. Because the floodings were viewed as having resulted from correctable flaws in levee design and from shortcomings in levee maintenance, confidence remained strong in the prospects for additional drainage and levee districts.

The drainage and levee districts probably received the most public attention in the 1920s, when creeks and river reoccupied a good part of the floodplain by breaching levee systems, including those of Beardstown and smaller communities, in 1922 and 1926–27. Levee breachings in 1922 affected districts with 75,600 acres of assessed land,

and in 1926 and 1927, districts with over 103,000 acres of assessed land were damaged by the river. Damaging runoff occurred in districts representing another 23,000 and 48,000 acres of assessed land in 1922 and 1926–27, respectively. The losses in property and crops were very heavy; some districts were kept out of production for months by standing water.

In view of the collapse in commodity prices that occurred in late 1921, no new drainage and levee districts were organized after 1922. Landowners were hard-pressed to pay the assessments for annual operating costs at the existing districts, let alone pay off the principal and interest due on drainage bonds. Delinquencies were common after 1926.

The drainage and levee districts were in such difficult financial straits by 1927 that the General Assembly provided funds for levee restoration. Before long, the U.S. Congress extended the jurisdiction of the Mississippi River Commission to such tributaries as the Illinois River, resulting in the involvement of the Corps of Engineers in the design and construction of flood control works in the 1930s and thereafter. The federal presence was solicited earnestly by a wide spectrum of political interests in Illinois, including the Association of Drainage and Levee Districts of Illinois, which drew its membership from the commissioners, engineers, and attorneys who served the districts of the lower Illinois Valley. Their attitude toward federal intervention was conditioned by the belief that only the federal government could rectify the perceived inequities resulting from the conduct of the Sanitary District of Chicago. As luck would have it, the outcome of a suit carried to the Supreme Court by the State of Wisconsin on behalf of maintaining water levels on the Great Lakes was a boon to the districts, which had been ineffectual in seeking removal of navigation dams and locks and in abating the growing volumes of sanitary water diverted through Chicago. Meanwhile, the districts had not consigned much of their land to floodway use or to storage of flood water. Indeed, land reclaimers constricted the river at several points; and they added to the flow of sediment into the system from the plains and bluffs by building diversion ditches and by preventing river tributaries from spreading across the floodplain. Each district gambled that relief from flood stages would come through piling up water over the residual floodplain and through levee breachings at other districts. Still, there was no obvious expression of rivalry among districts in the record.

By and large, the drainage and levee districts and their engineers and contractors accomplished the land drainage that they set out to do, although at much greater cost per assessed acre than anticipated, which was the rule in land drainage projects in the nation. Such endeavors increasingly removed segments of the floodplains from their natural functions in river systems. At the same time, land use development elsewhere in the watersheds, together with unpredictable occurrences of precipitation and rates of snow melt, delivered water and sediment into the river system in volumes made more extraordinary because of the modifications of the floodplain. The overarching outcome of nature's recurring challenges to urban and rural development behind artificially compartmentalized segments of floodplain was society's ultimate recognition that the provision of effective security for property and lives had to be approached systemwide in the watershed—an approach for which the operating framework could only be provided by collaborating state and national governments. Since central authorities had relegated the task of land drainage to the landowners and local authorities, who had accomplished the job along the lower Illinois River by extending settlement and revenue-producing activity to land previously viewed as having little value, there was an obligation to secure the achievement. The privately undertaken reclamation of the bottoms had destructive consequences for the wetlands and the gifts of nature, but the result is hard to gainsay. There was no alternative a century ago, given the state of public values and opinion, of technical knowledge, and of perceptions on the role of government, and given the complications attributable to the freewheeling interests of Chicago and of the landowners in the valley and to the behavior of nature. For that matter, none of the sectors that had open access to the resources of waterways or wetlands were given to exercising much restraint or compromise in the pursuit of well-being.

That the drainage and levee districts survived the travails created by nature and exacerbated by competing users of the Illinois River system was due to a conjunction of circumstances over which the districts had no control. The protection of navigation and hydroelectric power interests on the Great Lakes and the St. Lawrence River checked water diversions from Lake Michigan. The assumption by the United States government of responsibility for flood control work on the lower Mississippi River led ultimately to responsibility for similar endeavors on tributaries such as the Illinois River and in watersheds throughout the nation. Also, widespread flood damage in Illinois led to direct involvement by the State of Illinois in levee restoration work along the Mississippi and Ohio rivers, as well as along the Illinois River. State and federal leadership in restoring and improving levee systems after 1926–27 ended the era when the land drainage undertaking was left solely to private enterprise working through public corporations organized at the local level. Nevertheless, the districts were not freed of responsibility for the maintenance of the new generation of levees; and the maintenance and operation of collecting and outlet ditches and of pumping facilities remained their charge. The works maintained locally had become components in a flood control and drainage outlet system designed ultimately for the Illinois River basin. Federal intervention to resolve the district and landowner financial difficulties arising from the agricultural recession and prolonged by the Depression was yet another sign that the buoyant era of laissez-faire wetlands reclamation was over. At the same time, the viability of the drainage and levee district as a quasi-public, self-governing institution was preserved.

The entrepreneurs who developed drainage and levee districts in the lower Illinois Valley between the 1890s and 1920s added over 180,000 acres to the inventory of relatively secure farmland in Illinois. About 67 percent of the acreage was gained in the period 1905 to 1914, whereas 12 percent was leveed in the preceding 15 years and 23 percent between 1915 and 1924. Most of the 286 square miles of relatively secure bottoms were associated with 36 drainage and levee districts, which turned out to be resilient institutions. Few of the assessment districts went out of business. The investment in reclamation alone (i.e., excluding the cost of borrowed capital) probably aggregated more that $17 million by 1930. Additionally, the state expended over $1.3 million to restore and enlarge levees subsequent to 1927. Another $4.7 million were expended by the federal government in relocating and enlarging levees at the districts by mid-1944, the work being authorized by flood control acts passed in 1928, 1936, and 1938. The investment in farm improvements and flood-damage restorations was well over $5 million, or more than $26 per acre.[5] In any case, the process of securing arable land against runoff and overflow begun by individual settlers and pursued as a collaborative enterprise of local landowners between 1891 and 1924 could not be sustained without incorporating a systemic approach and using public funds. Standards of floodway and levee design, construction, and maintenance were formulated by federal and state agencies, and the enlarged defenses were largely paid for by the United States.

The circumstances leading to the transformation of the bottoms of the lower Illinois River into a distinct and highly productive agricultural landscape are barely a part of local tradition or historical record, and the relationship of the venture to the process of land development in the nation virtually is unknown anywhere. In essence, the half century of land drainage work carried out by the districts along the lower valley was contemporaneous with the irrigation development that was widespread in arid and subhumid lands and with widespread drainage of floodplains, wet prairies, and coastal marshes. This creation of arable land through irrigation and drainage grew in importance as the nation's reserve of land for rain-fed agriculture was all but exhausted. Although common manual tools and horse-drawn implements were used a great deal in shaping the works of irrigation and drainage, the scale adopted for ditch, canal, and levee systems in the late nineteenth century required the application of power machinery, notably the floating dredge and ditcher; the dragline appeared later. In this era of large-scale projects designed to transform high-risk and new cropland into relatively secure and very productive farms, pumps were introduced to lift and move irrigation water and to remove water from the former wetlands where gravity did not complete the drainage.

Among the nation's land drainage enthusiasts, there was a perception that the development of irrigation in the West was advantaged by direct federal support. This involvement of the United States government directly affected less than 10 percent of the aggregate of irrigated acreage (1919) in the West,[6] but there was no counterpart commitment to the improvement of wetlands in the eastern half of the country, which seemed inequitable to some people. Besides, the proponents of federal support to land drainage argued that the public health considerations (malaria in particular), the comparative costs of reclamation, and the closer proximity of the bulk of the nation's wetlands to the populous eastern half of the country warranted a commitment of public funds. The participants in large-scale land drainage projects for agriculture in the central Midwest might console themselves with pride in the primacy of achievement by collaborating individuals (most of them farmers), whereas institutional investors (such as railways) undertook drainage of the Mississippi River bottoms below Cairo to obtain lumber before having the cutover land transformed into farms, for which buyers were sought elsewhere. On the other hand, federal aid to the drainage of floodplains was not forthcoming because large corporate lumber, railway, and other moneyed interests would have been the primary beneficiaries.[7]

It is clear that geographically central to the drainage endeavor in the nation was the watershed of the Mississippi River and its major tributaries, among them the Illinois River. By the same token, the application of power machinery to the excavation of systems of drainage ditches was a major innovation of contractors with dipper dredges in Illinois, Indiana, and Ohio after 1883. The application of power excavators to levee building and ditching on the great floodplains of the region followed in the 1890s, the work of dipper dredges in the lower Illinois Valley being seminal. Some of the contractors subsequently engaged in similar projects on the bottoms of the Mississippi, Ohio, and Missouri rivers. Meanwhile, dredges had become essential to levee building and canal cutting in California and Louisiana. The adoption of pumps in the districts of the Illinois River was earlier and more essential than on most floodplains of the Midwest and middle Mississippi Basin. The relatively high cost of developing the pumping districts was, to begin with, related to the manner in which the works of navigation, metropolitan sanitation, and regional land drainage encroached upon a floodplain that had evolved adjacent a river of unusually low gradient.

The opportunities created for farmers by successful reclamations on the bottoms of the lower Illinois Valley appear to have drawn new owners and tenants from in-state, especially from areas adjacent the valley. Open recruitment of settlers was not a factor. The process appears to have been repeated elsewhere in the Midwest, such as Iowa and Missouri, as well as Arkansas and Mississippi. However, the migration by local people to the gravity drainage and pumping districts of the middle and lower Mississippi Valley appears to have accompanied and followed attempts by other regions of the country to attract colonists from the Midwest. Such recruitment of colonists was promoted in southern Louisiana, the Southeast, and Florida,[8] as it was by irrigation colonies in California and by vendors of wheat producing lands in the northern Great Plains and the Northwest. Thus, the settlement experience along the Illinois River was representative of the infilling end of the spectrum rather than of colonization. For the most part, the development of both irrigation and drainage districts was financed through the sale of bonds, the expectation being that bountiful crops and livestock would return a profit for the tenants and for the owners, who paid assessments to complete and operate the districts. Drainage district bonds were popular among conservative investors. By and large, the speculations succeeded where the project management was sound and where a high proportion of the drained land was committed promptly to cash-crop farming; but the combination of exceptional floods and the crash in commodity prices broke drainage districts across the nation in the 1920s. Many irrigation districts also defaulted on their bonds in the wake of the commodity crash. Land values plummeted. As Teele put it in 1924, "for the immediate future . . . there is no national need for reclaiming land" of either category.[9] A recovery was made in the late 1930s as a result of debt refinancing by the United States, bondholder concessions, and improvements in national economic conditions.

Thus, although there were general incentives nationwide to undertake land drainage on major floodplains around the turn of the twentieth century, the immediate inducement along the lower Illinois River was the presence of high water year-round. The distinctive local inducement resulted from a pyramiding effect of low dams and locks in the waterway, the arrival of waters diverted from Lake Michigan through the Chicago Sanitary and Ship Canal, and the building of levee and drainage systems that removed large areas of bottoms from the natural floodplain function. There were local circumstances that provided particular inducements to undertake the drainage of major floodplains elsewhere. For instance, the work of land drainage along the lower Mississippi River that faltered during and after the Civil War was revived when Congress determined that the Mississippi River Commission (established in 1879) should undertake measures to prevent destructive floods and the impairment of river-borne commerce. While the levees and other works were to benefit navigation, not reclamation, the strictures against the ancillary function softened after 1890;[10] and the commission's mandate was broadened by the Flood Control Act of 1917, as noted earlier.

That the act of 1917 also prescribed work that evolved into a comprehensive Sacramento River Flood Control Project was occasioned by the particular circumstances in northern California. They were akin to the circumstances along the lower Illinois River in that flood events were exacerbated by the unrestrained disposal of waste created by intensive economic activity, to the despair of landowners on the leveed floodplain. In the Sacramento Valley, however, the physical and political problems created by the debris of hydraulic mining were very much larger and more complex in their ramifications. Exceptional flows of water and enormous volumes of debris obliterated all signs of settlement from some tributary floodplains, choked principal waterways, flooded lowland towns and farms, and fouled tidal waters. To address the debris problem, Congress created the California Debris Commission (1893), which was modeled after the Mississippi River Commission. The agencies studied, monitored, and proposed remedial works in the interest of navigation. The experience in both areas guided the formulation of the Flood Control Act of 1917, which recognized that navigation and flood control problems were inseparable. Insofar as the Mississippi River Commission was concerned, the act of 1917 added to its purview the responsibility for such flood control measures as levee building as far upstream as Illinois. The Flood Control Act of 1928 brought the U.S. Corps of Engineers and the concept of integrated flood control measures well into the valley of the Illinois River and other tributaries of the Mississippi River, as it did in the Sacramento Valley and elsewhere.[11] Thus, decisions by the federal courts and Congress reflected a national sense that private developers of resources had an obligation to respect the security of the interests of other sectors while undertaking resource development. Laissez-faire was over insofar as the floodplains were concerned. Congressman Rainey's early alliance with irrigation enthusiasts was prescient.

This study of the lower valley of the Illinois River during the period 1890 to 1930 has focused on the institutions, technologies, and processes that shaped essential features of a settled landscape through intrusive volumes of water and widespread land drainage and clearing activities. In addition to documenting the marked changes in the nature of the valley and human pursuits and settlement therein, the study provided regional and national developmental contexts; and it related the local venture in floodplain modification to regional and national experiences in land reclamation. Thus, a contribution to understanding the valley's heritage was cast, as well, as a contribution to the comparative study of floodplain transformation into highly productive agricultural land.[12] This aspect of the national heritage in settlement geography and agricultural history is as much in need of recounting as is the heritage of institutions, technologies, and processes that added the last major segment of virgin and tenuously cropped land to the regional agricultural system, while causing the impoverishment and destruction of aquatic and wetlands ecosystems over a large area. In essence, intensive agriculture supplanted extensive land use activities, which continued on residual wetlands and water bodies that were modified by high-standing navigation pools wherein pollution and alluviation impaired the natural systems as never before.

# Notes
# Bibliography
# Index

# Notes

*The following abbreviations are used in citations of newspaper articles:*

| | |
|---|---|
| *CP* | *Carrollton Patriot* |
| *I-S* | *Illinoian-Star* (Beardstown) |
| *JR* | *Jerseyville Republican* |
| *MB* | *Meredosia Budget* |
| *MCD* | *Mason County Democrat* (Havana) |
| *ME* | *Morning Enterprise* (Beardstown) |
| *PCD* | *Pike County Democrat* (Pittsfield) |
| *PDT* | *Pekin Daily Times* |
| *PH-T* | *Peoria Herald-Transcript* |
| *PJ* | *Peoria Journal* |
| *RT* | *Rushville Times* |

## 1. The Physical Setting and General Context of Land Drainage

1. Map 1.1 is based on Illinois Rivers and Lakes Commission, *The Illinois River and Its Bottom Lands,* 2nd ed., by John W. Alvord and Charles B. Burdick (Springfield: Illinois State Journal Co., 1919), fig. 6 (cited hereafter as Alvord and Burdick, *Illinois River*). Depiction of valley floor materials is generalized from maps in Illinois Department of Registration and Education, Geological Survey, *Geology and Mineral Resources of the Beardstown, Glasford, Havana, and Vermont Quadrangles,* by Harold R. Wanless, Bulletin 82 (Urbana: 1957), 145 (cited hereafter as Wanless, *Geology and Mineral Resources*); and U.S. Army Corps of Engineers, St. Louis District, *Shallow Subsurface Geology, Geomorphology, and Limited Cultural Resource Investigations of the Meredosia Village and Meredosia Lake Levee and Drainage Districts, Scott, Morgan, and Cass Counties, Illinois,* by Edwin R. Hajic and David S. Lergh, Cultural Resources Management Report 17 (St. Louis: April 1985), app. C: "Outline of Late Wisconsinan and Holocene Geology of the Lower Illinois Valley," by Edwin R. Hajic, 137, 138 (cited hereafter as Hajic, "Outline of Geology").

2. It would be more precise to ascribe the valley widths to northern and southern sectors of 60 miles length, but the text is modified to facilitate integration with data based on river reaches, as in table 1.1.

3. Alvord and Burdick, *Illinois River,* 19; Lyman E. Cooley, *The Illinois River: Physical Relations and the Removal of the Navigation Dams* (Chicago: Clohesey & Co., 1914), 18–21, 32, 71, 91; Illinois Department of Purchases and Construction, Division of Waterways, *Flood Control Report,* by Jacob A. Harman, L. T. Berthe, M. Blanchard, and L. C. Craig (Springfield: Journal Printing Co., 1930), 28 (cited hereafter as Harman et al., *Flood Control Report*); U.S. Congress, House, Committee on Rivers and Harbors, *Illinois River, Ill.,* 72nd Cong., 1st sess., 1931, H. Doc. 182, 18; U.S. Department of the Interior, Geological Survey, *Geology and Mineral Resources of the Hardin and Brussels Quadrangles (in Illinois),* by William W. Rubey, Professional Paper 218 (Washington, D.C.: GPO, 1952), 115 (cited hereafter as Rubey, *Geology and Mineral Resources*).

4. U.S. Congress, House, Committee on Rivers and Harbors, *Survey, with Plans and Estimates of Cost, for a Navigable Waterway 14 Feet Deep from Lockport, Ill., by Way of Des Plaines and Illinois Rivers, and Thence by Way of the Mississippi River to St. Louis,* "Map of the Illinois and Des Plaines Rivers" (1:7,200), by J. W. Woermann, U.S. Army Corps of Engineers, 59th Cong., 1st sess., 1905, H. Doc. 263 (cited hereafter as Woermann, "Map"); Alvord and Burdick, *Illinois River,* 19–20; Cooley, *Illinois River,* 32; Association of Drainage and Levee Districts of Illinois, *First Annual Report of Transactions,* "Reclamation," by Lyman E. Cooley (n.p., 1911), 32–33 (cited hereafter as Cooley, "Reclamation"); Harman et al., *Flood Control Report,* 28; Rubey, *Geology and Mineral Resources,* 112, 124; Wanless, *Geology and Mineral Resources,* 19, 23, 27, 139; Hajic, "Outline of Geology," 150–52; Illinois Department of Registration and Education, State Museum, *Geomorphology of the Lower Illinois River Valley as a Spatial-Temporal Context for the Koster Archaic Site,* by Karl W. Butzer, Reports of Investigations 34 (Springfield, 1977), 1–60.

5. U.S. War Department, Chief of Engineers, *Report for the Year 1868* (Washington, D.C.: GPO, 1869), 252, 256–57; U.S. Treasury Department, Public Health Service, *A Study of the Pollution and Natural Purification of the Illinois River, II: The Plankton and Related Organisms,* by W. C. Purdy, Public Health Bulletin 198 (Washington, D.C.: GPO, November 1930), 1; U.S. Congress, House, Secretary of War, Chief of Engineers, *Annual Report, 1890,* "Improvement of the Illinois River, Illinois," 51st Cong., 2nd sess., 1890, H. Exec. Doc. 1, pt. 2:2, app. JJ3, 2443–44; Alvord and Burdick, *Illinois River,* 20; Harman et al., *Flood Control Report,* 52, 59–63, 80, 82–83.

6. S. G. Russell, "Russell's Early Recollections," *CP,* October 10, 1903, p. 1, cols. 2–3.

7. Wanless, *Geology and Mineral Resources,* 19–23, 143, 177; Hajic, "Outline of Geology," 134, 140.

8. Wanless, *Geology and Mineral Resources,* 16–19, 27, 139, 142; Hajic, "Outline of Geology," 140–52.

9. U.S. Congress, Senate, *Report from the Secretary of War, . . . Transmitting a Report of the Survey of the Kaskaskia and Illinois Rivers,* by Howard Stansbury, 25th Cong., 2nd sess., 1838, S. Doc. 272, v. 4, pp. 2–3 (cited hereafter as Stansbury, *Report*).

10. Stansbury, *Report,* 5. Table 1.1 and analysis are based on data in Alvord and Burdick, *Illinois River,* table 10.

11. Illinois Laboratory of Natural History, *The Plankton of the Illinois River, 1891–1899, with Introductory Notes upon the Hydrography of the Illinois River and Its Basin,* by Charles A. Kofoid, Bulletin 6:2 (Urbana, 1901–3), 95–574, especially 114–56.

12. Lewis M. Turner, "Grassland in the Floodplain of Illinois River," *American Midland Naturalist* 15 (1934), 770–80; Illinois Department of Registration and Education, State Museum, *Early Vegetation of the Lower Illinois Valley,* by April A. Zawacki and Glenn Hausfater, Reports of Investigations 17 (Springfield, 1969) (cited hereafter as Zawacki and Hausfater, *Early Vegetation*); Daniel H. Brush, *Growing Up with Southern Illinois, 1820 to 1861,* ed. Milo M. Quaife (Chicago: Lakeside Press, 1944).

13. Brush, *Growing Up,* 15–17; Zawacki and Hausfater, *Early Vegetation, 12; History of Greene and Jersey Counties, Illinois* (Springfield: Continental Historical Co., 1885), 366–68, 826–28, 839, 850; Capt. M. D. Massie, *Past and Present of Pike County, Illinois* (Chicago: S. J. Clarke Publishing Co., 1906), 80, 83; Judson F. Lee, "Transportation: A Factor in the Development of Northern Illinois Previous to 1860," *Journal of the Illinois State Historical Society* 10:1 (April 1917), 74–77.

14. Lewis M. Turner, "Plant Succession on Levees in the Illinois River Valley," *Transactions of the Illinois State Academy of Science* 24:2 (December 1931), 98–101; Illinois Laboratory of Natural History, *Forest Conditions in Illinois,* by R. Clifford Hall and O. D. Ingall, Bulletin 9:4 (Urbana: 1910–13), 197; Zawacki and Hausfater, *Early Vegetation,* 15.

15. Margaret B. Bogue, "The Swamp Land Act and Wet Land Utilization in Illinois, 1850–1890," *Agricultural History* 25:4 (October 1951), 170–72, 174, 179.

16. Benjamin Horace Hibbard, *A History of the Public Land Policies* (1924; reprint, Madison: University of Wisconsin Press, 1965), 269–88; Ray P. Teele, *The Economics of Land Reclamation in the United States* (Chicago: A. W. Shaw Co., 1927), 57–60, 88, 110–11; Gary Lane McDowell, "Local Agencies and Land Development by Drainage: The Case of 'Swampeast' Missouri" (Ph.D. diss., Columbia University, 1965), 94–101.

17. Although the writer reviewed the legislation, Illinois *Laws* (1879), 120 and 142, guidance was provided by the following studies: Illinois Department of Registration and Education, Geological Survey, *Engineering and Legal Aspects of Land Drainage in Illinois*, rev. ed., by George W. Pickels and Frank B. Leonard Jr., Bulletin 42 (Urbana, 1929), Part III: "Legal Problems" (cited hereafter as Pickels and Leonard, *Engineering*); Illinois Tax Commission, in cooperation with Works Progress Administration, *Drainage District Organization and Finance, 1879–1937*, Survey of Local Finance in Illinois, vol. 7 (Springfield, 1941); James E. Herget, "Taming the Environment: The Drainage District in Illinois," *Journal of the Illinois State Historical Society* 71:2 (May 1978), 107–18. Herget's paper is particularly useful in tracing the development of drainage law from the statutes of the Northwest Territory. Helpful too are the papers on Illinois drainage law published by Frank B. Leonard, "An Introduction to the Drainage Laws of Illinois," *Illinois Law Bulletin* 1:5 (April 1918), 227–47; and *University of Illinois Law Forum* 2 (summer 1960), ed. Harold W. Hannah. The papers in *Law Forum* provide an analysis of the recently enacted code, as well as trace the case law as it related to various types of districts.

18. Illinois *Laws* (1905), 197; Illinois *Laws* (1913), 299; Pickels and Leonard, *Engineering*, 263.

19. Pickels and Leonard, *Engineering*, 230–34, 237; Illinois Tax Commission, *Drainage District Organization*, 4, 39.

20. Harold W. Hannah, "History and Scope of Illinois Drainage Law," *University of Illinois Law Forum* 2 (summer 1960), 195–97.

21. Illinois Rivers and Lakes Commission, *Land Drainage in Illinois*, by Robert I. Randolph, Bulletin 4 (Springfield: Illinois State Journal Co., 1912), 10 (cited hereafter as Randolph, *Land Drainage*). For views on the law by contemporaries, see Association of Drainage and Levee Districts of Illinois, *Proceedings of Second Annual Meeting*, comments of Thomas Masters and Thomas Worthington (Beardstown: Enterprise Press, 1912), 57–63, and *Proceedings of the Third and Fourth Meetings*, "Advisability of a Recodification of Our Drainage Laws by a Commission of Experts," address by W. J. Graham (Winchester: Press of the Winchester Times, 1913), 7–19 (cited hereafter as Assoc. Drainage and Levee Dist., *Proceedings, 1912* and *Proceedings, 1913*, respectively).

22. Voluminous archival documentation of the organization and operation of the drainage and levee districts appears in University of Illinois, Water Resources Center, *Case Studies in Drainage and Levee District Formation and Development on the Floodplain of the Lower Illinois River, 1890s to 1930s*, by John Thompson, Special Report 017 (Urbana-Champaign, May 1989), 2, 8, 16, 57, 79 (cited hereafter as Thompson, *Case Studies*).

23. Illinois *Laws* (1879), 120; Thompson, *Case Studies*, 18, 64.

24. Illinois *Laws* (1879), 120–22, 124.

25. Illinois *Laws* (1879), 124–25, 128–29, 132.

26. Illinois *Laws* (1879), 125–28; Assoc. Drainage and Levee Dist., *First Annual Report*, "Organization of Districts—Engineering Features," address by C. W. Brown, 47–48.

27. Illinois *Laws* (1879), 128–29; Illinois *Laws* (1885), 74.

28. Thompson, *Case Studies,* 2, 5.

29. Illinois Tax Commission, *Drainage District Organization,* 10.

30. Assoc. Drainage and Levee Dist., *Proceedings, 1912,* "Address by Retiring President," by Jesse Lowe, 72 (cited hereafter as Lowe, "Address").

31. Illinois *Laws* (1879), 131; Thompson, *Case Studies,* 56, 58, 107.

32. Big Swan Drainage and Levee District Commissioners, "Ledger 1904–1931," Secretary, provisional minutes.

33. B. Henderson, "Land Drainage Needed in United States," *Reclamation and Farm Engineering* 8:5 (December 1925), 309–11; Robert W. Harrison, *Alluvial Empire: A Study of State and Local Efforts Toward Land Development in the Alluvial Valley of the Lower Mississippi River, Including Flood Control, Land Drainage, Land Clearing, Land Forming* (Little Rock: Pioneer Press, 1961), 261–63, 302–3.

34. Reprinted in "Reclaiming Bottom Land," *MB,* November 25, 1909, p. 1, col. 3.

35. Data for table 1.2 from University of Illinois, College of Agriculture, Agricultural Experiment Station, *Prices of Illinois Farm Products from 1866 to 1929,* by L. J. Norton and B. B. Wilson, Bulletin 351 (Urbana, July 1930), 495, 505, 530.

36. "The Loading of Wheat Barges: Snicarte," *MCD,* July 17, 1896, p. 5, col. 2; "Happenings in Illinois Cities and Villages," *I-S,* August 10, 1907, p. 3, col. 5; "The Nutwood Levee District," *CP,* July 29, 1909, p. 1, col. 5.

37. Lowe, "Address," 73.

38. Assoc. Drainage and Levee Dist., *Proceedings, 1913,* comments of H. J. Puterbaugh, 4.

39. Lowe, "Address," 73.

40. Lyman E. Cooley, *The Lakes and Gulf Waterway as Related to the Chicago Sanitary Problem . . . , a Preliminary Report* (Chicago: Press of John W. Weston, 1891), app. 13, 16–18; "Eldred," "Bluffdale," *CP,* September 5, 1902, p. 8, cols. 3, 5; U.S. Department of Agriculture, Office of Experiment Stations, *Annual Report of Irrigation and Drainage Investigations,* "Report of Drainage Investigations," by Charles G. Elliott, Bulletin 158 (Washington, D.C.: GPO, 1905), 713–14 (cited hereafter as Elliott, "Drainage Investigations"); Cleves H. Howell, "Reclamation of Agricultural Land by Diking" (C.E. thesis, University of Illinois, 1905), 14, 16; "$30,000 for Pegram Ranch," *CP,* October 20, 1905, p. 1, col. 3; "Scott County Drainage," *I-S,* September 7, 1906, p. 1, cols. 4–5; "Adding a Township," *CP,* October 5, 1906, p. 1, cols. 1–3; Joel E. Dunn, "Reclamation of Land by Levying" (C.E. thesis, University of Illinois, 1906), 18; Iowa State Drainage Association, *Proceedings of the First Annual Meeting,* "The Development of Agricultural Drainage in Illinois and Iowa," by Jacob A. Harman (n.p, 1908), 57 (cited hereafter as Harman, "Development of Drainage"); "Bottom Land $140 an Acre," *CP,* December 9, 1909, p. 1, col. 2; "A Big Wheat Yield," *CP,* July 22, 1909, p. 1, col. 6; "Big Land Transfer in Lacey Levee District," *MCD,* October 28, 1910, p. 1, col. 5; "More Reclamation Work," *RT,* July 25, 1912, p. 1, col. 5; "Land Deals Total $119,379.00," *RT,* August 15, 1912, p. 1, col. 2; "Eldred," *CP,* July 17, 1913, p. 8, col. 4; "Paid $24,800 for Drainage Farm," *RT,* October 9, 1913, p. 1, col. 2; Assoc. Drainage and Levee Dist., *Proceedings, 1913,* comments of Louis Lowenstein, 35; U.S. Department of Commerce, Bureau of the Census, *Fourteenth Census of the United States, 1920: Agriculture* (Washington, D.C.: GPO, 1922), vol. 6, pt. 1, 365, 376–85; Isham Randolph, "Results on Some Great Projects," *National Drainage Journal* 1:9 (February 1920), 2.

41. "Local News," *MCD,* January 17, 1896, p. 1, col. 4; "The Nutwood Levee District," *CP,* July 29, 1909, p. 1, col. 5; "New Drainage District," *PH-T,* July 29, 1910, p. 5, col. 1; "Want Permit to Drain Vast Area," July 31, 1910, p. 2, col. 7; Assoc. Drainage and Levee Dist., *First Annual Report,*

comments of Louis Lowenstein and Jesse B. Lowe, 50, 12; "Boast of Big Things," *CP*, September 14, 1911, p. 1, col. 6.

42. "Adding a Township," *CP*, October 5, 1906, p. 1, cols. 1–3.

43. Dunn, "Reclamation by Levying," 18; "Fabulous Prices for Farm Land," *PH-T*, June 15, 1906, p. 6, col. 2; "Land Reclaiming," *I-S*, August 31, 1906, p. 1, cols. 3–4.

44. Table 1.3 is from Elliott, "Drainage Investigations," 713–14; "Reclaiming Bottom Land," *MB*, November 25, 1909, p. 1, col. 3.

45. Elliott, "Drainage Investigations," 714; "Adding a Township," *CP*, October 5, 1906, p. 1, cols. 1–3; Harman, "Development of Drainage," 57–58; Alvord and Burdick, *Illinois River*, 63, table 11; Assoc. Drainage and Levee Dist., *Proceedings, 1913*, comments of Louis Lowenstein and Jacob A. Harman, 34, 36.

46. Randolph, *Land Drainage*, 3–4, 7–8; Woermann, "Map"; U.S. Congress, House, *Proceedings of a Conference of Governors*, "Conservation of Natural Resources," by Charles S. Deneen, 60th Cong., 2nd sess., 1909, H. Doc. 1425, 309–10, 313 (cited hereafter as Deneen, "Conservation"); Assoc. Drainage and Levee Dist., *First Annual Report*, comments of Frank W. DeWolf, 14–15; "Peorians Finish Survey of Vast Tract for State," *PH-T*, October 22, 1910, p. 3, col. 1.

47. Illinois Internal Improvements Commission, *The Lakes and Gulf Waterway* (Springfield: Phillips Bros., 1907), vi–ix; Assoc. Drainage and Levee Dist., *Proceedings, 1913*, "The Relation of the Rivers and Lakes Commission to Drainage," by Robert I. Randolph, 27–28.

48. Deneen, "Conservation," 313; Cooley, "Reclamation," 35.

49. Governor Deneen's denunciation of the perceived obstructionist behavior of the Association of Drainage and Levee Districts appears in Assoc. Drainage and Levee Dist., *Proceedings, 1913*, 64–66.

50. Assoc. Drainage and Levee Dist., *First Annual Report*, comments of Jesse Lowe, 12; Cooley, "Reclamation," 34.

51. Illinois Department of Public Works and Buildings, Division of Waterways, *Seventh Annual Report, July 1, 1923 to June 30, 1924* (Springfield: Illinois State Journal Co., 1925), 15.

## 2. The Construction and Spread of Drainage Districts

1. U.S. Department of Agriculture, Office of Experiment Stations, *Annual Report of Irrigation and Drainage Investigations*, "Report of Drainage Investigations," by Charles G. Elliott, Bulletin 158 (Washington, D.C.: GPO, 1905), 703 (cited hereafter as Elliott, "Drainage Investigations"). Generalizations are based on University of Illinois, Water Resources Center, *Case Studies in Drainage and Levee District Formation and Development on the Floodplain of the Lower Illinois River, 1890s to 1930s*, by John Thompson, Special Report 017 (Urbana-Champaign: May 1989) (cited hereafter as Thompson, *Case Studies*).

2. U.S. Congress, Senate, *Report of the Chief of Engineers, 1868*, "Report upon the Survey of the Illinois River," by Lt. Col. J. H. Wilson, 40th Cong., 1st sess., 1868, S. Exec. Doc. 16, 14.

3. Lyman E. Cooley, *The Lakes and Gulf Waterway as Related to the Chicago Sanitary Problem . . . , a Preliminary Report* (Chicago: Press of John W. Weston, 1891), app. 13, consisting of printed letters from residents of the valley.

4. Illinois Rivers and Lakes Commission, *The Illinois River and Its Bottom Lands*, 2nd ed., by John W. Alvord and Charles B. Burdick (Springfield: Illinois State Journal Co., 1919), 106 (cited hereafter as Alvord and Burdick, *Illinois River*).

5. Estimates based on measurements made of areas shown in U.S. Congress, House, *Survey*,

*with Plans and Estimates of Cost, for a Navigable Waterway 14 Feet Deep from Lockport, Ill., by Way of Des Plaines and Illinois Rivers, and Thence by Way of the Mississippi River to St. Louis,* "Map of the Illinois and Des Plains Rivers" (1:7,200), by J. W. Woermann, U.S. Army Corps of Engineers, 59th Cong., 1st sess., 1905, H. Doc. 263, sheets 29, 32 (cited hereafter as Woermann, "Map").

6. Description based on Woermann, "Map."

7. Alvord and Burdick, *Illinois River,* citing earlier sources, 27, 54.

8. "Frederick Department," *RT,* January 2, 1896, p. 1, col. 4, and January 29, 1896, p. 1, col. 4; "Lowlands Flooded," *RT,* January 7, 1897, p. 2, cols. 3–4; "Northeast Department," *RT,* May 26, 1898, p. 1, col. 3, May 25, 1899, p. 7, col. 5, April 19, 1900, p. 7, col. 2, and May 17, 1900, p. 7, col. 5; "Frederick Department," *RT,* May 28, 1899, p. 7, col. 5; "River Out of Banks," *RT,* July 10, 1902, p. 1, col. 4; "Coal Creek Levee Holds," *RT,* April 7, 1904, p. 1, col. 3; "Frederick," *RT,* April 7, 1904, p. 4, col. 3; "Land Reclamation Along the Illinois River," *I-S,* October 7, 1905, p. 3, cols. 1–5; "Looking Backward," *I-S,* February 7, 1906, p. 1, col. 5.

9. "Drainage Commissioners' Report," *RT,* September 23, 1897, p. 7, col. 4; "Frederick Department," *RT,* June 9, 1898, p. 2, col. 5, August 10, 1899, p. 7, col. 5, March 15, 1900, p. 7, col. 4, March 29, 1900, p. 7, col. 3, and July 4, 1901, p. 7, col. 6; "River Out of Banks," *RT,* July 10, 1902, p. 1, col. 4; "Frederick," *RT,* July 31, 1902, p. 7, col. 3, March 5, 1903, p. 7, col. 3, April 21, 1904, p. 4, col. 4, and May 5, 1904, p. 4, col. 4; "Looking Backward," *I-S,* February 7, 1906, p. 1, col. 5; "Making Big Improvements," *RT,* January 12, 1911, p. 1, col. 2; Charles A. E. Martin, ed., *Historical Encyclopedia of Illinois and History of Cass County* (Chicago: Munsell Publishing Co., 1915), 2:781; "Both Sides of the Bridge," *MCD,* November 1, 1901, p. 1, col. 5; "Contract Let for Grade West Havana Road," *MCD,* June 26, 1908, p. 1, col. 2; Peoria County Circuit Court, "Drainage Record—Pekin and La Marsh," vol. 1, Commissioners, "Assessment Roll," filed January 6, 1890, 22–27, "Assessment Roll, July 1, 1890 to July 1, 1891," n.d., 48–49, and "Assessment Roll . . . for 1900," n.d., 136–39; "Pekin—Work on La Marsh District," *PH-T,* November 20, 1905, p. 5, col. 1; "Volunteers Build Fulton Road," *MCD,* July 15, 1927, p. 1, col. 7, "Levee Road Is Now Driveable," October 28, 1927, p. 1, col. 3.

10. Elliott, "Drainage Investigations," 668–69.

11. "Land Reclaiming," *I-S,* August 31, 1906, p. 1, cols. 3–4.

12. Elliott, "Drainage Investigations," 713; Illinois Internal Improvements Commission, *Surface Water Supply of Illinois Central and Southern Portions, 1908–1910* (Springfield: Illinois State Journal Co., 1911), 229.

13. Iowa State Drainage Association, *Proceedings of the First Annual Meeting,* "The Development of Agricultural Drainage in Illinois and Iowa," by Jacob A. Harman (n.p., 1908), 56–57 (cited hereafter as Harman, "Development of Drainage").

14. Elliott, "Drainage Investigations," 713.

15. Descriptions based on Woermann, "Map."

16. Greene County Circuit Court, Hillview Drainage and Levee District files, "Order Approving and Confirming Commissioners' Report and Establishing District," filed June 6, 1906, 1–7; "Another Big Land Sale," *CP,* April 18, 1902, p. 1, col. 2; "Another Drainage District," *CP,* June 3, 1904, p. 1, col. 3; Elliott, "Drainage Investigations," 674–75; "Bit of History," *CP,* February 26, 1914, p. 1, cols. 1–2; Woermann, "Map," sheets 8, 9.

17. Thompson, *Case Studies,* 17–21, 80–81.

18. "Bottom Farmers at Work," *RT,* May 21, 1903, p. 1, col. 4.

19. Thompson, *Case Studies,* 19, 29, 42–43, 54, 81.

20. Peoria County Circuit Court, "Drainage Record—Pekin and La Marsh, 1," Jacob A.

Harman, "Report" (to Commissioners), filed August 31, 1905, 220–22 (cited hereafter as Harman, "Pekin and La Marsh Report"); "Hartwell District Will be Extended," *CP*, October 12, 1922, p. 8, col. 3; Thompson, *Case Studies*, 82.

21. Big Swan Drainage and Levee District, Commissioners, *Specifications for the Construction of Ditches and Levees* (Winchester: Press of the Winchester Times, 1904), 2, 7; Elliott, "Drainage Investigations," 694; Jersey County Circuit Court, Nutwood Drainage and Levee District, Commissioners' files, vol. 1, "Report of Engineer," by Jacob A. Harman, filed January 24, 1906, 21–23 (cited hereafter as Harman, "Report of Nutwood Engineer"); East Peoria Drainage and Levee District, Commissioners, "Complete Record" containing "Report of Engineer," by Jacob A. Harman, filed February 8, 1908, 83 (cited hereafter as Harman, "Report of East Peoria Engineer"); Brown County Circuit Court, "Drainage Record No. 1," Big Prairie Drainage and Levee District, Commissioners, "Report," filed May 8, 1916, containing "Exhibit G—Specifications for Proposed Levee," 56 (cited hereafter as Big Prairie Commissioners, "Report"); Charles E. Deleuw, "Problems in Design and Construction of Large Land Drainage Districts" (C.E. thesis, University of Illinois, 1912), 5; Howard C. Haungs, "Engineering Methods for the Reclamation of Overflowed Lands in Illinois" (C.E. thesis, University of Illinois, 1916), 27, 32, 36; "Reclaiming Bottom Land," *MB*, November 25, 1909, p. 1, cols. 3–4.

22. "Local Matters," *PCD*, November 9, 1904, p. 5, col. 3, and November 30, 1904, p. 5, col. 1; "Local and Other Matters," *MB*, April 25, 1907, p. 5, col. 4.

23. "Local Matters," *PCD*, July 4, 1906, p. 5, col. 2, January 23, 1907, p. 5, col. 3; "Local and Other Items," *MB*, September 13, 1906, p. 5, col. 2, October 4, 1906, p. 5, col. 1, January 24, 1907, p. 5, col. 1, May 16, 1907, p. 5, col. 3, and April 18, 1907, p. 7, col. 2; "McGee Creek on Rampage," *PCD*, January 23, 1907, p. 1, col. 3; "Man's Leg Cut Off by a Flying Sweep," *MB*, May 16, 1907, p. 1, col. 1; "Fatal Accident," *MB*, June 27, 1907, p. 2, col. 1; "Has Monster Boom," *PCD*, July 3, 1907, p. 1, col. 2; "A Man Drowned," *MB*, October 17, 1907, p. 5, col. 4; "McGee Levee and Drainage District," *MB*, October 24, 1907, p. 1, cols. 4–5; "Local and Other Items," *MB*, December 5, 1907, p. 5, col. 3, and January 30, 1908, p. 5, col. 3.

24. Elliott, "Drainage Investigations," 672–73; Joel E. Dunn, "Reclamation of Land by Levying" (C.E. thesis, University of Illinois, 1906), 13; Haungs, "Engineering Methods," 25; Deleuw, "Problems in Design and Construction," 6; "Reclaiming Bottom Land," *MB*, November 25, 1909, p. 1, cols. 3–4; Illinois Secretary of State, Archives, "Notes on Levee Districts, Illinois River Bottoms," by John W. Alvord and Charles B. Burdick [1914], entries 2 to 40 (cited hereafter as Alvord and Burdick, "Notes").

25. Elliott, "Drainage Investigations," 691; Cleves H. Howell, "Reclamation of Agricultural Land by Diking" (C.E. thesis, University of Illinois, 1905), 10; Illinois Secretary of State, Archives, Illinois Rivers and Lakes Commission, "Minute Book, 1911–1912," meeting of December 29, 1911, 67–68, 91; Alvord and Burdick, *Illinois River*, 56; Illinois Department of Registration and Education, Geological Survey, *Engineering and Legal Aspects of Land Drainage in Illinois*, rev. ed., by George W. Pickels and Frank B. Leonard Jr., Bulletin 42 (Urbana, 1929), 137 (cited hereafter as Pickels and Leonard, *Engineering*).

26. Big Prairie Commissioners, "Report," 57; Fulton County Circuit Court, "Thompson Lake Drainage and Levee District," "Petition to Form District," filed June 7, 1915, 6–9.

27. Elliott, "Drainage Investigations," 691; Harman, "Report of Nutwood Engineer," 7; Harman, "Development of Drainage," 57; Pickels and Leonard, *Engineering*, 260–62.

28. Elliott, "Drainage Investigations," 672, 692, 695, 700; Deleuw, "Problems in Design and Construction," 6; Harman, "Report of Nutwood Engineer," 7; Alvord and Burdick, "Notes,"

entries 2–40; Alvord and Burdick, *Illinois River,* 55–56, 61; Illinois Rivers and Lakes Commission, *Annual Report, 1916* (Springfield: Schnepp & Barnes, 1917), 21; Pickels and Leonard, *Engineering,* 265; Lewis M. Turner, "Plant Succession on Levees in the Illinois River Valley," *Transactions of the Illinois State Academy of Science* 24:2 (December 1931), 101–2.

29. Big Swan Commissioners, *Specifications,* 3, 4; Dunn, "Reclamation by Levying," 13; East Peoria Drainage and Levee District, Commissioners, "Minutes," entries for January 26, April 16, and October 17, 1912; Alvord and Burdick, *Illinois River,* 59–61; Pickels and Leonard, *Engineering,* 266.

30. Thompson, *Case Studies,* 20–22, 25–27, 61, 81, 108–9.

31. Thompson, *Case Studies,* 4; "Land Reclaiming," *I-S,* August 3, 1906, p. 1, cols. 3–4.

32. Elliott, "Drainage Investigations," 674, 704–5; Harman, "Report of Nutwood Engineer," 14; U.S. Department of Agriculture, *Land Drainage by Means of Pumps,* by S. M. Woodward, Bulletin 304 (Washington, D.C.: GPO, November 19, 1915), 15 (cited hereafter as Woodward, *Drainage by Pumps*); Haungs, "Engineering Methods," 12; Pickels and Leonard, *Engineering,* 268–69; Thompson, *Case Studies,* 27.

33. Thompson, *Case Studies,* 1, 76–77, 102.

34. Thompson, *Case Studies,* 4, 10–11, 22, 67; Association of Drainage and Levee Districts of Illinois, *Proceedings of the Third and Fourth Meetings* (Winchester: Press of the Winchester Times, 1913), comments of Louis Lowenstein, 48 (cited hereafter as Assoc. Drainage and Levee Dist., *Proceedings, 1913*).

35. "A Big Wheat Yield," *CP,* July 22, 1909, p. 1, col. 6; "The Nutwood Levee Problem," *CP,* September 30, 1915, p. 1, col. 5; Illinois General Assembly, Illinois Valley Flood Control Commission, *Hearings on the Causes and Control of Floods in the Illinois River Valley,* testimony by J. P. Kerr (n.p., 1929), 45.

36. U.S. Congress, House, Committee on Rivers and Harbors, *Hearings on the Subject of the Improvement of the Illinois and Mississippi Rivers, and the Diversion of Water from Lake Michigan into the Illinois River,* testimony by Henry T. Rainey, 68th Cong., 1st sess., 1924, pt. 2, 930 (cited hereafter as House Comm. R. and H., *Hearings, 1924*).

37. Thompson, *Case Studies,* 67, 81–82.

38. Deleuw, "Problems in Design and Construction," 26; Harman, "Report of Nutwood Engineer," 23; Harman, "Development of Drainage," 59; Haungs, "Engineering Methods," 17–18, 30; Woodward, *Drainage by Pumps,* 19; Association of Drainage and Levee Districts of Illinois, *Proceedings of Second Annual Meeting* (Beardstown: Enterprise Press, 1912), comments of C. W. Brown, J. A. Harman, and Jesse Lowe, 25–27.

39. Data for table 2.1 from the following: Big Swan Drainage and Levee District, Commissioners, "Ledger 1904–1931" (itemized statement of accounts); Big Prairie Commissioners, "Report," containing "Exhibit K—Statement of the Probable Cost . . . ," filed May 8, 1916, 63; Harman, "Report of East Peoria Engineer," 86; East Peoria Drainage and Levee District, Commissioners, "Minutes," meeting of September 9, 1908, 86; Fulton County Circuit Court, "East Liverpool Drainage and Levee District, 1," Commissioners, "Petition," containing "Report of Engineer," filed February 7, 1916, 64; Fulton County Circuit Court, "Lacey Drainage and Levee District, 1," Commissioners, "Report," filed April 23, 1896, 56; Fulton County Circuit Court, "Liverpool Drainage and Levee District, 1," Commissioners, "Report of Surveys," filed June 22, 1915, 81; Fulton County Circuit Court, "Thompson Lake Drainage and Levee District," Commissioners, "Petition for Second Assessment—Exhibit J," filed February 14, 1922; Greene County Circuit Court, Hillview Drainage and Levee District files, Engineer, "Report," filed April

25, 1906, 6; Schuyler County Circuit Court, "Coal Creek Drainage and Levee District," Commissioners, "Report," filed September 6, 1897, "Annual Report," filed September 5, 1898, 3–5, and "Kelly Lake Drainage and Levee District," Commissioners, "Petition for Additional Assessment," filed September 11, 1918, 2; Scott County Circuit Court, Scott County Drainage and Levee District files, Commissioners, "Report" containing "Engineer's Report," dated September 22, 1913.

40. Data for table 2.2 from Schuyler County Circuit Court, "Coal Creek Drainage and Levee District," Commissioners, "Report" containing "The Coal Creek Drainage and Levee District, Expenditures—Maintenance Fund," filed November 16, 1923, 1–4.

41. Data for table 2.3 from Big Swan Commissioners, "Ledger," 7–376.

42. Illinois Tax Commission, in cooperation with Works Projects Administration, *Drainage District Organization and Finance, 1879–1937,* Survey of Local Finance in Illinois, (Springfield, 1941), 7:13.

43. Pickels and Leonard, *Engineering,* 29–30.

44. Some examples are described in Thompson, *Case Studies,* 4, 32, 36–37, 40, 43, 50–51, 60, 85, 93.

45. Charles G. Elliott, *Engineering for Land Drainage* (New York: John Wiley & Sons, 1908), 3–4; Pickels and Leonard, *Engineering,* 29–30.

46. Assoc. Drainage and Levee Dist., *Proceedings, 1913,* address by Thomas Worthington, 50.

47. Thompson, *Case Studies,* 20–21, 60, 81–82, 85.

48. Deleuw, "Problems in Design and Construction," 2; Pickels and Leonard, *Engineering,* 30.

49. The first papers were Howell, "Reclamation by Diking" and Dunn, "Reclamation by Levying."

50. U.S. Congress, House, Committee on Flood Control, *Control of Floods on the Mississippi and Sacramento Rivers,* 64th Cong., 1st sess., 1916, H. Rept. 616, 4–7, 12.

51. Notable journal articles reprinted as a monograph were written by William Starling, *The Floods of the Mississippi River* (New York: Engineering News Publishing, 1897). A detailed article on levee building procedures and specifications based upon the experience of the Mississippi River Commission appeared as "Specifications for Levees," *Engineering News* 4:36 (September 1, 1887), 233–34; "Elliott, Charles Gleason," *National Cyclopaedia of American Biography* (New York: James T. White & Co., 1931), 21:287.

52. Harman, "Development of Drainage," 61.

53. Haungs, "Engineering Methods," 10; U.S. Congress, House, Committee on Rivers and Harbors, *Illinois River, Ill.,* 72nd Cong., 1st sess., 1931, H. Doc. 182, 45 (cited hereafter as House Comm. R. and H., *Illinois River, 1931*); R. A. Wheeler, "Report on the Status of Flood Control in Illinois," *Illinois Engineer* 12:4 (April 1936), 60; "The Nutwood Drainage and Levee District," *JR,* October 17, 1907, p. 1, cols. 1–4.

54. Ill. Valley Flood Control Comm., *Hearings,* 147; Assoc. Drainage and Levee Dist., *Proceedings, 1913,* "Construction and Maintenance of Levees," by E. J. Chamberlain, 41; Charles G. Elliott, "Drainage of Swamps," *Drainage and Farm Journal* 6:6 (June 1885), 19–25; "Elliot, Charles Gleason," 21:287.

55. The attitudes of Midwesterners toward the subject were noted a decade earlier by an early drainage engineer, William M. Whitten, "Drainage, Engineers and Engineering," in Association of County Surveyors and Civil Engineers of Indiana, *Proceedings, Sixth Annual Meeting, . . . 1886* (Indianapolis: Morning Star Publishing Co., 1886), 62–73.

56. Peoria County Circuit Court, "Drainage Record—Pekin and La Marsh, 1,"

Commissioners, "Petition for Additional Assessment," filed June 6, 1902, 157–58; Harman, "Pekin and La Marsh Report," 210, 214, 218; Elliott, "Drainage Investigations," 669–74; Haungs, "Engineering Methods," 11; Thompson, *Case Studies*, 54.

57. Data for tables 2.4 and 2.5 are from Illinois Department of Purchases and Construction, Division of Waterways, *Flood Control Report*, by Jacob A. Harman, L. T. Berthe, M. Blanchard, and L. C. Craig (Springfield: Journal Printing Co., 1930), 37–39 (cited hereafter as Harman et al., *Flood Control Report*); Pickels and Leonard, *Engineering*, 134.

58. For data on assessed and total acreages in the districts, see Harman et al., *Flood Control Report* 38–39, 59. However, the districts studied do not correspond with the complete list in the source.

59. Elliott, "Drainage Investigations," 693.

60. The more inclusive list of districts in Pickels and Leonard, *Engineering*, 134, reported the following:

| Sector | District | Area (ac.) | % |
|---|---|---|---|
| Upper | 9 | 37,380 | 19 |
| Middle | 11 | 29,200 | 15 |
| Lower | 19 | 129,950 | 66 |

Harman et al., *Flood Control Report*, 37, 59, agreed with the foregoing as to inclusive districts and areas for the upper and lower sectors but omitted 2 districts and 10,000 acres from the middle sector.

61. Turner, "Plant Succession on Levees," 94; House Comm. R. and H., *Hearings, 1924*, pt. 2, 197.

62. House Comm. R. and H., *Illinois River, 1931*, 47.

63. Ill. Valley Flood Control Comm., *Hearings*, testimony by John Goodell, J. W. Woermann, H. H. Fletcher, and G. Ratcliff, 31, 88, 136, 220–22; Illinois Department of Public Works and Buildings, Division of Waterways, *Sixteenth Annual Report, July 1, 1932 to June 30, 1933* (Springfield: Journal Printing Co., 1934), 15.

64. Contributing to the difficulty of estimating the cost is the diversity of experience among the districts in length of operating period, relative vulnerability to freshets and to the river, quality of management and maintenance, and strategy adopted by the commissioners in using assessments for construction or operations. Also, generalization is hampered by voids in the data.

65. This estimate of $16.25 million for 1930 is reasonably consistent with an estimate of $13.5 million made by the Association of Drainage and Levee Districts in 1923. House Comm. R. and H., *Hearings, 1924*, pt. 1, 197.

66. House Comm. R. and H., *Hearings, 1924*, pt. 1, 197; Alvord and Burdick, *Illinois River*, 13, table 11; Harman et al., *Flood Control Report*, 56.

67. In 1930, there were 7 districts with land values averaging $100–$105 per acre, 6 where the values were $90–$95, and 12 where the values were in the $70– $80 range. Six of the tracts had land in the $60 range and 2 in the $50 range. Illinois Planning Commission, *Report on the Lower Illinois River Basin* (Springfield, 1940), 26.

68. Harman et al., *Flood Control Report*, 38, 59.

69. U.S. Department of Agriculture, *Economic Status of Drainage Districts in the South in*

*1926*, by Roger D. Marsden and R. P. Teele, Technical Bulletin 194 (Washington, D.C.: GPO, October 1930), 3–6, 9, 27–29 (cited hereafter as Marsden and Teele, *Economic Status*).

70. Marsden and Teele, *Economic Status*, 4–6; C. E. Jacoby, "The Missouri River Valley Drainage Problems," in *Missouri's Swamp and Overflowed Lands and Their Reclamation*, by John Nolen, Report to the 47th Missouri General Assembly (Jefferson City, Mo.: Hugh Stephens Printing, 1913), 95–102; George R. Boyd, "The Refinancing of Drainage Districts," *Agricultural Engineering*, 13:10 (October 1932), 258.

71. The assertion is based on the writer's manuscript notes regarding production data in archives of the Marion Steam Shovel Co., the American Steel Dredge Co., and the Bay City Land Dredge Co. and reconstructed from trade publications of the Fairbanks Steam Shovel Co.

## 3. The Role of Pumps in Land Drainage

1. Illinois Secretary of State, Archives, "Notes on Levee Districts, Illinois River Bottoms," by John W. Alvord and Charles B. Burdick [1914], sheet B (cited hereafter as Alvord and Burdick, "Notes"); Iowa State Drainage Association, *Proceedings of the First Annual Meeting*, "The Development of Agricultural Drainage in Illinois and Iowa," by Jacob A. Harman (n.p., 1908), 55 (cited hereafter as Harman, "Development of Drainage").

2. U.S. Department of Agriculture, *Land Drainage by Means of Pumps*, by S. M. Woodward, Bulletin 304 (Washington, D.C.: GPO, November 19, 1915), 6 (cited hereafter as Woodward, *Drainage by Pumps*); S. H. McCrory, "Historical Notes on Land Drainage in the United States," *Proceedings of the American Society of Civil Engineers* 53:7 (September 1927), 1632; Robert W. Harrison, *Alluvial Empire: A Study of State and Local Efforts Toward Land Development in the Alluvial Valley of the Lower Mississippi River, Including Flood Control, Land Drainage, Land Clearing, Land Forming* (Little Rock: Pioneer Press, 1961), 262–76; John Thompson, "The Settlement Geography of the Sacramento–San Joaquin Delta, California" (Ph.D. diss., Stanford University, 1957), 277–81. Woodward's paper is the principal source of technical insights in this chapter. Nevertheless, the writer's search of district records, texts in drainage engineering (cited below), and other literature afforded fresh observations.

3. Woodward, *Drainage by Pumps*, 6–7; Alvord and Burdick, "Notes," entries 3, 11, 23.

4. "Electricity Versus Steam in Drainage Pumping," *Electrical World* 64:6 (August 8, 1914), 275.

5. U.S. Department of Agriculture, *Cost of Pumping for Drainage in the Upper Mississippi Valley*, by John G. Sutton, Technical Bulletin 327 (Washington, D.C.: GPO, October 1932), 94 (cited hereafter as Sutton, *Cost of Pumping*).

6. University of Illinois, Water Resources Center, *Case Studies in Drainage and Levee District Formation and Development on the Floodplain of the Lower Illinois River, 1890s to 1930s*, by John Thompson, Special Report 017 (Urbana-Champaign, May 1989), 19–21, 81 (cited hereafter as Thompson, *Case Studies*).

7. "New Drainage Ditch," *RT*, August 24, 1899, p. 1, col. 2.

8. Thompson, *Case Studies*, 21, 81.

9. Data for table 3.1 are from Woodward, *Drainage by Pumps*, 58; Illinois Rivers and Lakes Commission, *The Illinois River and Its Bottom Lands*, 2nd ed., by John W. Alvord and Charles B. Burdick (Springfield: Illinois State Journal Co., 1919), table 11 (cited hereafter as Alvord and Burdick, *Illinois River*); Green County Circuit Court, Hartwell Drainage and Levee District files, "Engineer's Report—1927," by Caldwell Engineering Co., filed July 15, 1927, table 1, "Drainage Plant Capacities of Illinois River"; Illinois Department of Purchases and Construction, Division

of Waterways, *Flood Control Report,* by Jacob A. Harman, L. T. Berthe, M. Blanchard, and L. C. Craig (Springfield: Journal Printing Co., 1930), 38–39; Illinois Department of Registration and Education, Geological Survey, *Engineering and Legal Aspects of Land Drainage in Illinois,* rev. ed., by George W. Pickels and Frank B. Leonard Jr., Bulletin 42 (Urbana, 1929), 120–34 (cited hereafter as Pickels and Leonard, *Engineering*); Sutton, *Cost of Pumping,* 4–5; and diverse district records. Where discrepancies arose in the data, the writer adopted the most plausible technical source.

10. "Good Road to River," *CP,* June 10, 1909, p. 1, col. 3.

11. "Boast of Big Things," *CP,* September 14, 1911, p. 1, col. 6; Thompson, *Case Studies,* 46.

12. "Good Road to River," *CP,* June 10, 1909, p. 1, col. 3.

13. "Both Sides of the Bridge," *MCD,* November 1, 1901, p. 1, col. 5; "Contract Let for Grade West Havana Road," *MCD,* June 26, 1908, p. 1, col. 2; Thompson, *Case Studies,* 17–18, 80.

14. Charles G. Elliott, *Engineering for Land Drainage* (New York: John Wiley & Sons, 1908), 3. A second text referred to in preparing general statements here and below on the task of the drainage engineer is Daniel W. Murphy, *Drainage Engineering* (New York: McGraw-Hill Book Co., 1920).

15. C. H. Kreiling, "Pumping Plants in Levee Districts," *Illinois Engineer* 5:5 (May 1929), 1; Woodward, *Drainage by Pumps,* 3–4; Elliott, *Engineering for Land Drainage,* 3, 8; Murphy, *Drainage Engineering,* v, 59.

16. Peoria County Circuit Court, "Drainage Record—Pekin and La Marsh, 1," Jacob A. Harman, "Report" (to Commissioners), filed August 31, 1905, 108 (cited hereafter as Harman, "Pekin and La Marsh Report"); Association of Drainage and Levee Districts of Illinois, *First Annual Report of Transactions,* "Maintenance of Districts—Commissioner's View Point," by Louis Lowenstein (n.p., 1911), 50–51 (cited hereafter as Lowenstein, "Maintenance of Districts"); Kreiling, "Pumping Plants" 1–2; Woodward, *Drainage by Pumps,* 3–4; Association of Drainage and Levee Districts of Illinois, *Proceedings of the Third and Fourth Meetings,* "Electric Power for Pumping Stations," by J. Paul Clayton (Winchester: Press of the Winchester Times, 1913), 85 (cited hereafter as Clayton, "Electric Pumping Stations"); Pike County Circuit Court, "Drainage Record, Book D," McGee Creek Drainage and Levee District, Commissioners, "Petition for Special Assessment . . . ," filed June term, 1908, 124 (cited hereafter as McGee Creek Dist., "Petition"); U.S. Department of Agriculture, *Yearbook of Agriculture, 1955: Water,* "The Use of Pumps for Drainage," by John G. Sutton (Washington, D.C.: GPO, 1955), 529.

17. "Electricity Versus Steam in Drainage Pumping," 275.

18. U.S. Department of Agriculture, Office of Experiment Stations, *Annual Report of Irrigation and Drainage Investigations,* "Report of Drainage Investigations," by Charles G. Elliott, Bulletin 158 (Washington, D.C.: GPO, 1905), 708; Pickels and Leonard, *Engineering,* 271–72; "McGee Creek Levee District," *MB,* March 27, 1913, p. 4, col. 1; "Local and Other Items," *MB,* November 26, 1914, p. 5, col. 4; Pike County Circuit Court, "Drainage Record, Book E," McGee Creek Drainage and Levee District, containing "Decree Ordering Assessment . . . ," filed April 11, 1913, 125–26, "Report and Order . . . ," filed July 3, 1917, 548–49; Jacob A. Harman, "Drainage of Areas Subject to Floods," *Illinois Engineer* 8:5 (May 1932), 3; Illinois Department of Public Works and Buildings, Division of Waterways, *Eighteenth Annual Report, July 1, 1934 to June 30, 1935,* "History of Agricultural Drainage in Illinois," by H. G. Potter (Springfield: Journal Printing Co., 1935), 63 (cited hereafter as Potter, "History of Drainage").

19. The conventional assumption was of discharge velocities at 10 f.p.s. through the pump. Woodward, *Drainage by Pumps,* 13–14; Kreiling, "Pumping Plants," 1; Harman, "Drainage of

Areas," 3; Pickels and Leonard, *Engineering,* 271–73; George W. Pickels, *Drainage and Flood Control Engineering* (New York: McGraw-Hill Book Co., 1925), 306.

20. Kreiling, "Pumping Plants," 1; John G. Sutton, "The Cost of Drainage Pumping," *Agricultural Engineering* 13:5 (May 1932), 124; U.S. Department of Agriculture, *Yearbook of Agriculture, 1955: Water,* "Outlet Ditches, Slopes, Banks, Dikes, and Levees," by John G. Sutton (Washington, D.C.: GPO, 1955), 521–22 (cited hereafter as Sutton, "Outlet Ditches").

21. The engine or motor for a 20-inch pump was apt to be 75 or 100 h.p., while 30-inch pumps usually had 150 h.p. power sources, and 36-inch pumps generally were linked to 250 h.p. units. The 60-inch pump required a 500 h.p. engine or motor. Sutton, "Cost of Drainage Pumping," 4–5; Clayton, "Electric Pumping Stations," 86; Pickels and Leonard, *Engineering,* 272, 274; Pickels, *Drainage and Flood Control,* 306; "Electricity Versus Steam in Drainage Pumping," 275.

22. Pickels and Leonard, *Engineering,* 271; Pickels, *Drainage and Flood Control,* 312–13; Clayton, "Electric Pumping Stations," 87.

23. Federal Land Bank of St. Louis, "South Beardstown Drainage and Levee District, Cass County, Illinois," by C. B. Schmeltzer (April 1934), 96, and "Spring Lake Drainage and Levee District, Tazewell County, Illinois," by F. H. Schreiner (July 14, 1945), 7.

24. Harman, "Pekin and La Marsh Report," 218–19; Woodward, *Drainage by Pumps,* 34.

25. "Finishing Up Drainage Work," *RT,* January 26, 1911, p. 1, col. 1; Lowenstein, "Maintenance of Districts," 54; McGee Creek Dist., "Petition," 124; "River News of Interest," *I-S,* April 9, 1914, p. 1, col. 2; Scott County Levee and Drainage District, Commissioners, "Ledger" (minutes book), March 3, 1917, 69; records obtained through courtesy of R. Edward Frost (Winchester).

26. Alvord and Burdick, "Notes," entries 3, 11, 23; Potter, "History of Drainage," 63; Sutton, "Cost of Drainage Pumping," 6; Clayton, "Electric Pumping Stations," 86; U.S. War Department, Chief of Engineers, *Annual Report, 1914* (Washington, D.C.: GPO, 1914), 2:2942; "Unified Public Utilities in Central Illinois," *Electrical World* 61:22 (May 31, 1913), 1146–48, 1151; "Electricity Replaces Steam in Drainage Pumping," *Electrical World* 66:2 (July 10, 1915), 87.

27. U.S. Department of Commerce, Bureau of the Census, *Sixteenth Census of the United States, 1940: Drainage of Agricultural Lands* (Washington, D.C.: GPO, 1942), 2, 134 (cited hereafter as Bureau of the Census, *Drainage of Agricultural Lands*). Good comparisons of the relative merits of the different power sources appear in Pickels, *Drainage and Flood Control,* 294–314; "Unified Public Utilities in Central Illinois," 1151–52; and Sutton, "Cost of Drainage Pumping," 6.

28. Clayton, "Electric Pumping Stations," 94–95.

29. Woodward, *Drainage by Pumps,* 55; Scott County Levee and Drainage District, Commissioners, "Ledger," minutes of October 4, 1913, November 18, 1913, May 6, 1916, July 1, 1916, September 2, 1916, March 5, 1918, 39, 41, 58, 60, 62, 84.

30. "Drainage District Started New Pumping Plant," *MCD,* March 7, 1930, p. 1, col. 2; Thompson, *Case Studies,* 56, 58, 107.

31. Bureau of the Census, *Drainage of Agricultural Lands,* 146–53; Sutton, "Cost of Drainage Pumping," 1.

32. Pickels and Leonard, *Engineering,* 54, 270; Sutton, "Cost of Drainage Pumping," 1, 2, 4–5; Woodward, *Drainage by Pumps,* 52.

33. U.S. Department of Commerce, Bureau of the Census, *Fourteenth Census of the United States, 1920: Irrigation and Drainage* (Washington, D.C.: GPO, 1922), 7:448–49 (cited hereafter as Bureau of the Census, *Irrigation and Drainage*); Alvord and Burdick, *Illinois River,* 61.

34. Bureau of the Census, *Irrigation and Drainage,* 358, 363, and *Drainage of Agricultural Lands,* 146–53; Sutton, "Outlet Ditches," 529–30.

## 4. Shapers of the Drainage Landscape

1. I. O. Baker, "Some of the Engineering Features of Illinois Drainage," *Association of Engineering Societies* 5:11 (September 1886), 426–29; "The Dredge Ditch in Drainage Work," *Drainage Journal* 23:10 (October 1901), 263. On the nature of antecedent drainage technology and work in the Grand Prairie, see Margaret B. Bogue, "The Swamp Land Act and Wet Land Utilization in Illinois, 1850–1890," *Agricultural History* 25:4 (October 1951), 169–80; Illinois Department of Registration and Education, Historical Library, *Patterns from the Sod: Land Use and Tenure in the Grand Prairie, 1850–1900,* by Margaret Beattie Bogue, Collections Series 34 (Springfield, 1959).

2. This and the following figures are from U.S. Department of Agriculture, Office of Experiment Stations, *Excavating Machinery Used for Digging Ditches and Building Levees,* by James O. Wright, Circular 74 (Washington, D.C.: GPO, 1907).

3. Baker, "Engineering Features of Drainage," 428; "Dredge Ditch in Drainage Work," 265.

4. Baker, "Engineering Features of Drainage," 428–29; A. W. Shaw, "The Inlet Swamp Drainage District," *Engineering News* 53:4 (January 26, 1905), 90; "Big Drainage Contract Let," I-S, January 20, 1906, p. 1, col. 4.

5. John Thompson, "Commemorating the Large Steam Dipper Dredges of the Panama Canal," *Nautical Research Journal* 43:2 (June 1998), 92–103.

6. S. H. McCrory, "Historical Notes on Land Drainage in the United States," *Proceedings of the American Society of Civil Engineers* 53:7 (September 1927), 1631. Contrary to McCrory's assertion, as will be reviewed in a forthcoming paper (John Thompson, "When Steam Dredges Transformed Wetlands to Cropland," *Illinois Steward* 10:3 [fall 2001]), this was not the first adoption of a steam dredge for land drainage in Illinois; however, the success was resounding. An account of subsequent use of dredges for drainage work in the upland wet prairies appears in Roger Andrew Winsor, "Artificial Drainage of East Central Illinois, 1820–1920" (Ph.D. diss., University of Illinois at Urbana-Champaign, 1975), 147–51. Earlier adoptions of the dipper dredge for levee building and ditching occurred in California. See Bucyrus-Erie Corp., "Excavating Machinery Shipment Ledger, 1882–1893," corporate archives, South Milwaukee, Wisconsin; A. H. Bell, "Drainage Districts and the Construction of Drainage Canals," *Engineering News and American Contract Journal* 15 (February 20, 1886), 113; I. O. Baker, "Dredges and Drainage," *Champaign County Gazette,* July 7, 1886, p. 6, cols. 3–4; John Thompson and Edward A. Dutra, *The Tule Breakers: The Story of the California Dredge* (Stockton, Calif.: Stockton Corral of Westerners, University of the Pacific, 1983) 37–41; Illinois Department of Business and Economic Development, *Inventory of Illinois Drainage and Levee Districts, 1971* (Springfield, 1971), 1:364.

7. Cass County Circuit Court, Hager Slough Special Drainage District, Commissioners, untitled report filed September 25, 1886, 1; "How Swamp Lands Are Reclaimed in Illinois," *Farmers' Review,* April 25, 1888, 260; Tazewell County Circuit Court, Hickory Grove Drainage and Levee District, Commissioners, "Annual Report, 1889," filed October 17, 1889, with appended "Expenditures," 2–3, and "Annual Report, 1891," filed September 7, 1891, 1; Tazewell County Treasurer, "Report for Hickory . . . ," filed December 1, 1897, 1–2; "Havana, Mason County's Metropolis," *PH-T,* December 29, 1905, p. 9, cols. 1–4; Marion Steam Shovel Co., "Machine Record No. 1," entries 41 and 96, not paginated. Records related to dredges sold, giving name of purchaser and delivery destination, are in the archives of Dresser Industries, Marion Power Shovel Co., Marion, Ohio.

8. U.S. Congress, House, *Survey, with Plans and Estimates of Cost, for a Navigable Waterway 14 Feet Deep from Lockport, Ill., by Way of Des Plaines and Illinois Rivers, and Thence by Way of the Mississippi River to St. Louis,* "Map of the Illinois and Des Plaines Rivers" (1:7,200), by J. W. Woermann, U.S. Army Corps of Engineers, 59th Cong., 1st sess., 1905, H. Doc. 263, sheets 8, 9.

9. University of Illinois, Water Resources Center, *Case Studies in Drainage and Levee District Formation and Development on the Floodplain of the Lower Illinois River, 1890s to 1930s,* by John Thompson, Special Report 017 (Urbana-Champaign, May 1989), 19, 25, 54, 78, 82, 109 (cited hereafter as Thompson, *Case Studies*).

10. Marion Steam Shovel Co., *Circular No. 21—Dredges* (n.p., ca. 1906), table showing "Ditching Dredges with Bank Spuds," 11.

11. F. G. Pulley, "A Levee Job of 2,260,000 Cubic Yards," *Contractors Review,* December 16, 1916, 131, 133; Marion Steam Shovel Co., *The Name and the Machine,* catalog 188 (Cleveland: Caxton Co., ca. 1918), 54–55.

12. "Another Big Dredge Coming," *MCD,* March 19, 1920, p. 1, col. 6; "New Dredge for Thompson Lake," *MCD,* June 4, 1920, p. 7, col. 3. The antecedent dredge type is described in Thompson and Dutra, *Tule Breakers.*

13. "Promoters, Will Use Big Dredge on New Railway," *PH-T,* October 25, 1912, p. 3, col. 7; "Reclaims Much Land," *PH-T,* November 18, 1912, p. 3, col. 3.

14. "Mammoth Dredge," *RT,* April 2, 1896, p. 1, col. 4; "Frederick Department," *RT,* November 12, 1896, p. 1, col. 3; "Drainage District News," *RT,* October 29, 1903, p. 1, col. 1; "Bluffdale," *CP,* September 8, 1905, p. 8, col. 3; "Eldred," *CP,* February 16, 1906, p. 8, col. 5; "Adding a Township," *CP,* October 5, 1906, p. 1, col. 2; "Big Land Transfer in Lacey Levee District," *MCD,* October 28, 1910, p. 1, col. 5; "Jesse Lowe, Big Works Contractor, Is Dead," *Contractor* 25:10 (May 10, 1918), 212.

15. Greene County Circuit Court, Hillview Drainage and Levee District, Commissioners, "Petition and Order Authorizing Contract to Finish Ditch Work," filed November 3, 1909; Scott County Circuit Court, "Big Swan Drainage and Levee District," Commissioners, "Petition," filed December 10, 1902.

16. "A Cross-Cut Excavating Machine for Drainage Ditches," *Engineering News* 54:10 (September 7, 1905), 250; "Adding A Township," *CP,* October 5, 1906, p. 1, cols. 2–3; Greene County Circuit Court, Hartwell Drainage and Levee District, Commissioners, "Petition for Additional Assessment," filed June 18, 1908, 3–4, 7; F. C. Austin Drainage Excavator Co., *Irrigation and Drainage Excavating Mach'y* (Chicago, n.d.), 2–3, 12.

17. "Cross-Cut Excavating Machine," 250; "Pekin News," *PH-T,* June 19, 1907, p. 7, col. 4; L. K. Sherman, "Drainage and Levee District Project," *Reclamation and Farm Engineering* 8:3 (September 1925), 212; Shaw, "Inlet Swamp District," 90; "Eldred," *CP,* December 20, 1923, p. 8, col. 7, May 15, 1924, p. 8, col. 2, and May 29, 1924, p. 8, col. 2; F. C. Austin Co., *Irrigation and Drainage,* 2–3.

18. McCrory, "Historical Notes on Land Drainage," 1631; "A New Style of Scraper Excavator," *Engineering News* 53:9 (March 2, 1905), 216–17. Among the contractors owning draglines were C. L. Jones, Nelson Cole & Sons. "Mammoth Pumping Plant," *JR,* December 3, 1908, p. 4, col. 1; "Hillview," *CP,* January 29, 1914, p. 8, col. 2; "Eldred," *CP,* January 27, 1916, p. 8, col. 3, February 10, 1916, p. 8, cols. 2–3, February 24, 1916, p. 8, col. 3, and March 2, 1916, p. 8, col. 4; "Dredge Starts in Seahorn District," *MCD,* March 26, 1920, p. 1, col. 5; Scott County Circuit Court, Scott County Drainage and Levee District files, "Proposal for Levee and Ditch Construction," by Federal Contracting Co., filed November 29, 1911; Greene County Circuit Court, *R. H. McWilliams and G. A. McWilliams v. Eldred Drainage and Levee District,* February term, 1918, no. 7354, "Record," 316.

19. "Local and Other Items," *MB,* September 3, 1914, p. 5, col. 4, and August 3, 1916, p. 5, col. 3.

20. "Local and Other Items," *MB,* December 10, 1914, p. 5, col. 4.

21. Fulton County Circuit Court, "Seahorn Drainage and Levee District, 1," "Engineer's Report," dated December 12, 1921, 220; Pulley, "Levee Job," 131, 133.

22. "Big Greene County Farm," *CP,* November 6, 1913, p. 2, cols. 1–2; "Drainage District Contracts," *CP,* January 20, 1916, p. 1, col. 2; "Drainage District," *RT,* August 2, 1917, p. 5, col. 1; Association of Drainage and Levee Districts of Illinois, *Proceedings of the Third and Fourth Meetings,* comments of Louis Lowenstein (Winchester: Press of the Winchester Times, 1913), 48 (cited hereafter as Assoc. Drainage and Levee Dist., *Proceedings, 1913*); Jersey County Circuit Court, "Drainage and Levee Record No. 1," Nutwood Drainage and Levee District, Commissioners, "Report for July 1, 1915–July 1, 1916," and "Report for July 1, 1917–July 1, 1918," 509, 546; Greene County Circuit Court, "Drainage Record 2," Eldred Drainage and Levee District, Commissioners, "Annual Financial Report," filed November 11, 1916, 100; Pike County Circuit Court, "Drainage Record, Book F," McGee Creek Drainage and Levee District, Commissioners, "Petition for a Special Assessment of $33,811.85," dated September 23, 1919, 474, 485–88; Scott County Levee and Drainage District, Commissioners, "Ledger," minutes of March 2, 1925, 278; Fulton County Circuit Court, "West Matanzas Drainage and Levee District, Record 2," Commissioners, "Annual Report," dated September 1, 1926, 23, 48, and "East Liverpool Drainage and Levee District, 1," Commissioners, "Petition for Leave to Purchase Drag Line," August 10, 1928, 237; Big Swan Drainage and Levee District, Commissioners, "Ledger 1904–1931," itemized statement of accounts, entry for October 6, 1926, 269 (obtained, along with other Scott County district records through the courtesy of R. Edward Frost, Winchester).

23. "Ditching with Dynamite," *PCD,* July 26, 1911, p. 7, col. 2; "Farming with Dynamite," *PCD,* April 24, 1912, p. 1, col. 4; "To Use Dynamite," *PCD,* May 29, 1912, p. 1, col. 4; Assoc. Drainage and Levee Dist., *Proceedings, 1913,* "The Uses for Explosives in Drainage Projects," by H. B. Owsley, 102–8.

24. Numerous statements of specifications and contracts were inspected. A model for checking contents was Big Swan Drainage and Levee District, Commissioners, *Specifications for the Construction of Ditches and Levee* (Winchester: Press of the Winchester Times, 1904).

25. Lost Creek Drainage and Levee District, Commissioners, minutes book, meeting of November 20, 1920, 172 (courtesy of Milton McClure Jr., Beardstown).

26. Observation based on Illinois Secretary of State, Archives, "Notes on Levee Districts, Illinois River Bottoms," by John W. Alvord and Charles B. Burdick [1914]; Association of Drainage and Levee Districts of Illinois, *First Annual Report of Transactions* (n.p., 1911), 67–71; Havana Public Library, "Sangamon River Outlet Organization," C. H. Kreiling letter to Col. C. Keller, District Engineer, Corps of Engineers, Chicago, July 5, 1941.

27. "Adding a Township," *CP,* October 5, 1906, p. 1, col. 2; "Hillview," *CP,* January 29, 1914, p. 8, col. 2, and March 26, 1914, p. 2, col. 1.

28. General observations are based on data contained in records of some 30 districts. The most accessible citations are in Thompson, *Case Studies,* 4, 7, 14, 20, 31, 34, 44, 45, 55, 59, 63, 85, 100. Representative primary sources include: East Peoria Drainage and Levee District, Commissioners, "Minutes," April 14, 1911, not paginated, and "Minute Book," August 10, 1923, not paginated (courtesy of Caterpillar Tractor Co., East Peoria); Fulton County Circuit Court, "Thompson Lake Drainage and Levee District," Commissioners, "Petition for Second Assessment—Exhibit J," filed February 14, 1922, "Statement of Probable Cost of Proposed Work, . . . as per Report of April 17, 1919," and "Statement . . . Nov. 4, 1921"; Pike County Circuit Court,

"Drainage Record, Book G," Valley City Drainage and Levee District, Commissioners, "Annual Financial Report," September 30, 1922, 451; and Illinois Department of Purchases and Construction, Division of Waterways, *Flood Control Report,* by Jacob A. Harman, L. T. Berthe, M. Blanchard, and L. C. Craig (Springfield: Journal Printing Co., 1930), 122.

29. "Funeral of A. V. Wills Is Held Here Today," *Pike County Republican,* October 4, 1939, p. 4, col. 4.

30. "Personal Mention," *PCD,* January 3, 1899, p. 3, col. 1, and May 23, 1899, p. 8, col. 1; "Local Matters," *PCD,* December 13, 1899, p. 5, col. 3, and July 4, 1900, p. 5, col. 1; "A. V. Wills & Sons," *PCD,* October 9, 1901, p. 8, col. 3; "Funeral of A. V. Wills . . ."; "Reclaiming Bottom Land," *MB,* November 25, 1909, p. 1, cols. 3–4.

31. "Personal Mention," *PCD,* January 24, 1899, p. 8, col. 1; "Local Matters," *PCD,* September 26, 1906, p. 5, col. 4, and November 14, 1906, p. 5, col. 2; "Has Monster Boom," *PCD,* July 3, 1907, p. 1, col. 2; "Funeral of A. V. Wills . . ."

32. "Local Matters," *PCD,* July 4, 1906, and November 14, 1906, p. 5, col. 2; "Has Monster Boom," *PCD,* July 3, 1907, p. 1, col. 2; "Local and Other Items," *MB,* January 30, 1908, p. 5, col. 3; "Big Dipper Dredge Moved," *MB,* June 17, 1909, p. 1, col. 5; Big Swan Drainage and Levee Dist., "Ledger 1904–1931," 7–12; Pike County Circuit Court, "Drainage Record, Book D," McGee Creek Drainage and Levee District, Commissioners, "Petition for Special Assessment . . . , Exhibit B," 101–2, filed June term, 1908, and "Annual Report, May 29, 1908–September 1, 1909, Exhibit A," 175; Schuyler County Circuit Court, "Crane Creek Drainage and Levee District, 1893–1918," Commissioners, "Record of Expenditures," November 30, 1912, and "Petition for Authority to Raise and Strengthen Levee," August 24, 1916.

33. Marion Steam Shovel Co., "Machine Record No. 1" and "Summary of Machine Record No. 2, Ditching Dredges," not paginated (cited hereafter as Marion Machine Records); "Personal Mention," *PCD,* January 24, 1899, p. 8, col. 1, and June 30, 1899, p. 8, col. 1; "Bad Cutting Affray," *PCD,* August 16, 1905, p. 1, col. 6; "Local Matters," *PCD,* November 14, 1906, p. 5, col. 2; "Big Dipper Dredge Moved," *MB,* June 17, 1905, p. 1, col. 5; "Dredge Moved," *MB,* June 17, 1909, p. 1, col. 5; "A. V. Wills & Sons," *Contractors Review* (December 2, 1916), 114; "Funeral of A. V. Wills . . ."; Schuyler County Circuit Court, "Crane Creek Drainage and Levee District," Commissioners, "Annual Report," filed July 25, 1923.

34. Greene County Circuit Court, *McWilliams v. Eldred,* testimony of Frank J. Traut, 550–51; "Gets a Big Contract," *I-S,* February 11, 1907, p. 1, col. 5; "Traut Before Grand Jury," *ME,* June 7, 1910, p. 1, cols. 3–4; "Concludes His Testimony to Jury," *ME,* June 8, 1910, p. 1, col. 3; "Browne and Four Others Indicted," *ME,* June 26, 1910, p. 1, col. 5; Scott County Levee and Drainage District, Commissioners, "Ledger," minutes of December 5, 1911, November 20, December 11, 13, 1912, 15, 27; "Pen Pictures of Prominent People," *I-S,* January 31, 1907, p. 1, cols. 3–5; Charles A. E. Martin, ed., *Historical Encyclopedia of Illinois and History of Cass County* (Chicago: Munsell Publishing Co., 1915), 2:969.

35. Illinois Secretary of State, Corporation Department, box 970-no. 55103, "Federal Contracting Co.," "Report of Commissioners . . . ," filed October 7, 1905, "Report of Subscription of Capital Stock," filed October 7, 1905, "[Petition to] Increase Stock," filed September 26, 1911, "Certificate of Cancellation of Charter," filed April 15, 1914; Marion Machine Records, No. 2; "Gets a Big Contract," *I-S,* February 11, 1907, p. 1, col. 5; "Loading the Machine," *I-S,* April 6, 1907, p. 7, col. 3; "The Nutwood Drainage and Levee District," *JR,* October 17, 1907, p. 1, cols. 1–4; "Work Progressing on Drainage Ditch," *PH-T,* December 13, 1907, p. 11, col. 1; "Dredge Closes Up Spring Lake," *PH-T,* June 2, 1910, p. 2, col. 4; "Secured Ditching Contract," *ME,* February 18,

1911, p. 1, col. 4; Eldred Drainage and Levee District, Commissioners, "Records," minutes of January 20, 1910 (courtesy of Jane Wilcox Villegas, Eldred); Scott County Levee and Drainage District, Commissioners, "Ledger," minutes of December 5, 1911, 15–97; East Peoria Drainage and Levee District, "Minutes," meeting of May 18, 1911, November 25, 1911, and October 17, 1912; Jersey County Circuit Court, "Drainage and Levee Record No. 1," Nutwood Drainage and Levee District, Commissioners, "Report of July 28, 1908 to July 1, 1909," 118–23; Brown County Circuit Court, "Drainage Record No. 1," Big Prairie Drainage and Levee District, Commissioners, "Disbursements, September 7, 1916 to January 6, 1917," 139–40; "Pen Pictures of Prominent People," *I-S,* January 31, 1907, p. 1, cols. 3–5; Charles M. Samson, comp., *Beardstown, Illinois, City Directory* (Bloomington: Pantagraph Printing and Stationery, 1915), 137, 187, 216; Martin, *Historical Encyclopedia,* 2:969; Howard C. Haungs, "Engineering Methods for the Reclamation of Overflowed Lands in Illinois" (C.E. thesis, University of Illinois, 1916), 34; *Farm Directory of Cass, Mason, Menard and Sangamon Counties, Illinois* (1917; reprint, Dixon, Ill.: Print Shop, 1984), 354. Traut is known to have been disposing of contracting equipment in 1919. Schuyler County Circuit Court, "Crane Creek Drainage and Levee District," Commissioners, "Annual Report," filed February 8, 1921.

36. Fairbanks Steam Shovel Co., *Dipper Dredge Catalogue* (n.p., ca. 1907), 33, and *Dipper Dredges* (Grand Rapids, Mich.: Dickinson Bros., 1912), 104; "A Bit of River News of Interest," *MB,* October 8, 1914, p. 1, col. 5; Haungs, "Engineering Methods," 36, 38.

37. "Eldred," *CP,* October 23, 1913, p. 8, col. 2; Schuyler County Circuit Court, "Big Lake Drainage and Levee District—1911–1920," Commissioners, "Annual Report," filed April 22, 1914; Mason County Circuit Court, *Knapp Island Gun Club v. M. J. O'Meara—Injunction,* "Answer of M. J. O'Meara," filed November 8, 1915, chancery no. 2214, box 772, 3; Brown County Circuit Court, "Drainage Record No. 1," Big Prairie Drainage and Levee District, Commissioners, "Report of Moneys Collected . . . and Expended," entries for October 21, 1916, to February 19, 1917, 139; Fulton County Circuit Court, "East Liverpool Drainage and Levee District, 1," Commissioners, "Agreement," December 26, 1917, 222, and "West Matanzas Drainage and Levee District, Record Book 1," Commissioners, "Agreement," January 17, 1918, and "Report," July 23, 1919, not paginated; Lost Creek Drainage and Levee District, Commissioners, minutes book, meeting of November 20, 1920, 171–75, 280, and Treasurer, "Reports, 1920–1934," 2; Schuyler County Circuit Court, "Crane Creek Drainage and Levee District," Commissioners, "Petition for Contract" and "Order," July 12, 1922, and Treasurer, "Report," July 25, 1923, not paginated; Peoria County Circuit Court, "Drainage Record—Pekin and La Marsh, 1," Commissioners, "Accounts," filed November 1, 1924, 100–103; "Dynamited Levee and Water Going Out Slowly," *CP,* November 4, 1926, p. 1, col. 4; Big Swan Drainage and Levee Dist., "Ledger 1904–1931," entry for March 10, 1927, 269.

38. "River News," *I-S,* September 17, 1917, p. 8, col. 4, September 18, 1917, p. 5, col. 3, September 20, 1917, p. 1, col. 5; W. H. Hoffman, *City Directory of Quincy, Illinois, 1930–1931* (Quincy: Hoffman City Directories, 1930), 370, 526; Artcraft Directory Publishers, *Directory of the City of Quincy,Illinois, 1938* (Quincy, April 1938), 17A, 297.

39. Marion Machine Records; Greene County Circuit Court, *McWilliams v. Eldred,* testimony of R. H. McWilliams, 310–11; "Frederick Department," *RT,* March 11, 1897, p. 1, col. 4, and July 1, 1897, p. 1, col. 4; "Otter Creek Valley," *MCD,* September 27, 1901, p. 1, col. 5, and October 11, 1901, p. 1, col. 5; "Big Drainage Contract Let," *I-S,* January 20, 1906, p. 1, col. 4; "Local and Other Items," *MB,* June 27, 1907, p. 5, col. 3; "Contract Let for Grade West Havana Road," *MCD,* June 26, 1908, p. 1, col. 2.

40. "Big Drainage Contract Let," *I-S*, January 20, 1906, p. 1, col. 4; "Begin Levee Soon," *CP*, January 27, 1910, p. 1, col. 4; Pike County Circuit Court, "Drainage Record, Book E," McGee Creek Drainage and Levee District, Commissioners, "Annual Report," filed December 24, 1915, 314–18; Big Swan Drainage and Levee Dist., "Ledger 1904–1931," entries for September 5, 1914, to January 13, 1916, 119–34; Schuyler County Circuit Court, "Crane Creek Drainage and Levee District," Commissioners, "Annual Report," January 1, 1919, and February 8, 1921; Morgan County Circuit Court, "Coon Run Drainage and Levee District," Commissioners, "Annual Report," filed July 7, 1924, 3, 8; "Electric-Hydraulic Dredge Work on Illinois Levees," *Engineering News-Record* 92:22 (May 29, 1924), 944–45; Illinois Department of Public Works and Buildings, Division of Waterways, *Ninth Annual Report, July 1, 1925 to June 30, 1926* (Springfield: Illinois State Journal Co., 1927), 76, 79 (cited hereafter as Ill. Div. Waterways, *Report, 1925–1926*).

41. Greene County Circuit Court, *McWilliams v. Eldred*, testimony of McWilliams, 310–11; "A Bit of River News of Interest," *MB*, October 8, 1914, p. 1, col. 5; "R. H. and G. A. McWilliams Awarded Drainage Contract in Arkansas," *Contractors Review* (December 30, 1916), 158; "McWilliams Dredging Co." (advertisement), *National Reclamation Magazine* 1:1 (December 1921), 25; R. L. Polk & Co., *Memphis City Directory, 1914* (Memphis, 1914), 919, *Memphis City Directory, 1926* (Memphis, 1926), 797, and *Memphis City Directory, 1929* (Memphis, 1929), 859.

42. "Big Concern to Do Business in Beardstown," *I-S*, June 12, 1915, p. 1, col. 6; "Work Completed on House Boat," *I-S*, August 18, 1915, p. 3, col. 1; "Inspecting Work," *I-S*, October 7, 1915, p. 5, col. 4; "Another Big Contract," *MCD*, May 14, 1920, p. 1, col. 2; "Edward Gillen Died in Racine," *MCD*, September 10, 1920, p. 1, col. 6; "Gillen Company Secures Large Contract," *MCD*, June 9, 1922, p. 1, col. 4; "River Front News," *MCD*, June 6, 1924, p. 1, col. 4; "Dredge Cook, Uncle of Well-Known Playwright," *MCD*, June 20, 1924, p. 1, col. 7; "Removing Rocks from River Bed," *MCD*, June 25, 1926, p. 1, col. 4; "Gillen Co. Sued Drainage District," *MCD*, June 30, 1926, p. 1, col. 2; "Gillen Co. Dredging for Super Plant," *MCD*, April 29, 1927, p. 1, col. 1.

43. "New Dredge for Thompson Lake," *MCD*, June 4, 1920, p. 7, col. 3; "River Rising; Bottoms Inundated," *MCD*, April 16, 1926, p. 1, col. 5; Mason County Probate Court, "Estate Papers for George William Gillen," box 219.

44. "Thieves Loot River Launch," *I-S*, August 10, 1915, p. 2, col. 2; "Work Completed on House Boat," *I-S*, August 18, 1915, p. 3, col. 1; "Gillen Company Lands $250,000 Drainage Contract," *I-S*, October 23, 1918, p. 1, col. 4; Pulley, "Levee Job," 131–33; "Thompson Lake Contract Let," *MCD*, June 20, 1919, p. 1, cols. 4–5; "Dredges Quit Work," *MCD*, December 5, 1919, p. 1, col. 1; "Gillen Company Secures Large Contract," *MCD*, June 9, 1922, p. 1, col. 4; "Gas Killed Geo. W. Gillen," *MCD*, January 9, 1925, p. 1, col. 6; "Gillen Co. Dredging for Super Plant," *MCD*, April 29, 1927, p. 1, col. 1.

45. Illinois Secretary of State, Corporation Department, box 806—no. 41404, "Pollard Goff and Co.," "Application for License to Form Corporation," filed September 22, 1899, "Certificate of Dissolution," filed February 16, 1907, and box 108—no. 8879, "Quincy Dredging and Towing Co.," "Statement of Incorporation," filed February 21, 1883, clerical note on expiration of corporation, filed March 26, 1913; "Pollard Estates Heavy Bond," *MCD*, April 20, 1906, p. 1, col. 3; Marion Machine Records.

46. "Rodgers' Dredger Sinks," *I-S*, October 14, 1909, p. 3, col. 5; Marion Machine Records; Mason County Circuit Court, "Mason and Tazewell Special Drainage District," Commissioners, "Contract," ca. May 1884, and "Long Branch Drainage District," letter from J. E. Rodgers, May 8, 1893; John Gresham, *Historical and Biographical Record of Douglas County, Illinois* (Logansport, Ind.: Wilson, Humphreys & Co., 1900), 139–40; Ruth Wallace Lynn, *Prelude to Progress: The*

*History of Mason County, Illinois, 1818–1968* (Havana: Mason County Board of Supervisors, 1968), 352; Morgan County Circuit Court, "Coon Run Drainage and Levee District," Commissioners, "Report," filed July 20, 1901, "Report," filed July 19, 1902, and "Annual Report," filed July 13, 1903; Ill. Div. Waterways, *Report, 1925–1926*, 76.

47. Piatt County Clerk, "Lake Fork Special Drainage District—No. 1," Commissioners, minutes of July 10, 1885, 99; Marion Machine Records; "Kilbourne-Land Improvements," *PH-T*, December 2, 1904, p. 2, cols. 1–2; "Dredge Will Go Through Bridge," *MCD*, July 27, 1906, p. 1, col. 3; "Big Dredge at Work," *MCD*, December 21, 1906, p. 1, col. 3; "High Water Delays Work by Dredges," *PH-T*, June 7, 1907, p. 10, col. 7; "Big Dredging Contract to Straighten the Sangamon," *MCD*, October 2, 1908, p. 1, col. 2.

48. Marion Machine Records; U.S. War Department, Secretary, *Annual Report, 1878*, vol. 2, pt. 2, "Improvement of Illinois River" (Washington, D.C.: GPO, 1878), BB5:1190; "Local Matters," *PCD*, June 12, 1907, p. 5, col. 1; Greene County Circuit Court, Hillview Drainage and Levee District, "Order to Make Second Additional Assessment," filed June 21, 1909, 2–4.

49. Fairbanks Steam Shovel Co., *Dipper Dredges*, 104; Peoria County Circuit Court, "Drainage District Book 1," Banner Special Drainage District, Commissioners, "Financial Report," November 9, 1915, 408; "River News," *MCD*, August 31, 1917, p. 10, col. 2, September 28, 1917, p. 1, col. 4, and November 9, 1917, p. 7, col. 4; Everett McKinley Dirksen, *The Education of a Senator* (Urbana: University of Illinois Press, 1998), 54, 56.

50. Federal Land Bank of St. Louis, "Report on Drainage and Levee Districts: Cass, Mason, Menard and Tazewell Counties, Illinois," containing "Chautauqua Drainage and Levee District . . .," by Hubert E. Goodell and Fred S. Morse, January 8, 1934, 41, 43.

51. Pike County Circuit Court, "Drainage Record, Book F," McGee Creek Drainage and Levee District, Commissioners, "Petition for a Special Assessment of $33,811.85," including "Exhibit C, Report of Engineer," September 23, 1919, 485–88.

52. U.S. Department of Agriculture, *Yearbook, 1918*, "The Drainage Movement in the United States," by S. H. McCrory (Washington, D.C.: GPO, 1919), 140, and *Yearbook of Agriculture, 1955: Water*, "The History of Our Drainage Enterprises," by Hugh H. Wooten and Lewis A. Jones (Washington, D.C.: GPO, 1955), 483–86.

53. Thompson and Dutra, *Tule Breakers*, 215, 247; John Thompson, "The Bay City Land Dredge and Dredge Works: Perspectives on the Machines of Land Drainage," *Michigan Historical Review* 12:2 (fall 1986), 24, 42.

54. *National Cyclopaedia of American Biography* (New York: James T. White & Co., 1931), 21:288. R. L. Polk & Co.'s *Memphis City Directories* for 1920, 1921, and 1928 record the presence of Elliott & Harmon.

55. Samuel P. Hays, *Conservation and the Gospel of Efficiency: The Progressive Conservation Movement, 1890–1920* (Cambridge: Harvard University Press, 1959), 228–29.

## 5. Challenges Engineered by Other Sectors

1. For guidance on the development of navigation and related issues, the writer relied particularly on several sources: U.S. Congress, House, Committee on Rivers and Harbors, *Hearings on the Subject of the Improvement of the Illinois and Mississippi Rivers, and the Diversion of Water from Lake Michigan into the Illinois River*, 68th Cong., 1st sess., 1924, pts. 1 and 2 (cited hereafter as House Comm. R. and H., *Hearings, 1924*). Well focused on the politics of the issue are: William B. Philip, "Chicago and the Down State: A Study of Their Conflicts, 1870–1934" (Ph.D. diss.,

University of Chicago, 1940), 159, 200–206, 212–28; and Robert A. Waller, "The Illinois Waterway from Conception to Completion, 1908–1933," *Journal of the Illinois State Historical Society* 65:2 (summer 1972), 126–27. Also see U.S. Army Corps of Engineers, Chicago District, *Those Army Engineers: A History of the Chicago District, U.S. Army Corps of Engineers,* by John W. Larson (Washington, D.C.: GPO, 1980), 190–92 (cited hereafter as Larson, *Those Army Engineers*). Relevant as well are: Illinois Board of Health, *Water Supplies of Illinois and the Pollution of Its Streams,* by John H. Rauch (Springfield, 1889), xvii; W. J. McGee, "Our Great Rivers," *World's Work* 13:4 (February 1907), 8576–78; Lakes-to-the-Gulf Deep Waterway Association, "Fourteen Feet Through the Valley," by Harry B. Hawes (St. Louis, ca. 1907), 9–13, 17–19; Association of Drainage and Levee Districts of Illinois, *First Annual Report of Transactions,* "Reclamation," by Lyman E. Cooley (n.p., 1911), 34–37 (cited hereafter as Cooley, "Reclamation"); and Samuel P. Hays, *Conservation and the Gospel of Efficiency: The Progressive Conservation Movement, 1890–1920* (Cambridge: Harvard University Press, 1959), 91–101, 108, 199.

2. Hays, *Conservation,* 11, 13–15, 91, 95, 100, 109, 252–53; U.S. Army Corps of Engineers, Office of History, *The Evolution of the 1936 Flood Control Act,* by Joseph L. Arnold (Washington, D.C.: GPO, 1988), 11–13; Marvin W. Block, "Henry T. Rainey of Illinois," *Journal of the Illinois State Historical Society* 65:2 (summer 1972), 144–45.

3. Illinois Department of Public Health, *The Rise and Fall of Disease in Illinois,* by Isaac D. Rawlings (Springfield: Phillips Bros., 1927), 2:402 (cited hereafter as Rawlings, *Rise and Fall*). The most comprehensive coverage in contemporary trade journals commenced as a 16-part series with "The Chicago Main Drainage Channel," *Engineering News* 33:20 (May 16, 1895), 314–16.

4. U.S. War Department, Secretary, *Annual Report on the Operations, 1870,* vol. 2, app. K2, "Improvement of Illinois River," 313–16; U.S. Congress, House, Army Chief of Engineers, *Annual Report for 1877,* 45th Cong., 2nd sess., 1877, Exec. Doc. 1, pt. 2:2, "Improving Illinois River," app. P5, 546–52; U.S. Congress, House, Secretary of War, Chief of Engineers, *Annual Report, 1880,* "Survey of Illinois River," 46th Cong., 3rd sess., 1880, Exec. Doc. 1, pt. 2:2, app. EE3, 1995–99; Illinois General Assembly, *Reports to, 1881,* vol. 3, Doc. D, "Report of Canal Commissioner" (Springfield: H. W. Rokker, 1881), 4–5; U.S. Congress, House, Committee on Rivers and Harbors, *Illinois River, Ill.,* 72nd Cong., 1st sess., 1931, H. Doc. 182, 15, 21–22 (cited hereafter as House Comm. R. and H., *Illinois River, 1931*).

5. U.S. War Department, Secretary, *Annual Report for the Year 1896,* pt. 1, 2:327; Cooley, "Reclamation," 35–36; House Comm. R. and H., *Illinois River, 1931,* 22; Larson, *Those Army Engineers,* 188–89; Philip, "Chicago and the Down State," 203–5.

6. "Farmers Want the Land," *RT,* August 14, 1902, p. 1, col. 3; "Dams and Damages," *CP,* August 22, 1902, p. 1, col. 6; Illinois Internal Improvements Commission, *The Lakes and Gulf Waterway* (Springfield: Phillips Bros., 1907), 28, 31; Illinois Department of Registration and Education, Geological Survey, *Geography of the Middle Illinois Valley,* by Harlan H. Barrows, Bulletin 15 (Urbana: Phillips Bros., 1910; reprint, 1925), 106; House Comm. R. and H., *Illinois River, 1931,* 48–49; Association of Drainage and Levee Districts of Illinois, *First Annual Report of Transactions,* comments of H. V. Teal (n.p., 1911), 4–5, and Cooley, "Reclamation," 34–36; W. G. Potter, "Improvement and Utilization of the Rivers of Illinois," *Western Society of Engineers Journal* 31:5 (May 1926), comments of Maj. R. W. Putnam, 213–14; Robert A. Waller, *Rainey of Illinois: A Political Biography, 1903–1934* (Urbana: University of Illinois Press, 1977), 57.

7. For a comprehensive account, see Louis P. Cain, *Sanitary Strategy for a Lakefront Metropolis* (DeKalb: Northern Illinois University Press, 1978); also see James C. O'Connell,

*Chicago's Quest for Pure Water,* Public Works Historical Society, Essay No. 1 (Washington, D.C.: Public Works Historical Society, 1976); Larson, *Those Army Engineers,* 185–206; and House Comm. R. and H., *Hearings, 1924,* "Statement," by George F. Barrett, 2:1003.

8. Illinois General Assembly, *Report to . . . , 1881,* 6; Illinois Board of Health, *Second Annual Report, 1879* (Springfield: H. W. Rokker, 1881), 114–18; New York State Engineer and Surveyor, *Annual Report for the Fiscal Year Ending September 30, 1891* (Albany: James B. Lynn, 1892), 475; Illinois Department of Purchases and Construction, Division of Waterways, *Flood Control Report,* by Jacob A. Harman, L. T. Berthe, M. Blanchard, and L. C. Craig (Springfield: Journal Printing Co., 1930), 52–53, 56 (cited hereafter as Harman et al., *Flood Control Report*); Larson, *Those Army Engineers,* 190, 193–94; House Comm. R. and H., *Hearings, 1924,* Barrett statement, 2:1012–13, 1176; Philip, "Chicago and the Down State," 203, 207; Waller, *Rainey of Illinois,* 57–65.

9. Cooley, "Reclamation," 32; Rawlings, *Rise and Fall,* 2:398–402, 404, 406; U.S. War Department, Chief of Engineers, *Annual Report, 1900* (Washington, D.C.: GPO, 1900), 1:42–43; Cain, *Sanitary Strategy,* 69–78, O'Connell, *Chicago's Quest,* 10–15; House Comm. R. and H., *Hearings, 1924,* Barrett statement, 2:1012, 1103–14; Philip, "Chicago and the Down State," 205–6, 215–18; Larson, *Those Army Engineers,* 193–94, 203.

10. Lyman E. Cooley, *The Lakes and Gulf Waterway as Related to the Chicago Sanitary Problem . . . , a Preliminary Report,* (Chicago: Press of John W. Weston, 1891) 8–9; Illinois Department of Public Works and Buildings, Division of Waterways, *Second Annual Report, July 1, 1918 to June 30, 1919* (Springfield: Illinois State Journal Co., 1920), 5, 7, and *Eleventh Annual Report, July 1, 1927 to June 30, 1928* (Springfield: Journal Printing Co., 1929), 41; University of Illinois, Water Survey, *Chemical and Biological Survey of the Waters of Illinois,* Bulletin 9 (Urbana, March 25, 1912), 38.

11. U.S. Congress, House, Committee on Rivers and Harbors, *Hearings on the Diversion of Waters from Lake Michigan,* 67th Cong., 2nd sess., 1922, 5, 14.

12. U.S. Congress, House, Secretary of War, Chief of Engineers, *Annual Report, 1890,* "Improvement of the Illinois River, Illinois," 51st Cong., 2nd sess., 1890, H. Exec. Doc. 1, pt. 2:2, app. JJ3, 2419–37, 2449–51; Association of Drainage and Levee Districts of Illinois, *Proceedings of the Third and Fourth Meetings,* comments of J. W. Woermann (Winchester: Press of the Winchester Times, 1913), 114–15 (cited hereafter as Assoc. Drainage and Levee Dist., *Proceedings, 1913*); Illinois Rivers and Lakes Commission, *The Illinois River and Its Bottom Lands,* 2nd ed., by John W. Alvord and Charles B. Burdick (Springfield: Illinois State Journal Co., 1919), 136 (cited hereafter as Alvord and Burdick, *Illinois River*); House Comm. R. and H., *Hearings, 1924,* Barrett statement, 2:1166; Harman et al., *Flood Control Report,* 57–58; Illinois Department of Public Works and Buildings, Division of Waterways, *Third Annual Report, July 1, 1919 to June 30, 1920* (Springfield: Illinois State Journal Co., 1921), 24, and *Fifth Annual Report, July 1, 1921 to June 30, 1922* (Springfield: Illinois State Journal Co., 1923), 95 (cited hereafter as Ill. Div. Waterways, *Report, 1919–1920* and *Report, 1921–1922,* respectively); "Illinois Fights for Health and Waterways," *National Reclamation Magazine* 3:5 (May 1924), 92–93; J. P. Kerr, "Chicago's Cesspool," *Outdoor America* 3:5 (December 1924), 36–37; Langdon Pearse, "The Sewage Treatment Program of the Sanitary District of Chicago," *Western Society of Engineers Journal* 31:7 (July 1926), 261–62; Rawlings, *Rise and Fall,* 2:404, 406; U.S. War Department, Chief of Engineers, *Annual Report, 1930,* "Illinois River, Ill." (Washington, D.C.: GPO, 1930), 1:1261; *Wisconsin et al. v. Illinois et al.,* 281 U.S. 179 (1930); Cain, *Sanitary Strategy,* 89–90, 96–97, 110–15, 118, 120; Waller, *Rainey of Illinois,* 87–88, 90; Larson, *Those Army Engineers,* 201, 203, 212–13; "Sanitary District Blames Delay on Chicago Tax Tangle," *Engineering News-Record* 109:20 (November 17, 1932), 602–3.

13. Alvord and Burdick, *Illinois River,* 28; Association of Drainage and Levee Districts of Illinois, *Proceedings of Second Annual Meeting,* comments of C. W. Brown and Baldwin Keech, (Beardstown: Enterprise Press, 1912), 29 (cited hereafter as Assoc. Drainage and Levee Dist., *Proceedings, 1912*); Illinois Natural History Survey, *Some Recent Changes in the Illinois River Biology,* by Stephen A. Forbes and Robert E. Richardson, Bulletin 13:6 (Urbana, April 1919), 139, 141 (cited hereafter as Forbes and Richardson, *Changes in River Biology*); Ill. Div. Waterways, *Report, 1921–1922,* 5; U.S. Congress, House, Committee on Flood Control, *Illinois River and Tributaries, Illinois—Flood Control,* 68th Cong., 1st sess., 1924, H. Doc. 276, 14; Jacob A. Harman, "Some Problems in Flood Control," *Transactions of the Illinois State Academy of Science* 24:2 (December 1931), 566.

14. Cooley, *Lakes and Gulf Waterway,* 11, 70–73, 78, app. 13, 17–19; Assoc. Drainage and Levee Dist., *Proceedings, 1912,* "Address by Retiring President," by Jesse Lowe, 11 (cited hereafter as Lowe, "Address"); Illinois General Assembly, Illinois Valley Flood Control Commission, *Hearings on the Causes and Control of Floods in the Illinois River Valley,* testimony and letters of Chris Kreiling, Allen T. Lucas, Charles W. Lebkuecher, Edward Krohe, William H. Hill, Charles E. Phelps, and F. W. Meyer (n.p., 1929), 183, 187, 205–6, 211, 214, 216, 255–56; Assoc. Drainage and Levee Dist., *First Annual Report,* comments of H. V. Teal, 5; U.S. War Department, Chief of Engineers, *Annual Report, 1925* (Washington, D.C.; GPO, 1925), 1:1394.

15. Optimum conditions for green plankton in the river prevailed at or above La Salle before 1900, about to Chillicothe in 1911, and about to Peoria in 1918. Forbes and Richardson, *Changes in River Biology,* 145. Among the descriptions of conditions along the river, see B. G. Merrill, "Where to Go Fishing," *Illinois Fish and Game Conservationist* 3:6 (June 1914), 11–12; House Comm. R. and H., *Hearings, 1924,* "Statement" by Hon. Henry T. Rainey, 2:788–89, and letters from the middle and upper Illinois Valley, 431–32. It should be added that the alleged pollution of the Mississippi River by the Sanitary District was cause for suit by the State of Missouri, which bill of complaint the U.S. Supreme Court dismissed in 1906. Waller, *Rainey of Illinois,* 88.

16. "Farmers Want the Land," *RT,* August 14, 1902, p. 1, col. 2.

17. "Farmers Want the Land," *RT,* August 14, 1902, p. 1, col. 2; Alvord and Burdick, *Illinois River,* 136; Assoc. Drainage and Levee Dist., *Proceedings, 1912,* "Report," by Committee on Resolutions, 45–46; House Comm. R. and H., *Hearings, 1924,* Rainey statement, 2:771; Harman et al., *Flood Control Report,* 57.

18. Ill. Valley Flood Control Comm., *Hearings,* testimony by Chester L. Whitnal, 204; "Local and Other Matters," *MB,* March 14, 1912, p. 7, cols. 1–2.

19. Ill. Valley Flood Control Comm., *Hearings,* testimony by Glenn Ratcliff, 87, and Charles E. Phelps, 216; Assoc. Drainage and Levee Dist., *Proceedings, 1912,* comments by C. W. Brown, 28.

20. House Comm. R. and H., *Hearings, 1924,* testimony by Hon. Guy L. Shaw, 2:826.

21. Lowe, "Address," 74–75; House Comm. R. and H., *Hearings, 1924,* Rainey statement, 2:768, 770, and Barrett statement, 2:1170–79; Philip, "Chicago and the Down State," 219–20; Waller, *Rainey of Illinois,* 63, 71–72; Larson, *Those Army Engineers,* 211–13.

22. "Chicago Sanitary District Has Large Force Here," *MCD,* November 15, 1908, p. 1, col. 5; "Report on Overflow Lands," *PJ,* March 4, 1909, p. 1, col. 2; Sanitary District of Chicago, *The Sanitary District of Chicago: History of Its Growth and Development,* by C. Arch Williams (Chicago: Sanitary District of Chicago, 1919), 79–83, 218–20 (cited hereafter as Williams, *Sanitary District of Chicago*); House Comm. R. and H., *Hearings, 1924,* testimony of L. A. Jarman, 1:163, and testimony of William L. Sackett, 2:934.

23. "Drainage Men Prepare to Fight," *PH-T,* August 24, 1906, p. 16, col. 5.

24. "Drainage Men Prepare to Fight," *PH-T,* August 24, 1906, p. 16, col. 5.

25. "Purport of Work Is Now Known," *PH-T,* April 13, 1907, p. 6, col. 1.

26. "Verdict of $750.00 Is Awarded," *PJ,* January 18, 1906, p. 12, col. 1; "Judge Green Says Verdict Is Wrong," *PJ,* February 11, 1906, p. 18, col. 3, p. 20, col. 6; "Court Is Scored for This Action," *PJ,* February 13, 1906, p. 3, col. 3; "May Settle All Drainage Litigation," *PJ,* February 16, 1906, p. 16, col. 7; House Comm. R. and H., *Hearings, 1924,* Jarman testimony, 1:165; Philip, "Chicago and the Down State," 220.

27. Williams, *Sanitary District of Chicago,* 82–83; Alvord and Burdick, *Illinois River,* 18; House Comm. R. and H., *Hearings, 1924,* Jarman testimony, 1:162–63, Barrett statement, 1:165, and Rainey statement, 2:804.

28. Alvord and Burdick, *Illinois River,* 133–34; "For a Clean River," reprint of editorial in *Henry Republican, MCD,* March 19, 1920, p. 1, col. 2.

29. House Comm. R. and H., *Hearings, 1924,* Sackett testimony, 2:934–35.

30. Ill. Valley Flood Control Comm., *Hearings,* statement by Spring Lake Drainage and Levee District, 75–76.

31. Ill. Valley Flood Control Comm., *Hearings,* testimony of Franklin L. Velde, 114.

32. House Comm. R. and H., *Hearings, 1924,* letter of Harry B. W. Kallista, 2:845; Philip, "Chicago and the Down State," 230.

33. Lowe, "Address," 73–75.

34. Philip, "Chicago and the Down State," 200–207, 212–14, 220–28, 302.

35. Assoc. Drainage and Levee Dist., *Proceedings, 1913,* "The Relation of the State Association to the National Drainage Congress," by Edmund T. Perkins, 22–24; John H. Nolen, *Missouri's Swamp and Overflowed Lands and Their Reclamation,* Report to the 47th Missouri General Assembly (Jefferson City, Mo.: Hugh Stephens Printing, 1913), 84, 120; Gary Lane McDowell, "Local Agencies and Land Development by Drainage: The Case of 'Swampeast' Missouri" (Ph.D. diss., Columbia University, 1965), 117; Waller, *Rainey of Illinois,* 56–57.

36. "Lowering the La Grange Dam," *MB,* September 28, 1905, p. 5, col. 1; "Federal Cooperation for Deep Waterways," *Steam Shovel and Dredge* 13:4 (April 1909), 370–75; Edward O. Phillips, "Illinois and the Deep Waterway," *Steam Shovel and Dredge* 13:7 (July 1909), 546–51; "To Remove Dams from Illinois River," *MB,* June 1, 1916, p. 1, col. 6; Ill. Valley Flood Control Comm., *Hearings,* comments of Dr. J. P. Kerry, 115–16; Waller, "The Illinois Waterway," 130–36; Robert P. Howard, *Illinois: A History of the Prairie State* (Grand Rapids, Mich.: William B. Eerdmans Publishing Co., 1972), 479–80.

37. Ill. Valley Flood Control Comm., *Hearings,* review of legislative history by Hon. Henry T. Rainey, 147–49, 151; U.S. War Department, Chief of Engineers, *Annual Report, 1929,* "Illinois River, Ill." (Washington, D.C.: GPO, 1929), pt. 1, 1473; House Comm. R. and H., *Hearings, 1924,* Barrett statement, 2:1170–76, 1178–79; Waller, *Rainey of Illinois,* 59–63, 86–87; Larson, *Those Army Engineers,* 211–12.

38. "Big Drainage Meeting in Beardstown Dec. 14," *MB,* December 6, 1923, p. 1, col. 3; "Chicago Urges Bill Authorizing Its Use of Lake Water," *Electrical World* 83:8 (February 23, 1924), 398; "New York Legislature Against Lake Michigan Diversion," *Electrical World* 83:12 (March 22, 1924), 589; "Supreme Court Hears Argument in Chicago Canal Case," *Electrical World* 84:25 (December 20, 1924), 1325; U.S. Congress, House, Committee on Rivers and Harbors, *Illinois River, Ill.,* 77th Cong., 2nd sess., 1942, H. Doc. 692, 10–11; Philip, "Chicago and the Down State," 229–30; Waller, *Rainey of Illinois,* 71, 73, 91; Larson, *Those Army Engineers,* 211, 214–18.

39. "Boulware Leaves for Drainage Trip," *PH-T,* September 9, 1907, p. 8, col. 3; Philip,

"Chicago and the Down State," 307; Ill. Div. Waterways, *Report, 1919–1920*, 24, and *Report, 1921–1922*, 95.

40. Ill. Valley Flood Control Comm., *Hearings*, testimony by John W. Beckwith, 117; Assoc. Drainage and Levee Dist., *Proceedings, 1913*, comments by Jesse Lowe, 59–61; House Comm. R. and H., *Hearings, 1924*, Rainey statement, 2:798, and Shaw testimony, 2:827.

41. Ill. Valley Flood Control Comm., *Hearings*, comments by Sen. Ben L. Smith, 120–21, and by Rainey, 146–47, 149, 153; Waller, *Rainey of Illinois*, 91–93.

## 6. Days of Reckoning for Reclamation

1. The source for tables 6.1 and 6.2 is Illinois Department of Purchases and Construction, Division of Waterways, *Flood Control Report*, by Jacob A. Harman, L. T. Berthe, M. Blanchard, and L. C. Craig (Springfield: Journal Printing Co., 1930), 65, 227 (cited hereafter as Harman et al., *Flood Control Report*). Illinois Department of Conservation, *Potential Conservation Areas along the Illinois River from Hennepin to Grafton*, by Jenkins, Merchant & Nankivil and W. B. Walraven (Springfield, May 1950), 34–37 (cited hereafter as Jenkins et al., *Survey*); U.S. Army Corps of Engineers, North Central and Lower Mississippi Valley Divisions, *Illinois River, Illinois, and Tributaries, Survey Report for Flood Control and Allied Water Uses* (n.p., April 1916) 1:16.

2. U.S. Congress, House, Committee on Rivers and Harbors, *Illinois River, Ill.*, 72nd Cong., 1st sess., 1931, H. Doc. 182, 18, 108 (cited hereafter as House Comm. R. and H., *Illinois River, 1931*).

3. Association of Drainage and Levee Districts of Illinois, *Proceedings of Second Annual Meeting*, comments of C. W. Brown (Beardstown: Enterprise Press, 1912), 28, 29.

4. Fulton County Circuit Court, "Kerton Valley Levee and Drainage District, 1," Commissioners, "Petition to Organize," filed ca. August 14, 1917, 1–2; University of Illinois, Water Resources Center, *Case Studies in Drainage and Levee District Formation and Development on the Floodplain of the Lower Illinois River, 1890s to 1930s*, by John Thompson, Special Report 017 (Urbana-Champaign: May 1989), 13, 54, 78 (cited hereafter as Thompson, *Case Studies*).

5. The effect of flood events at drainage and levee districts is more detailed and documented in Thompson, *Case Studies*.

6. Illinois General Assembly, Illinois Valley Flood Control Commission, *Hearings on the Causes and Control of Floods in the Illinois River Valley*, statement by John Goodell (n.p., 1929), 221; "River Highest since 1892," *I-S*, July 24, 1902, p. 1, col. 6.

7. Illinois Internal Improvements Commission, *The Lakes and Gulf Waterway* (Springfield: Philips Bros., 1907), 29; Harman et al., *Flood Control Report*, 36, 220–22, fig. 4; House Comm. R. and H., *Illinois River, 1931*, 33.

8. "Town Is Ruined," *I-S*, January 25, 1907, p. 1, col. 3.

9. The authoritative state agency data on the extent of flooding are modified by the writer to reflect data from district archives, which indicate that the Hartwell district was complete at the time. Harman et al., *Flood Control Report*, fig. 5; R. A. Wheeler, "Report on the Status of Flood Control in Illinois," *Illinois Engineer* 12:4 (October 17, 1907), 60; "Illinois River Falls; All Danger Is Past," *MB*, April 10, 1913, p. 4, cols. 4–5; "20,000 Homeless from Floods," *RT*, April 10, 1913, p. 1, cols. 1–2; "Naples Clear of Water; Citizens Return to Homes," *MB*, May 8, 1913, p. 1, col. 6.

10. Illinois Rivers and Lakes Commission, *The Illinois River and Its Bottom Lands*, 2nd ed., by John W. Alvord and Charles B. Burdick (Springfield: Illinois State Journal Co., 1919), 118 (cited hereafter as Alvord and Burdick, *Illinois River*). Data related to aggregate and district acreage are in assessable acres, as given in Harman et al., *Flood Control Report*, 38–39.

11. Alvord and Burdick, *Illinois River,* preface.

12. Alvord and Burdick, *Illinois River,* 85, 90–91, 93, 94; Harman et al., *Flood Control Report,* 52; "Pekin—Building Large Ditch," *PH-T,* December 30, 1903, p. 5, col. 3; "Kilbourne—Land Improvements," *PH-T,* December 2, 1904, sec. 2, p. 1, col. 2; U.S. Congress, House, Committee on Rivers and Harbors, *Sangamon River, Ill.,* 72nd Cong., 1st sess., 1932, H. Doc. 186, 36; Federal Land Bank of St. Louis, "Report on Drainage and Levee Districts: Cass, Mason, Menard and Tazewell Counties, Illinois," containing "Sangamon Outlet Drainage District, Preliminary Report," by Hubert E. Goodell and Fred S. Morse, January 8, 1934, 132 (cited hereafter as Goodell and Morse, "Sangamon Outlet").

13. Alvord and Burdick, *Illinois River,* 12, 16–17, 107, 114, 123, 127–29; Harman et al., *Flood Control Report,* 14–15; Illinois Department of Public Works and Buildings, Division of Waterways, *Fourth Annual Report, July 1, 1920 to June 30, 1921,* "Report," by M. G. Barnes (Springfield: Illinois State Journal Co., 1922), 2, and *Seventh Annual Report, July 1, 1923 to June 30, 1924* (Springfield: Illinois State Journal Co., 1925), 15 (cited hereafter as Ill. Div. Waterways, *Report, 1920–1921* and *Report, 1923–1924,* respectively); House Comm. R. and H., *Illinois River, 1931,* 33.

14. Two districts above Peoria were flooded, the Hennepin (org. 1909; 2,160 ac.) and Partridge (org. 1906; 5,500 ac.), near Chillicothe. Illinois River and Lakes Commission, *Annual Report, 1916,* "Illinois River Floods," by L. K. Sherman (Springfield: Schnepp & Barnes, 1917), 20–21; "Illinois River Rising Slowly; Gauge at 18 Ft.," *PDT,* January 24, 1916, p. 8, col. 5; "Illinois River Equals the Flood Record of 1904," *PDT,* January 25, 1916, p. 1, col. 2; "Floods Along Illinois River," *RT,* January 27, 1916, p. 1, col. 1; Illinois Department of Registration and Education, Geological Survey, *Engineering and Legal Aspects of Land Drainage in Illinois,* rev. ed., by George W. Pickels and Frank B. Leonard Jr., Bulletin 42 (Urbana, 1929), 117–19 (cited hereafter as Pickels and Leonard, *Engineering*).

15. Ill. Rivers and Lakes Comm., *Annual Report, 1916,* 21.

16. "Illinois River News," *MB,* February 3, 1916, p. 1, col. 1; "Crooked Creek Levee Break," *RT,* June 14, 1917, p. 1, cols. 1–2.

17. Data based on Harman et al., *Flood Control Report,* 38–39, 82, 215–17. Illinois Department of Public Works and Buildings, Division of Waterways, *Fifth Annual Report, July 1, 1921 to June 30, 1922* (Springfield: Illinois State Journal Co., 1923), 14; "Little Improvement in Flood Conditions," *PDT,* April 14, 1922, p. 3, col. 2; Jenkins et al., *Survey,* 36.

18. Ill. Div. Waterways, *Report, 1920–1921,* 1; Illinois Department of Public Works and Buildings, Division of Waterways, *Sixth Annual Report, July 1, 1922 to June 30, 1923* (Springfield: Illinois State Journal Co., 1924), 8, 10, 16, 20 (cited hereafter as Ill. Div. Waterways, *Report, 1922–1923*); House Comm. R. and H., *Illinois River, 1931,* 108; "Little Improvement in Flood Conditions," *PDT,* April 14, 1922, p. 3, col. 2.

19. "Flood Damages Grow," *RT,* April 27, 1922, p. 1, cols. 1–2.

20. Data for table 6.3 from House Comm. R. and H., *Illinois River, 1931,* 36–37, 39, 41. Discrepancies between this general table and tables 6.4–6.6, showing costs to the reclaimed districts, cannot be resolved.

21. Ill. Div. Waterways, *Report, 1922–1923,* 14, 17, 20; Harman et al., *Flood Control Report,* 14; "Many Pekin Homes Are Flooded," *PDT,* April 14, 1922, p. 1, cols. 1–2; "River at Meredosia Near 1858 Mark," *MB,* April 13, 1922, p. 1, col. 4; "Meredosia Lake Drainage Levee Breaks Wednesday," *MB,* April 20, 1922, p. 1, col. 2.

22. As many as 77,500 acres of leveed land may have flooded, raising the floodplain total to

175,000 acres. Ill. Div. Waterways, *Report, 1921–1922,* 16; Pickels and Leonard, *Engineering,* 136. Levee failure sequence is based on Harman et al., *Flood Control Report,* 53–54; and Jenkins et al., *Survey* 36.

23. "400 Are Homeless at Beardstown," *PDT,* April 11, 1922, p. 1, col. 1.

24. Data for table 6.4 from Illinois Department of Public Works and Buildings, Division of Waterways, *Eleventh Annual Report, July 1, 1927 to June 30, 1928* (Springfield: Journal Printing Co., 1929), 32–37. Levee breaks noted in Harman et al., *Flood Control Report,* 53–54; and district data in Thompson, *Case Studies.*

25. Ill. Div. Waterways, *Report, 1921–1922,* 16; "River Raises Three Points During Night," *PDT,* April 10, 1922, p. 1, col. 2; "Work Desperately to Save Dike," *PDT,* April 12, 1922, p. 1, col. 1; "Meredosia Lake Drainage Levee Breaks Wednesday," *MB,* April 20, 1922, p. 1, col. 2; "Flood Damages Grow," *RT,* April 27, 1922, p. 1, cols. 1–2.

26. Ill. Div. Waterways, *Report, 1920–1921,* 1–2, 5, and *Report, 1921–1922,* 1, 5, 20; House Comm. R. and H., *Illinois River, 1931,* 46.

27. Ill. Div. Waterways, *Report, 1921–1922,* 4, 20, and *Report, 1923–1924,* 18–29; Illinois, Department of Public Works and Buildings, Division of Waterways, *Twelfth Annual Report, July 1, 1928 to June 30, 1929* (Springfield: Journal Printing Co., 1930), 15.

28. U.S. Congress, House, Committee on Flood Control, *Illinois River and Tributaries, Illinois—Flood Control,* 68th Cong., 1st sess., 1924, H. Doc. 276, 1, 22–25; *U.S. Statutes at Large* 39:1 (1917): 948.

29. Illinois Department of Public Works and Buildings, Division of Waterways, *Tenth Annual Report, July 1, 1926 to June 30, 1927* (Springfield: Journal Printing Co., 1928), 25 (cited hereafter as Ill. Div. Waterways, *Report, 1926–1927*); Harman et al., *Flood Control Report,* 111.

30. Data for table 6.5 from Ill. Div. Waterways, *Report, 1926–1927,* 32–37; Harman et al., *Flood Control Report,* 54–56; and district data in Thompson, *Case Studies.* The text account is based on Ill. Div. Waterways, *Report, 1926–1927,* 24–26, 55–56, and *Report, 1927–1928,* 9, 26, 27; Ill. Valley Flood Control Comm., *Hearings,* testimony by W. H. Severns, L. Lowenstein, and H. G. Edwards, 36, 135, 191; Harman et al., *Flood Control Report,* 54–56, 212–13; "River Flood Will Break All Records," RT, October 6, 1926, p. 1, col. 1, p. 6, cols. 4–6; "River Flood Has Broken All Records," *RT,* October 13, 1926, p. 1, cols. 1–2, p. 5, cols. 1–3; "Pekin Wagon Bridge Closed at 2 P.M. Today," *PDT,* October 5, 1926, p. 1, col. 7, p. 3, cols. 5–6; "Banner District Levee Breaks; Floods Rush In," *PDT,* October 7, 1926, p. 1, col. 7; "River on Stand and the Levee Still Holding," *PDT,* October 11, 1926, p. 1, col. 3; "Levees Flooded 79,300 Acres," *MCD,* October 22, p. 1, col. 2.

31. Harman et al., *Flood Control Report,* 54–55, 85–86, 213.

32. Harman et al., *Flood Control Report,* 109; Ill. Div. Waterways, *Report, 1927–1928,* "Engineering Flood Relief—Illinois Division," by L. C. Craig, 96–97 (cited hereafter as Craig, "Engineering Flood Relief").

33. Ill. Div. Waterways, *Report, 1926–1927,* 56; Craig, "Engineering Flood Relief," 96; Harman et al., *Flood Control Report,* 214–15.

34. Data for table 6.6 from Ill. Div. Waterways, *Report, 1927–1928,* 32–37; Pickels and Leonard, *Engineering,* 113, 120–23; Harman et al., *Flood Control Report,* 42–43, 54–56; and district data in Thompson, *Case Studies.* Also see Craig, "Engineering Flood Relief," 96. The Crabtree area became a hunting preserve.

35. U.S. Army Corps of Engineers, Office of History, *The Evolution of the 1936 Flood Control Act,* by Joseph L. Arnold (Washington, D.C.: GPO, 1988), 18 (cited hereafter as Arnold, Evolution).

36. Illinois Tax Commission, in cooperation with Works Projects Administration, *Drainage District Organization and Finance, 1879–1937,* Survey of Local Finance in Illinois (Springfield: 1941), 7:26; Ill. Div. Waterways, *Report, 1926–1927,* 30, 33–35, and *Report, 1927–1928,* "Engineering Report," by L. D. Cornish, 66 (cited hereafter as Cornish, "Engineering Report").

37. Harman et al., *Flood Control Report,* 16, 22; Cornish, "Engineering Report," 71–74; Ill. Div. Waterways, *Report, 1928–1929,* 79; House Comm. R. and H., *Illinois River, 1931,* 43; U.S. Army Corps of Engineers, Chicago District, *Those Army Engineers: A History of the Chicago District, U.S. Army Corps of Engineers,* by John W. Larson (Washington, D.C.: GPO, 1980), 254 (cited hereafter as Larson, *Those Army Engineers*).

38. *U.S. Statutes at Large* 39:1 (1917): 948, 42:1 (1922): 1038, and 45:1 (1928): 534; Harman et al., *Flood Control Report,* 3, 13, 17, 22; U.S. Congress, House, Committee on Flood Control, *Control of Floods on the Mississippi and Sacramento Rivers,* 64th Congress, 1st sess., 1916, H. Rept. 616, 3–6, 24; House Comm. R. and H., *Illinois River, 1931,* 36, 71–72; Ill. Div. Waterways, *Report, 1927–1928,* "Flood Control a National Problem," memorial and petition to House Committee on Flood Control (November 17, 1927), by Gov. Len Small, 20–40; Illinois Department of Public Works and Buildings, Division of Waterways, *Fifteenth Annual Report, July 1, 1931 to June 30, 1932* (Springfield: Journal Printing Co., 1933), 110; Larson, *Those Army Engineers,* 254. See also Robert W. Harrison, *Alluvial Empire: A Study of State and Local Efforts Toward Land Development in the Alluvial Valley of the Lower Mississippi River, Including Flood Control, Land Drainage, Land Clearing, Land Forming* (Little Rock: Pioneer Press, 1961), 149–50, 205; Theodore M. Schad, "Evolution and Future of Flood Control in the United States," in William J. Hull and Robert W. Hull, eds., *The Origin and Development of the Waterways Policy of the United States* (Washington, D.C.: National Waterways Conference, 1967), 29–31; Arnold, *Evolution,* 11, 13–15; Jamie W. Moore and Dorothy P. Moore, *The Army Corps of Engineers and the Evolution of Federal Flood Plain Management Policy,* Natural Hazards Research and Applications Information Center, Special Publication 20 (Boulder: University of Colorado, 1989), 3–4; Gary Lane McDowell, "Local Agencies and Land Development by Drainage: The Case of 'Swampeast' Missouri" (Ph.D. diss., Columbia University, 1965), 134. California's mining debris and flood control experience is best rendered in Robert L. Kelley, *Gold vs. Grain* (Glendale, Calif.: Arthur H. Clark Co., 1959), and *Battling the Inland Sea* (Berkeley: University of California Press, 1989).

39. House Comm. R. and H., *Illinois River, 1931,* 72–75, 79; Arnold, *Evolution,* 16–17.

40. Harman et al., *Flood Control Report,* 116–17; Arnold, *Evolution,* 53–70, 79.

41. Illinois Department of Public Works and Buildings, Division of Waterways, *Sixteenth Annual Report, July 1, 1932 to June 30, 1933* (Springfield: Journal Printing Co., 1934), 35; House Comm. R. and H., *Illinois River, 1931,* 44; Larson, *Those Army Engineers,* 255–56.

42. Craig, "Engineering Flood Relief," 110–11; Ill. Valley Flood Control Comm., *Hearings,* statement by J. T. Rainey, 146–47; U.S. National Emergency Council, "Activities of Federal Agencies Operating in the State of Illinois," coordination meeting (mimeographed report, Chicago, November 7, 1935), 66–67; Illinois General Assembly, Legislative Flood Investigating Commission, *Report on Flood Situation in Illinois,* by Elliott & Porter, Consulting Engineers (Chicago, March 1947), 109–11, 114–17; Larson, *Those Army Engineers,* 256; Moore and Moore, *Army Corps,* 15.

43. U.S. Congress, House, Committee on Irrigation and Reclamation, *Loans for Relief of Drainage Districts,* "Statement" by J. P. Kerr (President, Association of Drainage and Levee Districts of Illinois), 71st Cong., 2nd sess., 1930, H.R. 11718, 70–71 (cited hereafter as House Comm. I. and R., *Loans for Relief*); House Comm. R. and H., *Illinois River, 1931,* 48; McDowell, "Local Agencies," 123–25.

44. University of Illinois, College of Agriculture, Agricultural Experiment Station, *Prices of Illinois Farm Products from 1866 to 1929,* by L. J. Norton and B. B. Wilson, Bulletin 351 (Urbana: July 1930), 495, 505, 530 (cited hereafter as Norton and Wilson, *Prices*).

45. Norton and Wilson, *Prices,* 495, 497; U.S. Congress, House, Committee on Rivers and Harbors, *Hearings on the Subject of the Improvement of the Illinois and Mississippi Rivers, and the Diversion of Water from Lake Michigan into the Illinois River,* statement by Henry T. Rainey, 68th Cong., 1st sess., 1924, pt 2:790 (cited hereafter as House Comm. R. and H., *Hearings, 1924*).

46. U.S. Congress, House, Committee on Irrigation and Reclamation, *Loans for Relief of Drainage Districts,* "Statement," by J. P. Kerr, 71st Cong., 2nd sess., 1930, H. Rept. 11718, 70–71; Charles L. Stewart, "Land Utilization in the Illinois River Basin," *Transactions of the Illinois State Academy of Science* 24:2 (1910–13), 559, 562–63; McDowell, "Local Agencies," 128 ff.

47. House Comm. I and R., *Loans for Relief,* "Statement," by Kerr, 71, 73, and "Statement," by William E. Hull (Member of Congress, Illinois), 11; Stewart, "Land Utilization," 558.

48. Willard D. Ellis, "Problems of Financing Land Reclamation," *Agricultural Engineering* 12:5 (May 1931), 168; Emil Schram, "Refinancing of Drainage, Levee, and Irrigation Districts," *Agricultural Engineering* 16:4 (April 1935), 151, 153; McDowell, "Local Agencies," 132–38; U.S. Department of Agriculture, *Yearbook of Agriculture, 1955: Water,* "The Use of Pumps for Drainage," by John G. Sutton (Washington, D.C.: GPO, 1955), 529; California Department of Public Works, Division of Water Resources, *Financial and General Data Pertaining to Irrigation, Reclamation, and Other Public Districts in California,* Bulletin 37 (Sacramento: California State Printing Office, 1931), 9–22.

49. The districts in most financial distress were Big Prairie and Liverpool (about 49% of total assessed), Valley City (about 32%), South Beardstown and Spankey (about 24%). Among the districts where defaults amounted to 5 to 16 percent of the assessment total and where flood related costs were high were McGee Creek ($1.3 million), Coal Creek ($.7 million), Scott County ($.6 million), Crane Creek and Meredosia Lake ($.4 million), and West Matanzas ($.3 million). Illinois Department of Public Works and Buildings, Division of Waterways, *Report on Drainage Districts, 1937* (n.p., 1937), 15–21.

50. Pickels and Leonard, *Engineering,* 137.

51. Illinois Natural History Survey, *Wildlife and Fishery Values of Bottomland Lakes in Illinois,* by Frank C. Bellrose and Clair T. Rollings, Biological Notes 21 (Urbana, June 1949), 4.

52. House Comm. R. and H., *Hearings, 1924,* statements by J. H. Cardes and J. L. Cook, 846–48; Havana Public Library, "Sangamon River Outlet Organization," C. H. Kreiling letter to Col. C. Keller, District Engineer, Corps of Engineers, Chicago, July 5, 1941; Illinois Tax Commission, *Drainage District Organization,* 105; Goodell and Morse, "Sangamon Outlet," 95.

53. Wheeler, "Report on Status," 61; House Comm. R. and H., *Illinois River, 1931,* 47; U.S. Congress, House, Committee on Rivers and Harbors, *Illinois River, Ill.,* 77th Cong., 2nd sess., 1942, H. Doc. 692, 23.

## 7. Agricultural Activity and Settlement in the 1920s

1. Maps 7.1 and 7.2 are reproduced from maps in the archives of the Illinois Natural History Survey, Champaign. Their scale is about 1 inch to 1.3 miles.

2. "Levee District Financed" (editorial), *JR,* August 22, 1907, p. 4, col. 1.

3. "A $1,000 Jury," *RT,* February 11, 1897, p. 7, col. 5.

4. "Adding a Township," *CP,* October 5, 1906, p. 1, col. 2; "Petition Was Filed," *I-S,* December

23, 1908, p. 1, col. 3; "City Dads in Session," *I-S*, June 18, 1909, p. 1, cols. 4–5; Brown County Circuit Court, "Drainage Record No. 1," Big Prairie Drainage and Levee District, "Petition for the Organization of . . . ," filed November 30, 1915, 1–2.

5. "Bit of History," *CP*, February 26, 1914, p. 1, cols. 1–2; Ralph W. Nauss, "Malaria in Illinois," *Illinois Health News* 8:6 (June 1922), 164–65, 188; Illinois Department of Public Health, *The Rise and Fall of Disease in Illinois*, by Isaac D. Rawlings (Springfield: Phillips Bros., 1927), 2:35, 380.

6. Illinois Department of Purchases and Construction, Division of Waterways, *Flood Control Report*, by Jacob A. Harman, L. T. Berthe, M. Blanchard, and L. C. Craig (Springfield: Journal Printing, 1930), 59 (cited hereafter as Harman et al., *Flood Control Report*); U.S. Department of Commerce, Bureau of the Census, *Sixteenth Census of the United States, 1940: Drainage of Agricultural Lands* (Washington, D.C.: GPO, 1942), 137–44. For comparative data on improved land within drainage enterprises, see U.S. Department of Agriculture, *Land Reclamation Policies in the United States*, by R. P. Teele, Bulletin 1257 (Washington, D.C.: GPO, August 23, 1924), 26.

7. "Land Reclaiming," I-S, August 31, 1906, p. 1, cols. 3–4; "Adding a Township," *CP*, October 5, 1906, p. 1, col. 1; "A Big Wheat Yield," *CP*, July 22, 1909, p. 1, col. 6; "The Nutwood Levee District," *CP*, July 29, 1909, p. 1, col. 5; "Eldred," *CP*, July 17, 1913, p. 8, col. 4; "Crops in Northern Greene," *CP*, August 21, 1913, p. 2, cols. 3–4; "150 Bushels per Acre," *CP*, November 26, 1914, p. 1, col. 2; "Wheat Crops Are Excellent," *I-S*, August 15, 1915, p. 5, col. 1; Harman et al., *Flood Control Report*, 40; U.S. Department of Commerce, Bureau of the Census, *Fourteenth Census of the United States, 1920: Agriculture* (Washington, D.C.: GPO, 1922), 6:1, 396 (cited hereafter as Bureau of the Census, *1920: Agriculture*).

8. "Farm Figures," *CP*, July 29, 1920, p. 1, col. 1. The introduction of silos is documented in: "Items from the Farm," *CP*, September 14, 1906, p. 1, col. 2; "Eldred," *CP*, July 15, 1909, p. 3, col. 3; "Building Second Silo," *MCD*, October 4, 1912, p. 1, col. 2.

9. "Farm Figures," *CP*, July 29, 1920, p. 1, col. 1; "Farm Acreage," *CP*, August 5, 1920, p. 1, col. 6.

10. "Bluffdale," *CP*, July 4, 1902, p. 8, col. 3, October 17, 1902, p. 8, col. 3, June 12, 1903, p. 8, col. 4, July 8, 1909, p. 8, col. 3, and June 13, 1912, p. 8, col. 1; "Eldred," *CP*, July 11, 1902, p. 8, col. 4; "Keach Ranch Is Sold," *CP*, July 17, 1903, p. 1, col. 5; "Keach Drainage District," *CP*, August 5, 1904, p. 1, col. 2; "The Keach Case Is Reversed," *CP*, October 21, 1904, p. 1, col. 6; "Adding a Township," *CP*, October 5, 1906, p. 1, col. 2; "A Big Wheat Yield," *CP*, July 22, 1909, p. 1, col. 6; "Bottom Land $140 an Acre," *CP*, December 9, 1909, p. 1, col. 2; "Ex-Vice President," *CP*, May 12, 1910, p. 1, col. 6; "Boast of Many Things," *CP*, September 14, 1911, p. 1, col. 6; "Cattle for Fairbanks Ranch," *CP*, January 30, 1913, p. 1, col. 5; "Crops in Northern Greene," *CP*, August 21, 1913, p. 2, cols. 3–4; "Big Green County Farm," reprint from *St. Louis Globe-Democrat*, in *CP*, November 6, 1913, p. 2, cols. 1–2; "Fairbanks Ranch Big Silos," *CP*, October 15, 1914, p. 1, col. 2.

11. "Fatal Accident," *PCD*, July 3, 1907, p. 1, col. 6; "Local Matters," *PCD*, August 24, 1907, p. 5, cols. 2–3; "Will Plow with Engine," *I-S*, October 9, 1908, p. 1, cols. 5–6; "Farming on Really Extensive Scale," *I-S*, July 31, 1910, p. 1, col. 3; "Boast of Big Things," *CP*, September 14, 1911, p. 1, col. 6; "Bought New Tractor," *MB*, June 6, 1912, p. 1, col. 6; "Bluffdale," *CP*, June 13, 1912, p. 8, col. 1; "The Big Bottom Ranches," *CP*, May 15, 1913, p. 1, col. 4; "Crops in Northern Greene," *CP*, August 21, 1913, p. 2, cols. 3–4; "Big Greene County Farm," *CP*, November 6, 1913, p. 2, cols. 1–2; "Thousands See Tractors Plow," *MCD*, April 4, 1919, p. 1, col. 6; "The Largest Wheat Harvest," *CP*, July 3, 1919, p. 2, col. 2; "Local and Other Items," *MB*, May 13, 1920, p. 5, col. 2; "Farm Figures," *CP*, August 5, 1920, p. 1, col. 6; Reynold M. Wik, "Steam Power on the American Farm, 1830–1880," *Agricultural History* 25:4 (October 1951), 185–86.

12. "Will Plow with Engine," *I-S*, October 9, 1908, p. 1, cols. 5–6; "A Big Wheat Yield," *CP*, July 22, 1909, p. 1, col. 6; "Farming on Really Extensive Scale," *I-S*, July 31, 1910, p. 1, col. 3; "Wheat Harvest Begins," *RT*, June 24, 1915, p. 1, col. 5; "Harvesting in the Drainage District," *I-S*, July 7, 1915, p. 2, cols. 1–2; U.S. War Department, Chief of Engineers, *Annual Report, 1930*, pt. 1, "Illinois River, Ill." (Washington, D.C.: GPO, 1930), 1263.

13. University of Illinois, Water Resources Center, *Case Studies in Drainage and Levee District Formation and Development on the Floodplain of the Lower Illinois River, 1890s to 1930s*, by John Thompson, Special Report 017 (Urbana-Champaign: May 1989), 75–76 (cited hereafter as Thompson, *Case Studies*); "Church at Eldred," *CP*, October 31, 1907, p. 1, cols. 1–2; "Boast of Big Things," *CP*, September 14, 1911, p. 1, col. 6.

14. Thompson, *Case Studies*, 19, 21, 22; "Frederick Department," *RT*, June 9, 1898, p. 2, col. 5, and August 10, 1899, p. 7, col. 5; "Frederick," *RT*, January 30, 1902, p. 7, col. 2, July 17, 1902, p. 7, col. 2, and July 31, 1902, p. 7, col. 3.

15. "150 Bushels per Acre," *CP*, November 26, 1914, p. 1, col. 2; "Prosperity in the Bottoms," *CP*, January 21, 1915, p. 1, col. 2.

16. "The Whiteside Ranch Owners," *CP*, April 25, 1902, p. 1, col. 2; "Farmers Who Move," *CP*, February 27, 1903, p. 1, col. 4.

17. Data for tables 7.1, 7.2, 7.3, and 7.4 from U.S. Department of Commerce, Bureau of the Census, "Fourteenth Census of the United States: 1920, Illinois" (microfilm T625), counties of Fulton (reel 369), Greene and Jersey (368), Mason (392), Morgan (396), Pike (401), Schuyler and Scott (408), and Tazewell (410) (cited hereafter as "Fourteenth Census: 1920, Illinois"). Townships and precincts (Scott Co.) are labeled on relevant sheets. All subsequent social and economic data for 1920 are derived from the above reels. About 13,700 individual tabulations are reflected in tables 7.3 and 7.4.

18. U.S. Department of Commerce, Bureau of the Census, *Fourteenth Census of the United States, 1920: Population* (Washington, D.C.: GPO, 1922), 3:244, 251; "Fourteenth Census: 1920, Illinois."

19. Bureau of the Census, *1920: Agriculture*, 6:1, 376–85.

20. Thompson, *Case Studies*, 21, 34.

21. "Boast of Big Things," *CP*, September 14, 1911, p. 1, col. 6; "Big Green County Farm," *CP*, November 6, 1913, p. 2, cols. 1–2; "Loaded 5 Cars of Rice Here," *MCD*, December 18, 1925, p. 1, col. 3; "Big Rice Fields in Greene County," *CP*, July 15, 1926, p. 1, col. 1; "Rice May Yield 100 Bu. Per Acre," *CP*, November 25, 1926; p. 1, col. 6; "More Than Half Corn Crop Has Been Gathered," *MCD*, January 14, 1927, p. 1, col. 3; "Eldred," *CP*, July 12, 1928, p. 8, col. 2.

22. U.S. War Department, Chief of Engineers, *Annual Report, 1912*, "Improvements of Illinois River, Ill." (Washington, D.C.: GPO, 1912), 2:2557, and *Annual Report, 1914*, "Improvement of Illinois River, Ill." (Washington, D.C.: GPO, 1914), 2:2942; "Hillview: Hustler," *CP*, June 13, 1912, p. 1, col. 4; "Hillview," *CP*, March 13, 1913, p. 8, col. 4; "Unified Public Utilities in Central Illinois," *Electrical World* 61:22 (May 31, 1913), 1146–48, 1151–52; "Electricity Replaces Steam in Drainage Pumping," *Electrical World* 66:2 (July 10, 1915), 87.

23. U.S. War Department, Chief of Engineers, *Annual Report, 1925*, "Illinois River, Ill." (Washington, D.C.: GPO, 1925), 1:1390; Thompson, *Case Studies*, 37–38, 46, 76.

24. University of Illinois, College of Agriculture, Agricultural Experiment Station and Extension Service, *Facts Assembled for Use of Committee on Mechanical Equipment, Drainage, and Farm Buildings*, Agricultural Adjustment Conferences, 1928–1929 (Urbana, October 1928), 12, 13, 15.

## 8. Harvesting Nature's Endowment

1. People gainfully employed in mussel digging and in shell factories were numerous in 1910, to judge from enumerator sheets, but not in 1900 and 1920. U.S. Department of Commerce, Bureau of the Census, "Twelfth Census of the United States: 1900, Illinois," microfilm T623: Brown Co., Versailles Twp. (reel 238), Calhoun Co., Harden and Kampsville villages (239); Cass Co., East and West Beardstown twps., Beardstown City (240); Fulton Co., Banner Twp. (302); Greene Co., Bluffdale Twp., Eldred Village; Mason Co., Bath, Havana, and Snicarte twps., Bath Village, and Havana City (303); Morgan Co., Meredosia Precinct and Tn. (332); Pike Co., Chambersberg, Flint, and Montezuma twps., Pearl and Valley City villages (337); Schuyler Co., Bainbridge, Browning, Frederick, and Hickory twps., Browning Village (344); Scott Co., Naples Precinct and Village (344) (cited hereafter as "Twelfth Census, 1900"); "Thirteenth Census of the United States: 1910, Illinois," microfilm T624: Calhoun Co., Crates, Gilead, and Hardin precincts, Hardin and Kampsville villages (reel 232); Cass Co., East and West Beardstown twps., Beardstown City (233); Morgan Co., Meredosia Precinct and Village (314) (cited hereafter as "Thirteenth Census, 1910"); "Fourteenth Census of the United States: 1920, Illinois," microfilm T625: Cass Co., Beardstown City (reel 301); Fulton Co., Waterford Twp. (369); Greene Co., Bluffdale, Patterson, and Woodville twps. (368); Mason Co., Bath Village (392); Morgan Co., Meredosia Village (396); Pike Co., Chambersberg and Flint twps., Pearl Village (401); Scott Co., Oxville Precinct and Naples Twp. (408); Schuyler Co., Bainbridge, Browning, Frederick, and Hickory twps. (408); Tazewell Co., Spring Lake Twp. (410) (cited hereafter as "Fourteenth Census, 1920"). A sense of employment patterns was obtained from the writer's interview with Everett Bull, Liverpool, November 23, 1992.

2. "Bluffdale," *CP*, October 17, 1902, p. 8, col. 3; "Local and Other Items," *MB*, December 5, 1907, p. 5, col. 3; "It Cost Two Lives," *CP*, October 22, 1908, p. 1, col. 6; "Twelfth Census, 1900," microfilm T623: Cass Co., Beardstown City (reel 240); Schuyler Co., Browning Village (344); "Fourteenth Census, 1920," microfilm T625: Fulton Co., Liverpool Tn. (reel 369); Pike Co., Flint Tn. (401); Schuyler Co., Hickory Tn. (408). Mason County Circuit Court, *Knapp Island Gun Club v. M. J. O'Meara—Injunction,* A. E. Schmoldt, "Answer," filed November 6, 1915, chancery no. 2214, box 772, 1–2; "Local and Other Items," *MB*, January 24, 1907, p. 5, col. 1, and April 1, 1915, p. 5, col. 3; "River News," *MCD,* January 25, 1907, p. 1, col. 4; "Pekin News," *PH-T,* February 23, 1907, p. 8, col. 4; Charles A. E. Martin, ed., *Historical Encyclopedia of Illinois and History of Cass County* (Chicago: Munsell Publishing Co., 1915), 2:952–53; Illinois Laboratory of Natural History, *Forest Conditions in Illinois,* by R. Clifford Hall and O. D. Ingall, Bulletin 9:4 (Urbana: 1910–13), 213, 215 (cited hereafter as Hall and Ingall, *Forest Conditions*); U.S. Congress, House, Committee on Rivers and Harbors, *Sangamon River, Ill.,* 72nd Cong., 1st sess., 1932, H. Doc. 186, 23.

3. "Bluffdale," *CP*, October 17, 1902, p. 8, col. 3; "The Water Ran Up Hill," *CP*, June 19, 1903, p. 1, col. 4; "It Cost Two Lives," *CP*, October 22, 1908, p. 1, col. 6; "Hardin Gets the Profit," *CP*, July 29, 1909, p. 1, col. 3; "Rail Rumors," *CP*, August 28, 1913, p. 1, col. 4; "Eldred," *CP*, March 6, 1913, p. 8, col. 4, December 21, 1916, p. 8, col. 2, July 3, 1919, p. 8, col. 3, and March 3, 1922, p. 8, col. 2; "Twelfth Census, 1900," microfilm T623: Cass Co., Beardstown City (reel 240); Schuyler Co., Browning Village (344); "Fourteenth Census, 1920," microfilm T625: Fulton Co., Liverpool Twp. (reel 369); Greene Co., Woodville Twp. (368); Pike Co., Flint Twp. (401); Schuyler Co., Hickory Twp. (408).

4. "Wood Butchers at Work," *CP*, February 5, 1914, p. 2, cols. 1–2.

5. "Wood Butchers at Work," *CP,* February 5, 1914, p. 2, cols. 1–2; "McGee Creek Levee and Drainage District Notice," *PCD,* January 18, 1917, p. 1, col. 2; "Turned Bottom Lands into Farms; 'Back to Nature' Says Conservation," *CP,* December 1, 1927, p. 1, cols. 3–4.

6. Hall and Ingall, *Forest Conditions,* 198; Lyman E. Cooley, *The Illinois River: Physical Relations and the Removal of the Navigation Dams* (Chicago: Clohesey & Co., 1914), 30; Illinois General Assembly, Illinois Valley Flood Control Commission, *Hearings on the Causes and Control of Floods in the Illinois River Valley,* statement by John Goodell (n.p., 1929), 220; Illinois Natural History Survey, *Some Recent Changes in the Illinois River Biology,* by Stephen A. Forbes and Robert E. Richardson, Bulletin 13:6 (Urbana, April 1919), 142 (cited hereafter as Forbes and Richardson, *Changes in River Biology*); Illinois Department of Purchases and Construction, Division of Waterways, *Flood Control Report,* by Jacob A. Harman, L. T. Berthe, M. Blanchard, and L. C. Craig (Springfield: Journal Printing Co., 1930), 105 (cited hereafter as Harman et al., *Flood Control Report*); Joan Devore, ed., *Meredosia Bicentennial Book, 1776–1976,* "Sawmill Operation," by Leona H. Lansink (Bluffs, Ill.: Jones Publishing, 1976), 53–54.

7. "Local Matters," *PCD,* January 4 and June 21, 1905, p. 5, col. 3; "Fur Buying on Increase," *MCD,* December 10, 1915, p. 13, col. 3; "Over 700 Muskrat Hides Marketed Here This Afternoon," *I-S,* December 7, 1916, p. 1, col. 2; "Local and Other Items," *MB,* December 14, 1916, p. 7, col. 4, and April 19, 1917, p. 5, col. 4; "Local Happenings," *MB,* February 15, 1923, p. 5, col. 3.

8. "Northeastern Department," *RT,* October 19, 1899, p. 7, col. 2.

9. C. Hallock, *The Sportsman's Gazetteer and General Guide* (New York: Forest and Stream Publishing Co., 1877), 46.

10. "Pekin," *PH-T,* March 14, 1902, p. 2, col. 2; "Duck Shooting Fair," *PH-T,* November 12, 1904, p. 8, col. 2; "Exodus of Hunters," *PH-T,* October 11, 1905, p. 7, col. 3; "Peoria Gunners Shoot Many Ducks," *PH-T,* March 9, 1906, p. 3, col. 4; "The Nutwood Drainage and Levee District," *JR,* October 17, 1907, p. 1, cols. 1–4; "Where to Go," *Field and Stream,* May 1913, 80, and August 1913, 414; P. E. Zartman, "Duck Hunting Reminiscences of Thirty Years Ago," *Outdoor America* 2:3 (October 1923), 68; "Turned Bottom Lands into Farms; 'Back to Nature' Says Conservation," *CP,* December 1, 1927, p. 1, cols. 3–4; "Bought 1900 Acre Hunting Preserve," *MCD,* October 19, 1928, p. 1, col. 3; "Duck Hunters Becoming Numerous," *MCD,* October 25, 1929, p. 1, col. 2; Eugene V. Connett, ed., *Wildfowling in the Mississippi Flyway,* "Illinois," by Kenneth H. Smith (New York: D. Van Nostrand, 1949), 248–51; Paul W. Parmalee and Forrest D. Loomis, *Decoys and Decoy Carvers of Illinois* (Dekalb, Ill.: Northern Illinois University Press, 1969), 10, 80, 488; Bull interview.

11. "At Spring Lake," *PH-T,* October 15, 1903, p. 5, col. 4; "Duck Hunting," *PH-T,* March 14, 1904, p. 2, col. 2; "Good Hunting," *PH-T,* March 16, 1904, p. 5, col. 5; "After Ducks," *PH-T,* March 19, 1904, p. 2, col. 3; "Duck Hunting," *PH-T,* September 1, 1904, p. 2, col. 3; "Down the River," *PH-T,* November 2, 1904, p. 2, col. 3; "Duck Marshes May Be Revived," *PH-T,* April 2, 1906, p. 2, col. 4; "Local Matters," *PCD,* July 4, 1906, p. 5, col. 3; "Local and Other Items," *MB,* March 1, 1908, p. 5, col. 1; "Keen Rivalry for Local Fishermen," *MB,* April 29, 1909, p. 3, col. 1; "Duck Hunters Becoming Numerous," *MCD,* October 25, 1929, p. 1, col. 2.

12. "The Evil of Spring Shooting," *American Field,* April 14, 1894, 344; "The Game-Hog or Pot-Hunter," *American Field,* December 1, 1894, 508; "Northeastern Department," *RT,* April 12 and 19, 1900, p. 7, col. 2; "Stop Fall Duck Shooting," *American Field,* March 1, 1902, 191; "Pekin," *PH-T,* March 20, 1902, p. 3, col. 3, and March 21, 1902, p. 6, col. 2; "Wagon Load of Ducks," *PH-T,* November 14, 1903, p. 2, col. 3; "Duck Hunting," *PH-T,* March 14, 1904, p. 2, col. 2, and September 1, 1904, p. 2, col. 3; "Hunters," *PH-T,* September 27, 1904, p. 3, col. 1.

13. "Peoria Gunners Shoot Many Ducks," *PH-T,* March 9, 1906, p. 3, col. 4.

14. "Peoria Gunners Shoot Many Ducks," *PH-T,* March 9, 1906, p. 3, col. 4; "Fields of Heavy Ice Rip Out Standing Timber," *PH-T,* January 24, 1910, p. 2, col. 1.

15. Illinois *Laws* (1903), 206; Illinois *Laws* (1905), 273; "Seize 4 Deer and 500 Ducks," *Illinois Fish and Game Conservationist* 2:10 (November 1913), 13; "Busy at Havana," *Illinois Fisherman and Hunter* 2:6 (January 1914), 10; Bull interview.

16. Illinois General Assembly, *Reports to, 1881,* "Annual Report of Fish Commission" (Springfield: H. W. Rokker, 1881), 7 (cited hereafter as Ill. Fish Comm., "Report, 1881").

17. Illinois General Assembly, *Reports to, 1885,* 2, "Report of Fish Commission, 1884" (Springfield: H. W. Rokker, 1885), 18 (cited hereafter as Ill. Fish Comm., "Report, 1884"); Illinois Board of Fish Commissioners, *Report, from October 1, 1894, to September 30, 1896* (Springfield: Phillips Bros., 1897), 8, and *Report, from October 1, 1902, to September 30, 1904* (Springfield: Illinois State Journal Co., 1905), 22, and *Report, from October 1, 1904, to September 30, 1906* (n.p., n.d.), 6–7 (cited hereafter as Ill. Fish Comm., *Report, 1894–1896, Report, 1902–1904,* and *Report, 1904–1906,* respectively); S. P. Bartlett, "The Value of the Carp as a Food Product of Illinois Waters," *Transactions of the American Fisheries Society* 29 (1900), 82, and "Fish Waste, Past and Present," *Transactions of the American Fisheries Society* 47:1 (December 1917), 22.

18. Ill. Fish Comm., "Report, 1881," 3, 7, 9–12, "Report, 1884," 18–21, and *Report, 1894–1896,* 8.

19. Ill. Fish Comm., "Report, 1884," 21.

20. Ill. Fish Comm., "Report, 1881," 3, 9–10, 12, and "Report, 1884," 12; Illinois General Assembly, *Reports to, 1890,* "Report of Board of Fish Commissioners" (Springfield: H. W. Rokker, 1890), 4; U.S. Department of Commerce, Commissioner of Fisheries, *Annual Report for the Fiscal Year Ended June 30, 1915* (Washington, D.C.: GPO, 1915), 26 (cited hereafter as U.S. Comm. Fisheries, *Report, 1915*).

21. U.S. Department of Commerce, Commission of Fish and Fisheries, *Report for the Year Ending June 30, 1901,* "Statistics of Fisheries of the Mississippi River," by C. H. Townsend (Washington, D.C.: GPO, 1902), 679 (cited hereafter as Townsend, "Statistics of Fisheries"); Ill. Fish Comm., *Report, 1902–1904,* 19; Illinois Board of Fish Commissioners, *Report, from October 1, 1906, to September 30, 1908* (n.p., n.d.), 16 (cited hereafter as Ill. Fish Comm., *Report, 1906–1908*); Illinois Internal Improvements Commission, *The Lakes and Gulf Waterway* (Springfield: Phillips Bros., 1907), 32; S. P. Bartlett "The Decrease of the Coarse Fish and Some of Its Causes," *Transactions of the American Fisheries Society* 41 (1912), 200; Bartlett, "Fish Waste," 22, 25.

22. Ill. Fish Comm., "Report, 1881," 5, and "Report, 1884," 10–12; Bartlett, "Value of the Carp," 80–81; U.S. Department of Commerce and Labor, Bureau of the Census, in cooperation with Bureau of Fisheries, *Fisheries of the United States, 1908,* Special Report (Washington, D.C.: GPO, 1911), 50, 116 (cited hereafter as Bureau of Census, *Fisheries, 1908*); "Carp Culture in Illinois," and related articles, *Prairie Farmer,* April 24, 1886, 260–61.

23. The earlier annual totals of record are understood to be understatements, numerous commercial fishermen operating independently from the dealers' association that compiled production data. Illinois Board of Fish Commissioners, *Report, from October 1, 1896, to September 30, 1898* (Springfield: Phillips Bros., 1899), 8, *Report, from October 1, 1898 to September 30, 1900* (Springfield: Phillips Bros., 1900), 12, and *Report, from October 1, 1900, to September 30, 1902* (n.p., n.d.), 13–14 (cited hereafter as Ill. Fish Comm., *Report, 1896–1898, Report, 1898–1900,* and *Report, 1900–1902,* respectively); Ill. Fish Comm., *Report, 1902–1904,* 1, and *Report, 1904 to 1906,* 5; "Fishermen Howl for Lower Water," *PH-T,* October 21, 1907, p. 3, col. 3; "Goes to

Beardstown," *I-S,* November 1, 1909, p. 8, col. 3; Illinois Rivers and Lakes Commission, *The Illinois River and Its Bottom Lands,* 2nd ed., by John W. Alvord and Charles B. Burdick (Springfield: Illinois State Journal Co., 1919), 64–65, 71, 81 (cited hereafter as Alvord and Burdick, *Illinois River*); Stephen A. Forbes, "The Native Animal Resources of the State," *Transactions of the Illinois State Academy of Science* 5 (1912), 41–42; Forbes and Richardson, *Changes in River Biology,* 139, 148–49, 151–52; U.S. Department of Commerce, Bureau of Fisheries, *Fisheries Industries of the United States, 1923,* by Oscar E. Sette, Document 976 (Washington, D.C.: GPO, 1925), 222 (cited hereafter as Sette, *Fishery Industries, 1923*); David H. Thompson, "The Fishing Industry of Illinois River," *Transactions of the Illinois State Academy of Science* 24:2 (December 1931), 592, 593; Fred R. Jelliff, "The State-Wide Menace of Stream Pollution in Illinois," *Outdoor America* 2:10 (May 1924), 19; U.S. Congress, House, Committee on Rivers and Harbors, *Hearings on the Subject of the Improvement of the Illinois and Mississippi Rivers, and the Diversion of Water from Lake Michigan into the Illinois River,* 68th Cong., 1st sess., 1924, pt. 2, 955 (cited hereafter as House Comm. R. and H., *Hearings, 1924*). A well-documented review of the long-term experience with pollution in the Illinois River appears in R. T. Oglesby, C. A. Carlson, and J. A. McCann, eds., *River Ecology and Man,* "Man and the Illinois River," by William C. Starrett (New York: Academic Press, 1972), 148–52.

24. (Untitled), *PCD,* December 2, 1898, p. 5, col. 3; "Seining Objectionable Fish from the Illinois River," *American Field,* February 1, 1902, 98; "Fish Industry of Illinois," *RT,* February 25, 1904, p. 1, col. 6; "Where are the Bass," *PH-T,* September 11, 1904, p. 8, col. 6; "Lakes Overstocked with German Carp," *PCD,* July 25, 1906, p. 3, col. 4; "Fishermen Howl for Lower Water," *PH-T,* October 21, 1907, p. 3, col. 3; Ill. Fish Comm., *Report, 1906–1908,* 71; Illinois Laboratory of Natural History, *The Fishes of Illinois,* by Stephen A. Forbes and Robert E. Richardson (Danville: Illinois Printing Co., 1908), 267–69 (cited hereafter as Forbes and Richardson, *Fishes*); Forbes and Richardson, *Changes in River Biology,* 151; Martin, *History of Cass County,* 2:808; Alvord and Burdick, *Illinois River,* fig. 25, 67; Bartlett, "Fish Waste," 23; U.S. Bur. Fisheries, *Fisheries Industries,* 223–24; D. H. Thompson, "Fishing Industry," 593, 594; "Illinois Fish Law," *Illinois Fisherman* 1:3 (October 1912), 12; (untitled), *Illinois Fisherman* 1:10 (May 1913), 7, 8; "Trammel Nets Doing the Same Old Illegal Work," *Illinois Fish and Game Conservationist* 2:10 (November 1913), 12; (untitled), *Illinois Fisherman and Hunter* 2:7 (February 1914), 8; Bull interview.

25. (Untitled), *PCD,* April 25, 1899, p. 5, col. 2; "Northeastern District," *RT,* October 12, 1899, p. 7, col. 2; "Fish Industry of Illinois," *RT,* February 25, 1904, p. 1, col. 6; "In a Mass of Fish," *PCD,* August 21, 1901, p. 2, col. 2; "Lake Overstocked with German Carp," *PCD,* July 25, 1906, p. 3, col. 4; "Three Mason County Cities . . . ," *PH-T,* October 4, 1907, p. 7, col. 1, and p. 11, cols. 1–2; E. Jack, "The Ghetto Lives on Carp from Illinois River Nets," *PH-T,* December 15, 1912, p. 20, cols. 1–7; "Illinois River Fisheries," *MB,* October 23, 1913, p. 4, cols. 1–2; "Shipping Live Fish," *Illinois Fisherman* p. 1, col. 9 (April 1913), not paginated; Ill. Fish Comm., *Report, 1896–1898,* 5, 7, *Report, 1902–1904,* 19–21, and *Report, 1906–1908,* 49; Illinois Game and Fish Conservation Commission, *Annual Report for the Fiscal Year 1913–14* (Springfield: Illinois State Journal Co., 1914), 118 (cited hereafter as Ill. Game and Fish Comm., *Report, 1913–1914*); Alvord and Burdick, *Illinois River,* 64, 80. Data for map 8.2 from Ill. Fish Comm., *Report, 1906–1908.*

26. "Northeastern District," *RT,* October 12, 1899, p. 7, col. 2, and February 15, 1900, p. 7, col. 2; "Illinois Fish Industry," *RT,* October 24, 1901, p. 4, col. 3; "Fish Industry of Illinois," *RT,* February 25, 1904, p. 1, col. 6; "Lakes Overstocked with German Carp," *PCD,* July 25, 1906, p. 3, col. 4; "Ship Carp to Boston," *MB,* February 14, 1907, p. 1, col. 2; E. Jack, "The Ghetto Lives on Carp from Illinois," *PH-T,* December 15, 1912, p. 20, cols. 1–7; "Fishermen Make Large Haul,"

*MCD*, January 9, 1916, p. 1, col. 4; "Send Fresh Fish to Philadelphia," *MB*, February 27, 1913, p. 4, col. 2; Ill. Fish Comm., *Report, 1902–1904*, 18–21; Forbes and Richardson, *Fishes*, cxvii–cxviii, and *Changes in River Biology*, 151; Alvord and Burdick, *Illinois River*, 65, 66, 71–73; Bartlett, "Fish Waste," 23–24.

27. "Northeast District," *RT*, August 11, 1898, p. 7, col. 2, and November 16, 1899, p. 7, col. 2; S. P. Bartlett, "Discussion on Carp," *Transactions of the American Fisheries Society* 30 (1901), 131; Ill. Fish Comm., *Report, 1898–1900*, 1, 7, and *Report, 1902–1904*, 20; "Fish Industry of Illinois," *RT*, February 25, 1904, p. 1, col. 6; "Fish Company Organized," *PCD*, July 5, 1905, p. 3, col. 5; "Raising Fish," *PCD*, October 18, 1905, p. 3, col. 1; "River News," *I-S*, September 14, 1917, p. 8, col. 4; Devore, *Meredosia*, 46–47; D. H. Thompson, "Fish Industry," 594–95.

28. "Frederick Department," *RT*, February 14, 1901, p. 7, col. 6; "Feeding Carp for Market," *RT*, January 23, 1902, p. 1, col. 3; Cleves H. Howell, "Reclamation of Agricultural Land by Diking" (C.E. thesis, University of Illinois, 1905), 15; Joel E. Dunn, "Reclamation of Land by Levying" (C.E. thesis, University of Illinois, 1906), 10; Alvord and Burdick, *Illinois River*, 14.

29. "Will Get 400,000 Pounds of Fish," *MB*, December 12, 1907, p. 1, col. 2; East Peoria Drainage and Levee District, Commissioners, "Minutes," entry for November 25, 1911; "Local and Other Items," *MB*, February 6, 1913, p. 5, col. 3; Scott County Levee and Drainage District, Commissioners, "Ledger" (minutes book), entry for November 18, 1913, 41; Illinois Department of Agriculture, *First Annual Report* (July 1, 1917, to June 30, 1918), "Division of Game and Fish," by R. F. Bradford (Springfield: Illinois State Journal Co., 1918), 4, and *Second Annual Report* (July 1, 1918, to June 30, 1919), "Division of Game and Fish," by R. F. Bradford (Springfield: Illinois State Journal Co., 1920), 11 (cited hereafter as Bradford, "Division of Game and Fish, 1917–1918," and "Division of Game and Fish, 1918–1919," respectively); Illinois Department of Agriculture, *Seventh Annual Report* (July 1, 1923, to June 30, 1924), "Division of Game and Fish," by W. J. Stratton and S. B. Roach (Springfield: Illinois State Journal Co., 1925), 21.

30. Ill. Fish Comm., *Report, 1902–1904*, 20, and *Report, 1904–1906*, 10; Bartlett, "Fish Waste," 24.

31. Illinois Board of Health, *Second Annual Report, 1879*, "Sanitary Problems of Chicago, Past and Present," by John H. Rauch (Springfield: H. W. Rokker, 1881), 108–17, *Ninth Annual Report*, "Sanitary Survey of Beardstown," by Henry Ehrhardt (Springfield: Springfield Printing, 1889), 22–25, and *Report of the Sanitary Investigations of the Illinois River and Its Tributaries* (Springfield: Phillips Bros., 1901), xxviii, 4, 100; A. N. Talbot, "Sewage Disposal," *Proceedings of the Fourth Annual Meeting, Illinois Society of Engineers and Surveyors* (Bloomington, Ill., January 23–25, 1889), 55; Sanitary District of Chicago, *Reports of International Waterways Commission Concerning the Chicago Diversion and Terms of Treaty*, "Dilution Ratios in Chicago Drainage Canal," by Lyman E. Cooley, app. 1 (Chicago, n.d.), 68; Bartlett, "Decrease of Coarse Fish," 200.

32. Illinois Department of Public Works and Buildings, Division of Waterways, *Second Annual Report, July 1, 1918, to June 30, 1919* (Springfield: Illinois State Journal Co., 1920), 29, *Third Annual Report, July 1, 1919, to June 30, 1920*, "Report," by Stephen A. Forbes (Springfield: Illinois State Journal Co., 1921), 25, 26, 36–37, and *Seventh Annual Report, July 1, 1923 to June 30, 1924* (Springfield: Illinois State Journal Co., 1925), 12 (cited hereafter as Ill. Div. Waterways, *Report, 1918–1919; Report, 1919–1920;* and *Report, 1923–1924*, respectively); Illinois Natural History Survey, *Changes in the Bottom and Shore Fauna of the Middle Illinois River and Its Connecting Lakes since 1913–1915 as a Result of the Increase, Southward, of Sewage Pollution*, by Robert E. Richardson, Bulletin 14:4 (Urbana, 1921), 74–75, and *The Bottom Fauna of the Middle Illinois River, 1913–1915*, by Robert E. Richardson, Bulletin 17:12 (Urbana, December 1928), 470–71; Stephen A. Forbes, "Sewage Pollution of the Illinois River," *Outdoor America* 3:5 (December

1924), 35–36; Illinois Water Survey, *Illinois River Studies, 1925–1928,* by C. S. Boruff and A. M. Buswell, Bulletin 28 (Urbana, 1929), 7, 8, 48, 54.

33. Ill. Fish Comm., *Report, 1898–1900,* 11, and *Report, 1902–1904,* 32; Illinois Department of Public Works and Buildings, Division of Waterways, *First Annual Report, July 1, 1917 to June 30, 1918* (Springfield: Illinois State Journal Co., 1918), 3–4 (cited hereafter as Ill. Div. Waterways, *Report, 1917–1918*), and *Eighth Annual Report, July 1, 1924 to June 30, 1925* (Springfield: Illinois State Journal Co., 1926), 51; Ill. Div. Waterways, *Report, 1918–1919,* 29–31, and *Report, 1923–1924,* 12–14; "First Annual Convention of the Izaak Walton League of America," *Izaak Walton League Monthly* 1:9 (May 1923), 478; "League Opens Campaign to Save the Illinois River," *Outdoor America* 3:4 (November 1924), 11; W. G. Potter, "Improvement and Utilization of the Rivers of Illinois," *Western Society of Engineers Journal* 31:5 (May 1926), 204; Illinois Association of Sanitary Districts, *Eighth Annual Report, 1931–1932,* "Dedication Address of the Greater Peoria Sewage Treatment Work . . . ," by Henry B. Ward (Urbana, 1932), 19, and *Ninth Annual Report, 1932–1933,* "Sanitary Districts in Illinois," insert by Illinois Department of Public Health, Division of Sanitary Engineering (n.p., May 1933), 34; S. B. Locke, "Developments in the Clean Streams Campaign," *Transactions of the American Fisheries Society* 62 (1932), 355, 359; Illinois Department of Public Works and Buildings, Division of Waterways, *Fifteenth Annual Report, July 1, 1931 to June 30, 1932,* "Stream Pollution and Sanitation," by M. C. Sjoblom (Springfield: Journal Printing Co., 1933), 118.

34. An instructive general perspective on the development of the mussel fishing industry in the Midwest appears in Philip V. Scarpino, *Great River: An Environmental History of the Upper Mississippi, 1890–1950* (Columbia: University of Missouri Press, 1985), 80–113. Also for details about the Illinois River, see Townsend, "Statistics of Fisheries," 677; U.S. Department of Commerce, Bureau of Fisheries, *The Mussel Resources of the Illinois River,* by Ernest Danglade, Document 804 (Washington, D.C.: GPO, 1914) (cited hereafter as Danglade, *Mussel Resources*); Bureau of Census, *Fisheries, 1908,* 116; U.S. Comm. Fisheries, *Report, 1915,* 63–65; Sette, *Fishery Industries, 1923,* 222–26; Ill. Fish Comm., *Report, 1904–1906,* 17–18; Illinois Natural History Survey, *A Survey of the Mussels (Unionacea) of the Illinois River: A Polluted Stream,* by William C. Starrett, Bulletin 30:5 (Urbana, February 1971), 351, 354, 363 (cited hereafter as Starrett, *Survey of Mussels*); and "The Mussell-Shell Industry of Illinois," *Illinois Fish Conservation Society News-Letter* 1:3 (March 15, 1912), not paginated.

35. Danglade, *Mussel Resources,* 7–10; Scarpino, *Great River,* 84.

36. Townsend, "Statistics of Fisheries," 678; Danglade, *Mussell Resources,* 8, 20–23, 28–29; Herbert H. Vertrees, *Pearls and Pearling* (Columbus, Ohio: A. R. Harding, 1913), 108–34; Scarpino, *Great River,* 90–91. The crowfoot bar consisted of a 10- to 15-foot-long pipe, .5 to .75 inch in diameter, from which multiple short lines were strung, each with a number of hooks that were fashioned from heavy-gauge wire so as to produce four barbs and an appearance reminiscent of a crow's foot. As a rule, two bars were carried by a small boat. The drifting craft dragged lines and hooks parallel to the alignment usually taken by feeding bivalves. Upon contact with a barb, the bivalves grasped firmly enough that numbers of them were retrieved with each cycling of a bar across the bed. U.S. Department of Commerce, Commissioner of Fisheries, *Annual Report for the Fiscal Year Ended June 30, 1916* (Washington, D.C.: GPO, 1916), 50.

37. Danglade, *Mussel Resources,* 8, 28, 32; Scarpino, *Great River,* 83, 87, 95; "Thirteenth Census, 1910, Illinois" (microfilm T624:233, 314).

38. Danglade, *Mussel Resources,* 14, 17ff., 31; "Illinois Fish Law," *Illinois Fisherman* 1:3 (October 1912), 12; Scarpino, *Great River,* 81, 110–13.

39. Ill. Fish Comm., *Report, 1896–1898,* 16; (untitled), *PCD,* January 13, 1899, p. 5, col. 3, and March 3, 1899, p. 5, col. 2; "Thompson Lake Hunting Preserve," *MCD,* September 6, 1901, p. 4, col. 4; "Stop Fall Duck Shooting," *American Field,* March 1, 1902, 191; "Grand Island Lodge," *American Field,* June 28, 1902, 599–600; "Immense Drainage Scheme," *PH-T,* November 2, 1903, p. 8, col. 4; "'Freak' Measures Are Introduced," *I-S,* February 13, 1907, p. 9, cols. 1–3; "Hunting Clubs Hard Hit," *PCD,* May 10, 1895, p. 6, col. 2; "Fishermen Are Enjoined," *MB,* February 24, 1910, p. 5, col. 3; "Illinois News Briefly Told," *MB,* February 6, 1913, p. 7, col. 2; "Lake Now Exclusive Property," *I-S,* August 19, 1914, p. 3, col. 3; "Local News," *MB,* August 19, 1920, p. 5, col. 4; "Club House Burns," *Illinois Fish and Game Conservation Society and Warden's Journal* 3:8 (August 1914), 6; "Illinois Lake Purchased by Club," *Illinois Fish and Game Conservation Society and Warden's Journal* 3:9 (September 1914), 12; "Turned Bottom Lands into Farms; 'Back to Nature,' Says Conservation," *CP,* December 1, 1927, p. 1, cols. 3–4; Illinois Secretary of State, Corporation Department, "Duck Island Hunting and Fishing Club," "Certificate" (of incorporation); Illinois Natural History Survey, *Waterfowl Hunting in Illinois: Its Status and Problems,* by Frank C. Bellrose Jr., Biological Notes 17 (Urbana, March 1944), 4, 16, 31 (cited hereafter as Bellrose, *Waterfowl Hunting*).

40. "A Cry for Water," *MCD,* September 27, 1901, p. 1, col. 4; "Work of Shooting Duck Ponds for Duck Shooting," *I-S,* September 19, 1916, p. 1, col. 6; Edward Cave, "Down at Grand Island by Proxy," *Outers' Book-Recreation,* March 1918, 208; Smith, "Illinois," in Connett, *Wildfowling,* 251.

41. "Fishing and Hunting Notice," *I-S,* September 16, 1909, p. 7, cols. 4–6; "New Drainage District," *RT,* February 9, 1911, p. 1, cols. 1–2; Illinois General Assembly, House, Submerged and Shore Lands Investigating Committee, *Report to the Governor and 47th General Assembly* (Springfield: Illinois State Journal Co., 1911), 1:166 (cited hereafter as Submerged and Shore Lands Comm., *Report*).

42. (Untitled), *PCD,* March 3, 1899, p. 5, col. 2; "Northeastern District," *RT,* November 29, 1900, p. 7, col. 2; "Local Matters," *PCD,* January 31, 1900, p. 5, col. 3, June 13, 1900, p. 5, col. 3, and August 22, 1900, p. 5, col. 4; "Thompson Lake Hunting Preserve," *MCD,* September 6, 1901, p. 4, col. 4; "Grand Island Lodge," *American Field,* March 8, 1902, 214, and June 28, 1902, 599–600; "Industrious Pekin," *PH-T,* March 17, 1905, sec. 3, p. 4, col. 2; "A Day Spent at Sni E'Carte," *PCD,* July 25, 1906, p. 1, col. 4; Illinois Secretary of State, Corporation Department, "Pike County Fishing Club," "Rice Lake Hunting Club," "Thompson Lake Rod and Gun Club," "Clear Lake Outing Club," ""Knapp Island Gun Club," certificates of incorporation; Peoria County Circuit Court, "Drainage Record—Pekin and LaMarsh, 1," Pekin and LaMarsh Drainage and Levee District, Commissioners, "Assessment Roll . . . for 1900," n.d., 136; "Illinois News Briefly Told," *I-S,* August 19, 1914, p. 3, col. 3; "New Drainage District," *RT,* February 9, 1911, p. 1, cols. 1–2; Mason County Circuit Court, *Knapp Island Gun Club v. M. J. O'Meara—Injunction,* filed January 6, 1915, chancery no. 2214, box 772, 1, 3; "Let Contract for Club House," *RT,* July 19, 1917, p. 1, col. 3; Cave, "Down at Grand Island," 208; Gus Munch, "Illinois River Duck Chatter," *Outers' Book-Recreation,* March 1919, 148; U.S. Congress, House, *Survey, with Plans and Estimates of Cost, for a Navigable Waterway from Lockport, Ill., by Way of Des Plaines and Illinois Rivers and Thence by Way of the Mississippi River to St. Louis,* "Map of the Illinois and Des Plaines Rivers, . . . " (1:7,200), by J. W. Woermann, U.S. Army, Corps of Engineers, 59th Cong., 1st sess., 1905, H. Doc. 263, sheets 10, 27, 30, 32, 33.

43. (Untitled), *PCD,* January 13, 1899, p. 5, col. 3.

44. Munch, "Illinois River Duck Chatter," 147; Aldo Leopold, *Report on a Game Survey of the North Central States,* for the Sporting Arms and Ammunition Manufacturers Institute

(Madison: Democrat Printing Co., 1931), 208–9; Ralph F. Bradford, "A Few Remarks on Conservation in the Illinois Valley," *Transactions of the Illinois State Academy of Science* 24:2 (December 1931), 589.

45. "Drainage Plan to Cost Much," *PH-T,* September 30, 1903, p. 8, col. 5; "State Legislative Investigating Committee Makes Tour of River and Lakes," *MCD,* September 3, 1909, p. 1, cols. 1–2.

46. "Will Organize," *MB,* October 18, 1906, p. 1, col. 1. Also see "Fishing Club Fight in Illinois House," *MB,* April 11, 1907, p. 1, col. 1.

47. "Stop Fall Duck Shooting," *American Field,* March 1, 1902, 191; "Drainage Plan to Cost Much," *PH-T,* September 30, 1903, p. 8, col. 5; "Will Organize," *MB,* October 18, 1906, p. 1, col. 1; "Charge That There Is a Duck Trust," *MB,* April 29, 1909, p. 1, col. 3; "Thompson Lake Ruling by the Rivers and Lakes Commission," *MCD,* May 28, 1915, p. 1, cols. 6–7, and p. 2, cols. 3–5; "Corn vs. Carp, Which Will Win?", *MCD,* July 2, 1915, p. 1, cols. 2–3; Munch, "Illinois River Duck Chatter," 148; Zartman, "Duck Hunting Reminiscences," 68–69; Leopold, *Report on a Game Survey,* 207–8; C. W. E. Eifrig, "Taking Inventory," *Audubon Annual Bulletin* 22 (1932), 26–27; Smith, "Illinois," in Connett, *Wildfowling,* 249, 251–52; Bellrose, *Waterfowl Hunting,* 3–4.

48. "Thompson Lake Belongs to State of Illinois," *MCD,* November 12, 1909, p. 1, col. 1; "Fishermen Win Suit," *RT,* October 20, 1910, p. 1, col. 4.

49. "Spring Lake Controversy Heard This Week," *MCD,* September 17, 1909, p. 1, col. 3.

50. Attributed to the *Mason County Democrat* in "Would Limit Game Preserves," *PH-T,* March 11, 1907, p. 2, col. 1; "Federal Injunction Halts Governor Deneen and Party," *MCD,* September 10, 1909, p. 1, col. 1; *Who Was Who* (Chicago: A. N. Marquis Co., 1943), 1:892.

51. *Schulte v. Warren et al.,* 218 Illinois Reports (1906), 108–9, 116.

52. "Schulte Sells Clear Lake," *MCD,* March 2, 1906, p. 1, col. 5; "Hunting Clubs Hard Hit," *PCD,* May 10, 1895, p. 6, col. 2; "'Freak' Measures Are Introduced," *I-S,* February 13, 1907, p. 9, cols. 1–3; "Law to Abolish Hunting Clubs," *MB,* February 14, 1907, p. 1, col. 3; "Fishing Club Fight in Illinois House," *MB,* April 11, 1907, p. 1, col. 1; Submerged and Shore Lands Comm., *Report,* 168–69.

53. Attributed to the *Mason County Democrat* in "Would Limit Game Preserves," *PH-T,* March 11, 1907, p. 2, col. 1.

54. "Thompson Lake Hunting Preserve," *MCD,* September 6, 1901, p. 4, col. 4; Illinois Secretary of State, Corporation Department, "Thompson Lake Rod and Gun Club," "Record of Dissolution," filed January 2, 1909; "Thompson Lake Belongs to State of Illinois," *MCD,* November 12, 1909, p. 1, col. 1; "Famous Case Is Now Ended," *MB,* October 20, 1910, p. 1, col. 4; "Fishermen Win Suit," *RT,* October 20, 1910, p. 1, col. 4; Submerged and Shore Lands Comm., *Report,* 169–70.

55. "Fishermen Win Suit," *RT,* October 20, 1910, p. 1, col. 4; Bartlett, "Decrease of Coarse Fish," 197; Fulton County Circuit Court, Thompson Drainage and Levee District files, "Petition to form District," filed June 7, 1915, 1.

56. "Corn vs. Carp Interests All," *MCD,* July 23, 1915, p. 1, col. 3. An account of the reclamation of the Thompson Lake Drainage and Levee District appears in University of Illinois, Water Resources Center, *Case Studies in Drainage and Levee District Formation and Development on the Floodplain of the Lower Illinois River, 1890s to 1930s,* by John Thompson, Special Report 017 (Urbana-Champaign, May 1989), 98–101 (cited hereafter as Thompson, *Case Studies*).

57. "Stay Off Thompson Lake," *MCD,* February 24, 1922, p. 1, col. 5; "Thompson Lake Trespassers Fined," *MCD,* April 11, 1924, p. 1, col. 5.

58. J. M. Entwistle, "To Save Fish and Game," *RT,* January 21, 1915, p. 4, cols. 4–5. As early as

1912, outright purchase was urged upon the state. "Thompson Lake at Stake," *Illinois Fish Conservation Society News-Letter* 1:11 (October 1912), not paginated; (untitled), *Illinois Fisherman* 1:5 (December 1912), 13.

59. "Havana, Mason County's Metropolis," *PH-T,* December 29, 1905, p. 9, cols. 1–4; "Quiver Beach Attracts Many Summer Visitors," *MCD,* July 28, 1911, p. 1, col. 1; Bartlett, "Decrease of Coarse Fish," 197; "Liverpool Improves," *MCD,* July 28, 1911, p. 1, col. 4; "A Popular Resort," *Illinois Fisherman* 1:3 (October 1912), 4; "For Sale," *Illinois Fisherman* 1:8 (March 1913), not paginated; "Illinois Warden's Action Relieves Health Conditions" *Illinois Fish and Game Conservation Society and Warden's Journal* 3:8 (August 1914), 4; "Matanzas Beach a New Resort," *MCD,* February 26, 1915, p. 1, col. 5; "Summer Visitors Invade Havana," *MCD,* June 26, 1925, p. 1, col. 7; "Corn vs. Carp; Farms vs. Resort," *MCD,* July 30, 1915, p. 5, col. 4; "The Illinois Is Chicago Sewer," *CP,* February 12, 1925, p. 1, col. 7; "Havana Mecca for Pleasure Seekers," *MCD,* August 3, 1928, p. 1, col. 2; Alvord and Burdick, *Illinois River,* 64; House Comm. R. and H., *Hearings, 1924,* 199, 825.

60. Illinois Rivers and Lakes Commission, *Land Drainage in Illinois,* by Robert I. Randolph, Bulletin 4 (Springfield: Illinois State Journal Co., 1912), 22–23 (cited hereafter as Randolph, *Land Drainage*). Accounts of the reclamation of the tracts appear in Thompson, *Case Studies,* 3–5, 59–61, 95–97, 106–8.

61. "Immense Drainage Scheme," *PH-T,* November 2, 1903, p. 8, col. 4; "Fishermen to Meet Here Sunday," *PH-T,* November 4, 1903, p. 5, col. 2; "Will Improve Land," *PH-T,* December 16, 1903, p. 8, col. 7; "The Large Drainage District," *PH-T,* January 5, 1904, p. 5, col. 7; "Save Spring Lake," *Illinois Fish Conservation Society News-Letter* 1:3 (March 15, 1912), not paginated; "Famous Resort Gone," *Illinois Fisherman* 1:8 (March 1913), not paginated; Submerged and Shore Lands Comm., *Report,* 158, 166.

62. "Drainage Plan to Cost Much," *PH-T,* September 30, 1903, p. 8, col. 6; "Spring Lake Case Now Being Heard," *PH-T,* February 4, 1910, p. 8, cols. 1–3; "Old Rivermen in Spring Lake Case," *PH-T,* February 4, 1910, p. 16, col. 5; "Rendered Decision," *I-S,* April 7, 1910, p. 1, col. 6; *People v. Spring Lake District,* 253 Illinois Reports 479 (1912).

63. "Duck Marshes May Be Ruined," *PH-T,* April 2, 1906, p. 2, col. 4; "Fight Not Apparent," *I-S,* August 30, 1909, p. 1, col. 4; "Spring Lake Controversy Heard This Week," *MCD,* September 17, 1909, p. 1, col. 3; "Cannot Close Lake," *I-S,* November 13, 1909, p. 5, cols. 4–5; "State Officials File Suit in Equity," *PH-T,* February 4, 1910, p. 8, cols. 1–3; "Grants Injunction to People of State," *PH-T,* February 6, 1910, p. 16, cols. 1–2; "Spring Lake Commissioners Answer Attorney General," *PH-T,* February 21, 1910, p. 3, col. 1; Submerged and Shore Lands Comm., *Report, 168; People v. Spring Lake District,* 253 Illinois Reports 479 (1912); *Spring Lake Drainage District v. Stead,* 263 Illinois Reports 247 (1914); Illinois Secretary of State, Archives, Illinois Rivers and Lakes Commission, "Minute Book, 1912–1913," meetings of March 31, May 17, and August 21, 1913, 138–43, 152, 186, and "Minute Book, 1914," meeting of September 21, 1914, 106; Randolph, *Land Drainage,* 22–28; "Famous Resort Gone," *Illinois Fisherman* 1:8 (March 1913), not paginated; Illinois Department of Registration and Education, Geological Survey, *Engineering and Legal Aspects of Land Drainage in Illinois,* rev. ed., by George W. Pickels and Frank B. Leonard Jr., Bulletin 42 (Urbana, 1929), 121 (cited hereafter as Pickels and Leonard, *Engineering*).

64. Ill. Fish Comm., *Report, 1898–1900,* 12, *Report, 1902–1904,* 1, and *Report, 1904–1906,* 5.

65. Ill. Fish Comm., *Report, 1898–1900,* 1, 2, 6, 10–11, *Report, 1902–1904,* 3, 7–10, 20, 32, *Report, 1904–1906,* 6–11, 19–20, 33–42, and *Report, 1906–1908,* 8, 71–89. Lengthiest page references are to reports by wardens. "Misleading Legislation" and "Conditions That Confront Us," *Illinois Fish*

*Conservation Society News-Letter* 1:3 (March 15, 1912), not paginated; W. E. Meehan, "The Relationship between State Fish Commission and Commercial Fishermen," *Illinois Fish Conservation Society News-Letter* 1:5 (April 1912), not paginated; "Fishermen Accused of Arson," *Illinois Fish Conservation Society News-Letter* 2:8 (September 1913), 14.

66. Bartlett, "Decrease of Coarse Fish," comments by Prof. H. B. Wood, University of Illinois, 203; "The Proposed Illinois Law," *American Field*, February 16, 1895, 147–49; "The Old Illinois Game Law Stands," *American Field*, June 22, 1895, 578; "Important Convention," *American Field*, April 26, 1902, 1; Illinois Game Commission, *First Annual Report* (Springfield: Phillips Bros., 1900), 11–12; "Fishermen, Hunters Organize, Would Preserve Haunts," *MCD*, February 24, 1911, p. 1, col. 5; Forbes, "Native Animal Resources," 42–44; "Commercial Fishermen Organize," *Illinois Fish Conservation Society News-Letter* 2:7 (June 1, 1913), not paginated; B. G. Merrill, "Report of Committee on Membership and Publicity," *Illinois Fish Conservation Society News-Letter* 2:8 (September 1913), 6–7; "State News Comments on Work of Sportsmen's Organizations," *Illinois Fish Conservation Society News-Letter* 2:10 (November 1913), 11–12; "Would Take a Backward Step," *Illinois Fish Conservation Society News-Letter* 3:2 (February 1914), 5; "Fisheries of the Illinois River," *Illinois Fish and Game Conservationist* 3:2 (February 1914), 7–8; "Federal Regulations and the Fish and Game Conservation Society of Illinois," *Illinois Fish and Game Conservationist* 3:3 (March 1914), 7–8; "Would Have Game Preserves Revert to the People," *Illinois Fish Conservation Society News-Letter* 3:5 (May 1914), 10; "The Attack on the Federal Bird Laws" and "Sportsmen, Congressmen, Audubonites and Others," *Audubon Bulletin* (winter 1916–17), 9–13, 26–27; Mary Drummond, "The Formative Years of the Illinois Audubon Society," *Audubon Bulletin* (fall 1920), 4–5.

67. Nat H. Cohen, "Concerning Fish Laws in Illinois," *Transactions of the American Fisheries Society* 30 (1901), 134; Ill. Fish Comm., *Report, 1906–1908*, 5; "Work of the Illinois Fish Commission," *Illinois Fish Conservation Society News-Letter* 1:5 (April 15, 1912), not paginated; "Commission Wages Successful War Against Gars," *Illinois Fish Conservation Society News-Letter* 2:1 (January 1913), 1–4; H. Wheeler Perce, "Fair Play," *Illinois Fish Conservation Society News-Letter* 2:2 (February 1913), 1; "Sen. Beall Introduced Society's Bills in Senate," *Illinois Fish Conservation Society News-Letter* 2:4 (April 1913), 1–2; "Governor Dunne's Special Message on Fish and Game Department," *Illinois Fish Conservation Society News-Letter* 2:5 (May 1913), 7–9; "Synopsis of Fish Law," *Illinois Fish Conservation Society News-Letter* 2:7 (August 1913), not paginated; "Exterminating the Gar," *Illinois Fish Conservation Society News-Letter* 3:3 (March 1914), 4–5; "Urges Consolidation," *Illinois Fisherman* 1:10 (May 1913), not paginated; Ill. Game and Fish Comm., *Report, 1913–1914*, 116, 118, 120, 129; Illinois Game and Fish Conservation Commission, *Third Annual Report, for the Fiscal Year 1915–1916* (Springfield: Illinois State Journal Co., 1916), 43; "The Illinois Sportsmen's League," *Audubon Bulletin* (fall 1921), 4–5; Will H. Dilg, "A Message to Rod and Gun Clubs," *Izaak Walton League Monthly* 1:2 (September 1922), 55, and "First Annual Convention of the Izaak Walton League of America," *Izaak Walton League Monthly* 1:9 (May 1923), 478; "Brief News of the Chapters . . . ," *Outdoor America* 2:9 (April 1924), 44; Paul D. Barnes, "Illinois Convention," *Izaak Walton League Monthly* 5:9 (April 1927), 54–55; Leopold, *Report on a Game Survey*, 259, 261.

68. Dilg, "Message," 54; "League Wins Bass Fight in Illinois," *Izaak Walton League Monthly* 1:11 (July 1923), 603; "Turned Bottom Lands into Farms . . . ," *CP*, December 1, 1927, p. 1, cols. 2–3.

69. Ill. Fish Comm., *Report, 1906–1908*, 19.

70. Robert G. Hays, *State Science in Illinois: The Scientific Surveys, 1850–1978* (Carbondale: Southern Illinois University Press, 1980), 44–56; Sanitary District of Chicago, *Report of Stream*

*Examination, Chemic and Bacteriologic of the Waters Between Lake Michigan at Chicago and the Mississippi River at St. Louis, for the Purpose of Determining Their Condition and Quality Before and After the Opening of the Drainage Channel*, by Arthur W. Palmer (Chicago: Blakely Printing Co., December 1902); Illinois Laboratory of Natural History, *Biological Investigations on the Illinois River,* "The Work of the Illinois Biological Station," by Stephen A. Forbes (Urbana, 1910), 1 (cited hereafter as Forbes, "Biological Station").

71. Forbes, "Biological Station," 1; Illinois Laboratory of Natural History, *Biological Investigations on the Illinois River,* "The Investigation of a River System in the Interest of Its Fisheries," by Stephen A. Forbes, 12–13.

72. Submerged and Shore Lands Comm., *Report,* 6–10, 15, 17; Randolph, *Land Drainage,* 5, 7.

73. Illinois Rivers and Lakes Commission, *Annual Report, 1912* (Springfield: Illinois State Journal Co., 1913), 3–9, 26–30, and *Annual Report, 1913–1914* (Chicago: Fred Klein Co., 1914), 3–13, 24–25.

74. Alvord and Burdick, *Illinois River,* preface; Forbes and Richardson, *Changes in River Biology,* 148.

75. Ill. Game and Fish Comm., *Report, 1913–1914,* 140.

76. Ill. Fish Comm., *Report, 1902–1904,* 23; "Thompson Lake Ruling by the Rivers and Lake Commission," *MCD,* May 28, 1915, p. 1, cols. 6–7; Alvord and Burdick, *Illinois River,* 131.

77. Bradford, "Division of Game and Fish, 1917–1918," 5, and "Division of Game and Fish, 1918–1919," 7–8; Ill. Div. Waterways, *Report, 1917–1918,* 3; "Legislation," *Audubon Bulletin* (winter, 1917–18), 8; Ernest Ludlow Bogart and John Mabry Mathews, *Centennial History of Illinois: The Modern Commonwealth, 1893–1918,* vol. 4 (Chicago: A. C. McClurg & Co., 1922), 153; Illinois Department of Conservation, *Preliminary Report . . . Fiscal Year Beginning July 1, 1928* (Springfield: Journal Printing Co., 1929), 1, 23; "Facts Concerning the State Sanitary Water Board," *Illinois Conservation* 1:3 (fall 1936), 9–10.

78. Forbes and Richardson, *Changes in River Biology,* 139–55; Illinois Natural History Survey, *Duck Food Plants of the Illinois River Valley,* by Frank C. Bellrose Jr., Bulletin 21:8 (Urbana, August 1941), 237–41, and *The Fate of Lakes in the Illinois River Valley,* by Frank C. Bellrose, S. P. Havera, F. L. Paveglio Jr., and D. W. Steffeck, Biological Notes 119 (Urbana, April 1983), 6–7, 9–10 (cited hereafter as Bellrose et al., *Fate of Lakes*); Frederick C. Lincoln, "The Mississippi Flyway," in Connett, *Wildfowling,* 11–12; Illinois Department of Public Works and Buildings, Division of Waterways, in cooperation with Department of Conservation, *Report for Recreational Development, Illinois River Backwater Areas* (Springfield, 1969), 33–34 (cited hereafter as Ill. Div. Waterways, *Report for Recreational Development*); Donald W. Steffeck, Fred L. Paveglio Jr., Frank C. Bellrose, and Richard E. Sparks, "Effects of Decreasing Water Depths on the Sedimentation Rate of Illinois River Bottomland Lakes," *Water Resources Bulletin* 16:3 (June 1980), 554.

79. Alvord and Burdick, *Illinois River,* 14–15, 18, 133–34; Harman et al., *Flood Control Report,* 22, 84–85, 123; Illinois Department of Public Works and Buildings, Division of Waterways, *Sixteenth Annual Report, July 1, 1932 to June 30, 1933,* "Supervision of Rivers and Lakes," by W. G. Potter (Springfield: Illinois State Journal Co., 1934), 15; Pickels and Leonard, *Engineering,* 137.

80. U.S. Congress, House, Committee on Rivers and Harbors, *Illinois River, Ill.,* 77th Cong., 2nd sess., 1942, H. Doc. 692, 19, 22–23, 27–83; Starrett, *Survey of Mussels,* 156, 159, 161; Bellrose et al., *Fate of Lakes,* 23–26.

81. Leopold, *Report on a Game Survey,* 202; U.S. Department of Agriculture, Bureau of Biological Survey, *Report of the Chief, 1935* (Washington, D.C.: GPO, 1935), 25, 27, and *The*

*Survey* 18:1 (January 1937), 19; Illinois Department of Conservation, Division of Technical Services, *Land and Water Report* (Springfield, June 30, 1984), 4.

82. In 1950, the merits for conservation and flood control were especially well regarded for the Spring Lake–Clear Lake, Banner Special, East Liverpool, Thompson Lake, Big Prairie, Hartwell, and Keach districts and for the northerly Hennepin district (2,900 acres), which is not in the study area. Illinois Department of Conservation, *Potential Conservation Areas along the Illinois River from Hennepin to Grafton,* by Jenkins, Merchant & Nankivil and W. B. Walraven (Springfield, May 1950), 77–79; Ill. Div. Waterways, *Report for Recreational Development,* 77–79, 98–99.

83. Meandered lakes in which title to the bed is vested in the state are, from north to south: Pekin and Spring (Tazewell County); Clear, Mud, Liverpool, Quiver, and Matanzas (Mason County); and Meredosia (Cass and Morgan counties). Illinois Department of Public Works and Buildings, Division of Waterways, *Meandered Lakes in Illinois* (n.p., 1912), 16–20, 37–42, 56–60.

## 9. Retrospect

1. U.S. Congress, House, Committee on Rivers and Harbors, *Hearings on the Subject of the Improvement of the Illinois and Mississippi Rivers, and the Diversion of Water from Lake Michigan into the Illinois River,* "Memorandum of Information by Drainage and Levee Districts of Illinois," and testimony by Guy L. Shaw, 68th Cong., 1st sess., 1924, 1:119–203, 2:826 (cited hereafter as House Comm. R. and H., *Hearings, 1924*).

2. U.S. Department of Commerce, Bureau of the Census, *Census of Agriculture, 1959: Drainage of Agricultural Lands* (Washington, D.C.: Government Printing Office, 1961), 4:12.

3. Gary Lane McDowell, "Local Agencies and Land Development by Drainage: The Case of 'Swampeast' Missouri" (Ph.D. diss., Columbia University, 1965), 13, 89. McDowell's remarkably comprehensive review of the national drainage experience anticipated much of this writer's separate inquiry.

4. Benjamin Whipple Palmer, *Swamp Land Drainage with Special Reference to Minnesota,* Bulletin of the University of Minnesota, Studies in the Social Sciences 5 (Minneapolis: University of Minnesota, 1915), 21, 96, 101.

5. Illinois Department of Purchases and Construction, Division of Waterways, *Flood Control Report,* by Jacob A. Harman, L. T. Berthe, M. Blanchard, and L. C. Craig (Springfield: Journal Printing Co., 1930), 44–45; Illinois General Assembly, Legislative Flood Investigating Commission, *Report on Flood Situation in Illinois,* by Elliott & Porter, Consulting Engineers (Chicago, March 1947), 114–17; House Comm. R. and H., *Hearings, 1924,* 196–97; U.S. Department of Agriculture, *Land Reclamation Policies of the United States,* by R. P. Teele, Bulletin 1257 (Washington, D.C.: GPO, August 23, 1924), 7–9 (cited hereafter as Teele, *Land Reclamation Policies*).

6. Teele, *Land Reclamation Policies,* 14.

7. "To Reclaim Swamps," *American Contractor,* November 2, 1907, 74; "Swamp Land Reclamation," *American Contractor,* May 16, 1914, 103; McDowell, "Local Agencies and Land Development," 117, 129–34.

8. Teele, *Land Reclamation Policies,* 27, 30–31; Robert W. Harrison and Walter M. Kollmorgan, "Past and Present Drainage Reclamations in the Coastal Marshlands of the Mississippi River Delta," *Journal of Land and Public Utility Economics* 23:3 (August 1947), 306–10. More comprehensive is the statement in Robert W. Harrison, *Alluvial Empire: A Study of State and Local Efforts*

*Toward Land Development in the Alluvial Valley of the Lower Mississippi River, Including Flood Control, Land Drainage, Land Clearing, Land Forming* (Little Rock: Pioneer Press, 1961), 262.

9. Teele, *Land Reclamation Policies,* 35.

10. Theodore M. Schad, "Evolution and Future of Flood Control in the United States," in William J. Hull and Robert W. Hull, eds., *The Origin and Development of the Waterways Policy of the United States* (Washington, D.C.: National Waterways Conference, 1967), 28; E. W. Gould, *Fifty Years on the Mississippi; or, Gould's History of River Navigation* (St. Louis: Nixon-Jones Printing, 1889), 226–29.

11. Harrison, *Alluvial Empire,* 132–33, 148–54; Hull and Hull, *Origin and Development of Policy,* ix–xi; Schad, "Evolution and Future of Flood Control," 28; U.S. Congress, House, *Laws of the United States Relating to the Improvement of Rivers and Harbors from August 11, 1790 to March 4, 1913,* 62nd Cong., 3rd sess., 1913, H. Doc. 1491, 3:1703–6; U.S. Congress, House, Committee on Flood Control, *Control of Floods on the Mississippi and Sacramento Rivers,* 64th Cong., 1st sess., 1916, H. Rept. 616, 24–28, 36–43.

12. For that matter, comparisons may be drawn with the origins, effects, and resolution of problems arising from competing uses of water and land beyond the floodplain of first-order streams like the Colorado River, as in the conceptually rich contributions of various scholars in Gary D. Weatherford and F. Lee Brown, eds., *New Courses for the Colorado River: Major Issues for the Next Century* (Albuquerque: University of New Mexico Press, 1986).

# Bibliography

*This bibliography is divided into the following major sections: Archival Materials; Government Publications; Books and Trade Catalogs; Journals, Proceedings, and Pamphlets; and Dissertations and Theses.*

## Archival Materials

### United States


U.S. Department of Commerce. Bureau of the Census. "Twelfth Census of the United States: 1900, Illinois." Census Schedule Microfilm T623: Brown, Calhoun, Cass, Fulton, Greene, Mason, Morgan, Pike, Schuyler, and Scott counties.

———. "Thirteenth Census of the United States: 1910, Illinois." Census Schedule Microfilm T624: Calhoun, Cass, and Morgan counties.

———. "Fourteenth Census of the United States: 1920, Illinois." Census Schedule Microfilm T625: Cass, Fulton, Green, Mason, Morgan, Pike, Schuyler, Scott, and Tazewell counties.


### Illinois


Illinois Secretary of State. Archives. "Notes on Levee Districts, Illinois River Bottoms." By John W. Alvord and Charles B. Burdick. [1914]

———. Illinois Rivers and Lakes Commission. "Minute Book, 1911–1912."

———. "Minute Book, 1912–1913."

———. "Minute Book, 1914."

Illinois Secretary of State. Corporation Department. "Clear Lake Outing Club."

———. "Duck Island Hunting and Fishing Club."

———. "Federal Contracting Company."

———. "Knapp Island Gun Club."

———. "Pike County Fishing Club."

———. "Pollard Goff and Company."

———. "Quincy Dredging and Towing Company."

———. "Rice Lake Hunting Club."

———. "Thompson Lake Rod and Gun Club."


### Illinois Counties


Brown County Circuit Court. "Drainage Record No. 1."

Cass County Circuit Court. "Hager Slough Special Drainage District."

Fulton County Circuit Court. "East Liverpool Drainage and Levee District, 1."

———. "Kerton Valley Levee and Drainage District, 1."

———. "Lacey Drainage and Levee District, 1."

———. "Liverpool Drainage and Levee District, 1."

———. "Seahorn Drainage and Levee District, 1."
———. "Thompson Lake Drainage and Levee District."
———. "West Matanzas Drainage and Levee District, Record Book 1."
———. "West Matanzas Drainage and Levee District, Record 2."
Greene County Circuit Court. "Drainage Record 2."
———. Hartwell Drainage and Levee District files.
———. Hillview Drainage and Levee District files.
———. *R. H. McWilliams and G. A. McWilliams v. Eldred Drainage and Levee District.* February Term, 1918. No. 7354.
Jersey County Circuit Court. "Drainage and Levee Record No. 1."
———. Nutwood Drainage and Levee District files.
Mason County Circuit Court. *Knapp Island Gun Club v. M. J. O'Meara.* Chancery No. 2214. Box 772.
———. "Long Branch Drainage District."
———. "Mason and Tazewell Special Drainage District."
Mason County Probate Court. "Estate Papers for George William Gillen." Box 219.
Morgan County Circuit Court. "Coon Run Drainage and Levee District."
Peoria County Circuit Court. "Drainage District Book 1."
———. "Drainage Record—Pekin and LaMarsh." Vol. 1.
Piatt County Circuit Court. "Lake Fork Drainage District, No. 1."
Pike County Circuit Court. "Drainage Record, Book D."
———. "Drainage Record, Book E."
———. "Drainage Record, Book F."
———. "Drainage Record, Book G."
Schuyler County Circuit Court. "Big Lake Drainage and Levee District—1911–1920."
———. "Coal Creek Drainage and Levee District."
———. "Crane Creek Drainage and Levee District."
———. "Crane Creek Drainage and Levee District, 1893–1918."
———. "Kelley Lake Drainage and Levee District."
Scott County Circuit Court. "Big Swan Drainage and Levee District."
———. Files of Scott County Drainage and Levee District.
Tazewell County Circuit Court. Files of Hickory Grove Drainage and Levee District.

### City of Havana

Havana Public Library. "Sangamon River Outlet Organization."

### Drainage and Levee Districts

Big Swan Drainage and Levee District. Commissioners. "Ledger 1904–1931."
———. *Specifications for the Construction of Ditches and Levees.* Winchester: Press of the Winchester Times, 1904.
East Peoria Drainage and Levee District. Commissioners. "Complete Record."
———. "Minute Book."
———. "Minutes [August 15, 1907 to January 31, 1921]."
Eldred Drainage and Levee District. Commissioners. "Records."
Lost Creek Drainage and Levee District. Commissioners. Book of Minutes.

———. Treasurer. "Reports, 1920–1934."
Scott County Levee and Drainage [*sic*] District. Commissioners. "Ledger."

### Corporate

Bucyrus-Erie Corporation [South Milwaukee, Wisconsin]. "Excavating Machinery Shipment Ledger, 1882–1893."
Federal Land Bank of St. Louis. "Report on Drainage and Levee Districts: Cass, Mason, Menard and Tazewell Counties, Illinois." By Hubert E. Goodell and Fred S. Morse. January 1934.
———. "South Beardstown Drainage and Levee District, Cass County, Illinois." By C. B. Schmeltzer. April 1934.
———. "Spring Lake Drainage and Levee District, Tazewell County, Illinois." By F. H. Schreiner. July 14, 1945.
Marion Steam Shovel Company. "Machine Record No. 1." Dresser Industries, Marion Power Shovel Company, Marion, Ohio.
———. "Summary of Machine Record No. 2, Ditching Dredges" Dresser Industries, Marion Power Shovel Company, Marion, Ohio.

## Government Publications

### United States

U.S. Army Corps of Engineers. Chicago District. *Those Army Engineers: A History of the Chicago District, U.S. Army Corps of Engineers.* By John W. Larson. Washington, D.C.: GPO, 1980.
U.S. Army Corps of Engineers. North Central and Lower Mississippi Valley Divisions. *Illinois River, Illinois, and Tributaries, Survey Report for Flood Control and Allied Water Uses.* Vol. 1. N.p., April 1916.
U.S. Army Corps of Engineers. Office of History. *The Evolution of the 1936 Flood Control Act.* By Joseph L. Arnold. Washington, D.C.: Government Printing Office, 1988.
U.S. Army Corps of Engineers. St. Louis District. *Shallow Subsurface Geology, Geomorphology, and Limited Cultural Resource Investigations of the Meredosia Village and Meredosia Lake Levee and Drainage Districts, Scott Morgan, and Cass Counties, Illinois.* By Edwin R. Hajic and David S. Lergh. Cultural Resources Management Report 17. St. Louis, April 1985.
U.S. Congress. House. *Laws of the United States Relating to the Improvement of Rivers and Harbors from August 11, 1790 to March 4, 1913.* 62nd Cong., 3rd sess., 1913. H. Doc. 1491. Vols. 1–3.
———. *Proceedings of a Conference of Governors.* 60th Cong., 2nd sess., 1909. H. Doc. 1425.
U.S. Congress. House. Committee on Flood Control. *Control of Floods on the Mississippi and Sacramento Rivers.* 64th Cong., 1st sess., 1916. H. Rept. 616.
———. *Illinois River and Tributaries, Illinois—Flood Control.* 68th Cong., 1st sess., 1924. H. Doc. 276.
U.S. Congress. House. Committee on Irrigation and Reclamation. *Loans for Relief of Drainage Districts.* 71st Cong., 2nd sess., 1930. H. Rept. 11718.
U.S. Congress. House. Committee on Rivers and Harbors. *Survey, with Plans and Estimates of Cost, for a Navigable Waterway 14 Feet Deep from Lockport, Ill., by Way of Des Plaines and*

*Illinois Rivers, and Thence by Way of the Mississippi River to St. Louis.* By Mississippi River Commission . . . and U.S. Army Corps of Engineers. 59th Cong., 1st sess., 1905. H. Doc. 263.
———. *Hearings on the Diversion of Waters from Lake Michigan.* 67th Cong., 2nd sess., 1922.
———. *Hearings on the Subject of the Improvement of the Illinois and Mississippi Rivers, and the Diversion of Water from Lake Michigan into the Illinois River.* 68th Cong., 1st sess., 1924. Pts. 1 and 2.
———. *Illinois River, Ill.* 72nd Cong., 1st sess., 1931. H. Doc. 182.
———. *Sangamon River, Ill.* 72nd Cong., 1st sess., 1932. H. Doc. 186.
———. *Illinois River, Ill.* 77th Cong., 2nd sess., 1942. H. Doc. 692.
U.S. Congress. House. Secretary of War. Chief of Engineers. *Annual Report for 1877.* "Improving Illinois River." 45th Cong., 2nd sess., 1877. Exec. Doc. 1. Pt. 2:2.
———. *Annual Report, 1880.* "Survey of Illinois River." 46th Cong., 3rd sess., 1880. Exec. Doc. 1. Pt. 2:2.
———. *Annual Report, 1890.* "Improvement of the Illinois River, Illinois." 51st Cong., 2nd sess., 1890. H. Exec. Doc. 1. Pt. 2:2.
U.S. Congress. Senate. *Report from the Secretary of War, . . . Transmitting a Report of the Survey of the Kaskaskia and Illinois Rivers.* By Howard Stansbury. 25th Cong., 2nd sess., 1838. S. Doc. 272.
———. *Report of the Chief of Engineers, 1868.* "Report upon the Survey of the Illinois River." By Lt. Col. J. H. Wilson. 40th Cong., 1st sess., 1868. S. Exec. Doc. 16.
———. Committee on Irrigation and Reclamation. *Loans for Relief of Drainage Districts.* 71st Cong., 3rd sess., 1930. S. Rept. 4123.
U.S. Department of Agriculture. *Land Drainage by Means of Pumps.* By S. M. Woodward. Bulletin 304. Washington, D.C.: GPO, November 19, 1915.
———. *Yearbook, 1918.* Washington, D.C.: GPO, 1919.
———. *Land Reclamation Policies in the United States.* By R. P. Teele. Bulletin 1257. Washington, D.C.: GPO, August 23, 1924.
———. *Economic Status of Drainage Districts in the South in 1926.* By Roger D. Marsden and R. P. Teele. Technical Bulletin 194. Washington, D.C.: GPO, October 1930.
———. *Cost of Pumping for Drainage in the Upper Mississippi Valley.* By John G. Sutton. Technical Bulletin 327. Washington, D.C.: GPO, October 1932.
———. *Yearbook of Agriculture, 1955: Water.* Washington, D.C.: GPO, 1955.
———. Bureau of Biological Survey. *Report of the Chief,* 1935. Washington, D.C.: GPO, 1935.
———. *The Survey* 18:1 (January 1937).
U.S. Department of Agriculture. Office of Experiment Stations. *Annual Report of Irrigation and Drainage Investigations.* "Report of Drainage Investigations." By Charles G. Elliott. Bulletin 158. Washington, D.C.: GPO, 1905.
———. *Excavating Machinery Used for Digging Ditches and Building Levees.* By James O. Wright. Circular 74. Washington, D.C.: GPO, 1907.
U.S. Department of Commerce. Bureau of Fisheries. *The Mussell Resources of the Illinois River.* By Ernest Danglade. Document 804. Washington, D.C.: GPO, 1914.
———. *Fisheries Industries of the United States, 1923.* By Oscar E. Sette. Document 976. Washington, D.C.: GPO, 1925.
U.S. Department of Commerce. Bureau of the Census. *Fourteenth Census of the United States, 1920: Agriculture.* Vol. 6:1. Washington, D.C.: GPO, 1922.
———. *Fourteenth Census of the United States, 1920: Irrigation and Drainage.* Vol. 7. Washington, D.C.: GPO, 1922.

———. *Fourteenth Census of the United States, 1920: Population.* Vol. 3. Washington, D.C.: GPO, 1922.
———. *Sixteenth Census of the United States, 1940: Drainage of Agricultural Lands.* Washington, D.C.: GPO, 1942.
———. *Census of Agriculture, 1959: Drainage of Agricultural Lands.* Vol. 4. Washington, D.C.: GPO, 1961.
U.S. Department of Commerce. Commissioner of Fisheries. *Annual Report for the Fiscal Year Ended June 30, 1915.* Washington, D.C.: GPO, 1915.
———. *Annual Report for the Fiscal Year Ended June 30, 1916.* Washington, D.C.: GPO, 1916.
U.S. Department of Commerce. Commission of Fish and Fisheries. *Report for the Year Ending June 30, 1901.* Washington, D.C.: GPO, 1902.
U.S. Department of Commerce and Labor. Bureau of the Census, in cooperation with Bureau of Fisheries. *Fisheries of the United States, 1908.* Special Report. Washington, D.C.: GPO, 1911.
U.S. Department of the Interior. Geological Survey. *Geology and Mineral Resources of the Hardin and Brussels Quadrangles (in Illinois).* By William W. Rubey. Professional Paper 218. Washington, D.C.: GPO, 1952.
U.S. National Emergency Council. "Activities of Federal Agencies Operating in the State of Illinois." Coordination Meeting. Chicago, November 7, 1935. Mimeographed.
*U.S. Statutes at Large* 39:1 (1917): 948.
*U.S. Statutes at Large* 42:1 (1922): 1038.
*U.S. Statutes at Large* 45:1 (1928): 534.
U.S. Treasury Department. Public Health Service. *A Study of the Pollution and Natural Purification of the Illinois River, II: The Plankton and Related Organisms.* By W. C. Purdy. Public Health Bulletin 198. Washington, D.C.: GPO, November 1930.
U.S. War Department. Chief of Engineers. *Report for the Year 1868.* Washington, D.C.: GPO, 1869.
———. *Annual Report, 1900.* Vol. 1. Washington, D.C.: GPO, 1900.
———. *Annual Report, 1912.* Pt. 2. Washington, D.C.: GPO, 1912.
———. *Annual Report, 1914.* Pt. 2. Washington, D.C.: GPO, 1914.
———. *Annual Report, 1925.* Pt. 1. Washington, D.C.: GPO, 1925.
———. *Annual Report, 1929.* Pt. 1. Washington, D.C.: GPO, 1929.
———. *Annual Report, 1930.* Pt. 1. Washington, D.C.: GPO, 1930.
U.S. War Department. Secretary. *Annual Report on the Operations, 1870.* Vol. 2. Washington, D.C.: GPO, 1870.
———. *Annual Report, 1878.* Washington, D.C.: GPO, 1878.
———. *Annual Report for the Year 1896.* Pt. 1. Vol. 2. Washington, D.C.: GPO, 1896.

## California

California Department of Public Works. Division of Water Resources. *Financial and General Data Pertaining to Irrigation, Reclamation, and Other Public Districts in California.* Bulletin 37. Sacramento: California State Printing Office, 1931.

## Illinois

Illinois Board of Fish Commissioners. *Report, from October 1, 1894, to September 30, 1896.* Springfield: Phillips Bros., 1897.

———. *Report, from October 1, 1896, to September 30, 1898*. Springfield: Phillips Bros., 1899.
———. *Report, from October 1, 1898, to September 30, 1900*. Springfield: Phillips Bros., 1900.
———. *Report, from October 1, 1900, to September 30, 1902*. N.p., n.d.
———. *Report, from October 1, 1902, to September 30, 1904*. Springfield: Illinois State Journal Co., 1905.
———. *Report, from October 1, 1904, to September 30, 1906*. N.p., n.d.
———. *Report, from October 1, 1906, to September 30, 1908*. N.p., n.d.
Illinois Board of Health. *Second Annual Report, 1879*. Springfield: H. W. Rokker, 1881.
———. *Ninth Annual Report*. Springfield: Springfield Printing, 1889.
———. *Water Supplies of Illinois and the Pollution of Its Streams*. By John H. Rauch. Springfield, 1889.
———. *Report of the Sanitary Investigations of the Illinois River and Its Tributaries*. Springfield: Phillips Bros., 1901.
Illinois Department of Agriculture. *First Annual Report* (July 1, 1917, to June 30, 1918). Springfield: Illinois State Journal Co., 1918.
———. *Second Annual Report* (July 1, 1918, to June 30, 1919). Springfield: Illinois State Journal Co., 1920.
———. *Seventh Annual Report* (July 1, 1923, to June 30, 1924). Springfield: Illinois State Journal Co., 1925.
Illinois Department of Business and Economic Development. *Inventory of Illinois Drainage and Levee Districts, 1971*. Vol. 1. Springfield, 1971.
Illinois Department of Conservation. *Preliminary Report . . . Fiscal Year Beginning July 1, 1928*. Springfield: Journal Printing Co., 1929.
———. *Potential Conservation Areas along the Illinois River from Hennepin to Grafton*. By Jenkins, Merchant & Nankivil and W. B. Walraven. Springfield, May 1950.
———. Division of Technical Services. *Land and Water Report*. Springfield, June 30, 1984.
Illinois Department of Public Health. *The Rise and Fall of Disease in Illinois*. Vol. 2. By Isaac D. Rawlings. Springfield: Phillips Bros., 1927.
Illinois Department of Public Works and Buildings. Division of Waterways. *Meandered Lakes in Illinois*. N.p., 1912. Mimeographed.
———. *First Annual Report, July 1, 1917 to June 30, 1918*. Springfield: Illinois State Journal Co., 1918.
———. *Second Annual Report, July 1, 1918 to June 30, 1919*. Springfield: Illinois State Journal Co., 1920.
———. *Third Annual Report, July 1, 1919 to June 30, 1920*. Springfield: Illinois State Journal Co., 1921.
———. *Fourth Annual Report, July 1, 1920 to June 30, 1921*. Springfield: Illinois State Journal Co., 1922.
———. *Fifth Annual Report, July 1, 1921 to June 30, 1922*. Springfield: Illinois State Journal Co., 1923.
———. *Sixth Annual Report, July 1, 1922 to June 30, 1923*. Springfield: Illinois State Journal Co., 1924.
———. *Seventh Annual Report, July 1, 1923 to June 30, 1924*. Springfield: Illinois State Journal Co., 1925.
———. *Eighth Annual Report, July 1, 1924 to June 30, 1925*. Springfield: Illinois State Journal Co., 1926.

———. *Ninth Annual Report, July 1, 1925 to June 30, 1926.* Springfield: Illinois State Journal Co., 1927.

———. *Tenth Annual Report, July 1, 1926 to June 30, 1927.* Springfield: Journal Printing Co., 1928.

———. *Eleventh Annual Report, July 1, 1927 to June 30, 1928.* Springfield: Journal Printing Co., 1929.

———. *Twelfth Annual Report, July 1, 1928 to June 30, 1929.* Springfield: Journal Printing Co., 1930.

———. *Fifteenth Annual Report, July 1, 1931 to June 30, 1932.* Springfield: Journal Printing Co., 1933.

———. *Sixteenth Annual Report, July 1, 1932 to June 30, 1933.* Springfield: Journal Printing Co., 1934.

———. *Eighteenth Annual Report, July 1, 1934 to June 30, 1935.* Springfield: Journal Printing Co., 1935.

———, in cooperation with Department of Conservation. *Report on Drainage Districts, 1937.* N.p., 1937.

———. *Report for Recreational Development, Illinois River Backwater Areas.* Springfield, 1969.

Illinois Department of Purchases and Construction. Division of Waterways. *Flood Control Report.* By Jacob A. Harman, L. T. Berthe, M. Blanchard, and L. C. Craig. Springfield: Journal Printing Co., 1930.

Illinois Department of Registration and Education. Geological Survey. *Geography of the Middle Illinois Valley.* By Harlan H. Barrows. Bulletin 15. Urbana: Phillips Bros., 1910. Reprint, 1925.

———. *Engineering and Legal Aspects of Land Drainage in Illinois.* Rev. ed. By George W. Pickels and Frank B. Leonard Jr. Bulletin 42. Urbana, 1929.

———. *Geology and Mineral Resources of the Beardstown, Glasford, Havana, and Vermont Quadrangles.* By Harold R. Wanless. Bulletin 82. Urbana, 1957.

Illinois Department of Registration and Education. Historical Library. *Patterns from the Sod: Land Use and Tenure in the Grand Prairie, 1850–1900.* By Margaret Beattie Bogue. Collections Series 34. Springfield, 1959.

Illinois Department of Registration and Education. State Museum. *Early Vegetation of the Lower Illinois Valley.* By April A. Zawacki and Glenn Hausfater. Reports of Investigations 17. Springfield, 1969.

———. *Geomorphology of the Lower Illinois River Valley as a Spatial-Temporal Context for the Koster Archaic Site.* By Karl W. Butzer. Reports of Investigations 34. Springfield, 1977.

Illinois Game and Fish Conservation Commission. *Annual Report for the Fiscal Year 1913–14.* Springfield: Illinois State Journal Co., 1914.

———. *Third Annual Report, for the Fiscal Year 1915–1916.* Springfield: Illinois State Journal Co., 1916.

Illinois Game Commission. *First Annual Report.* Springfield: Phillips Bros., 1900.

Illinois General Assembly. *Reports to, 1881.* Springfield: H. W. Rokker, 1881.

———. *Reports to, 1885.* Springfield: H. W. Rokker, 1885.

———. *Reports to, 1890.* Springfield: H. W. Rokker, 1890.

———. House. Submerged and Shore Lands Investigating Committee. *Report to the Governor and 47th General Assembly.* Springfield: Illinois State Journal Co., 1911.

Illinois General Assembly. Illinois Valley Flood Control Commission. *Hearings on the Causes and Control of Floods in the Illinois River Valley.* N.p., 1929.

Illinois General Assembly. Legislative Flood Investigating Commission. *Report on Flood Situation in Illinois.* By Elliott & Porter, Consulting Engineers. Chicago, March 1947.
Illinois Internal Improvements Commission. *The Lakes and Gulf Waterway.* Springfield: Phillips Bros., 1907.
———. *Surface Water Supply of Illinois Central and Southern Portions, 1908–1910.* Springfield: Illinois State Journal Co., 1911.
Illinois Laboratory of Natural History. *The Plankton of the Illinois River, 1891–1899, with Introductory Notes upon the Hydrography of the Illinois River and Its Basin.* By Charles A. Kofoid. Bulletin 6:2. Urbana, 1901–3.
———. *The Fishes of Illinois.* By Stephen A. Forbes and Robert E. Richardson. Danville: Illinois Printing Co., 1908.
———. *Biological Investigations on the Illinois River.* Urbana, 1910.
———. *Forest Conditions in Illinois.* By R. Clifford Hall and O. D. Ingall. Bulletin 9:4. Urbana, 1910–13.
Illinois *Laws.* 1879.
Illinois *Laws.* 1885.
Illinois *Laws.* 1903.
Illinois *Laws.* 1905.
Illinois *Laws.* 1913.
Illinois Natural History Survey. *Some Recent Changes in the Illinois River Biology.* By Stephen A. Forbes and Robert E. Richardson. Bulletin 13:6. Urbana, April 1919.
———. *Changes in the Bottom and Shore Fauna of the Middle Illinois River and its Connecting Lakes since 1913–1915 as a Result of the Increase, Southward, of Sewage Pollution.* By Robert E. Richardson. Bulletin 14:4. Urbana, 1921.
———. *The Bottom Fauna of the Middle Illinois River, 1913–1915.* By Robert E. Richardson. Bulletin 17:12. Urbana, December 1928.
———. *Duck Food Plants of the Illinois River Valley.* By Frank C. Bellrose Jr. Bulletin 21:8. Urbana, August 1941.
———. *Waterfowl Hunting in Illinois: Its Status and Problems.* By Frank C. Bellrose Jr. Biological Notes 17. Urbana, March 1944.
———. *Wildlife and Fishery Values of Bottomland Lakes in Illinois.* By Frank C. Bellrose and Clair T. Rollings. Biological Notes 21. Urbana, June 1949.
———. *A Survey of the Mussels (Unionacea) of the Illinois River: A Polluted Stream.* By William C. Starrett. Bulletin 30:5. Urbana, February 1971.
———. *The Fate of Lakes in the Illinois River Valley.* By Frank C. Bellrose, S. P. Havera, F. L. Paveglio Jr., and D. W. Steffeck. Biological Notes 119. Urbana, April 1983.
Illinois Planning Commission. *Report on the Lower Illinois River Basin.* Springfield: 1940.
Illinois Rivers and Lakes Commission. *Land Drainage in Illinois.* By Robert I. Randolph. Bulletin 4. Springfield: Illinois State Journal Co., 1912.
———. *Annual Report, 1912.* Springfield: Illinois State Journal Co., 1913.
———. *Annual Report, 1913–1914.* Chicago: Fred Klein Co., 1914.
———. *Annual Report, 1916.* Springfield: Schnepp & Barnes, 1917.
———. *The Illinois River and Its Bottom Lands.* 2nd ed. By John W. Alvord and Charles B. Burdick. Springfield: Illinois State Journal Co., 1919.
Illinois Tax Commission, in cooperation with Works Progress Administration. *Drainage*

*District Organization and Finance, 1879–1937*. Survey of Local Finance in Illinois. Vol. 7. Springfield, 1941.

Illinois Water Survey. *Chemical and Biological Survey of the Waters of Illinois*. Bulletin 9. Urbana, March 25, 1912.

———. *Illinois River Studies, 1925–1928*. By C. S. Boruff and A. M. Buswell. Bulletin 28. Urbana, 1929.

University of Illinois. College of Agriculture. Agricultural Experiment Station. *Prices of Illinois Farm Products from 1866 to 1929*. By L. J. Norton and B. B. Wilson. Bulletin 351. Urbana, July 1930.

University of Illinois. College of Agriculture. Agricultural Experiment Station and Extension Service. *Facts Assembled for Use of Committee on Mechanical Equipment, Drainage, and Farm Buildings*. Agricultural Adjustment Conferences, 1928–1929. Urbana, October 1928.

University of Illinois. Water Resources Center. *Case Studies in Drainage and Levee District Formation and Development on the Floodplain of the Lower Illinois River, 1890s to 1930s*. By John Thompson. Special Report 017. Urbana-Champaign, May 1989.

### New York

New York State Engineer and Surveyor. *Annual Report for the Fiscal Year Ending September 30, 1891*. Albany: James B. Lynn, 1892.

### Sanitary District of Chicago

Sanitary District of Chicago. *Reports of International Waterways Commission Concerning the Chicago Diversion and Terms of Treaty*. Chicago, n.d.

———. *Report of Stream Examination, Chemic and Bacteriologic of the Waters Between Lake Michigan at Chicago and the Mississippi River at St. Louis, for the Purpose of Determining Their Condition and Quality Before and After the Opening of the Drainage Channel*. By Arthur W. Palmer. Chicago: Blakely Printing Co., December 1902.

———. *The Sanitary District of Chicago: History of Its Growth and Development*. By C. Arch Williams. Chicago: Sanitary District of Chicago, 1919.

## Books and Trade Catalogs

Artcraft Directory Publishers. *Directory of the City of Quincy, Illinois, 1938*. Quincy, April 1938.

Bogart, Ernest Ludlow, and John Mabry Mathews. *Centennial History of Illinois: The Modern Commonwealth, 1893–1918*. Vol. 4. Chicago: A. C. McClurg & Co., 1922.

Brush, Daniel H. *Growing Up with Southern Illinois, 1820 to 1861*. Edited by Milo M. Quaife. Chicago: Lakeside Press, 1944.

Cain, Louis P. *Sanitary Strategy for a Lakefront Metropolis*. DeKalb: Northern Illinois University Press, 1978.

Connett, Eugene V., ed. *Wildfowling in the Mississippi Flyway*. New York: D. Van Nostrand, 1949.

Cooley, Lyman E. *The Lakes and Gulf Waterway as Related to the Chicago Sanitary Problem . . . , a Preliminary Report*. Chicago: Press of John W. Weston, 1891.

———. *The Illinois River: Physical Relations and the Removal of the Navigation Dams.* Chicago: Clohesey & Co., 1914.

Devore, Joan, ed. *Meredosia Bicentennial Book, 1776–1976.* Meredosia Junior Women's Club. Bluffs, Ill.: Jones Publishing Co., 1976.

Dirksen, Everett McKinley. *The Education of a Senator.* Urbana: University of Illinois Press, 1998.

Elliott, Charles G. *Engineering for Land Drainage.* New York: John Wiley & Sons, 1908.

Fairbanks Steam Shovel Company. *Dipper Dredge Catalogue.* N.p., ca. 1907.

———. *Dipper Dredges.* Grand Rapids, Mich.: Dickinson Bros., 1912.

*Farm Directory of Cass, Mason, Menard and Sangamon Counties, Illinois.* Chicago: Farmers' Review, 1917. Reprint, Dixon, Ill.: Print Shop, 1984.

F. C. Austin Drainage Excavator Company. *Irrigation and Drainage Excavating Mach'y.* Chicago, n.d.

Gould, E. W. *Fifty Years on the Mississippi; or, Gould's History of River Navigation.* St. Louis: Nixon-Jones Printing, 1889.

Gresham, John. *Historical and Biographical Record of Douglas County, Illinois.* Logansport, Ind.: Wilson, Humphreys & Co., 1900.

Hallock, C. *The Sportsman's Gazeteer and General Guide.* New York: Forest and Stream Publishing Co., 1877.

Harrison, Robert W. *Alluvial Empire: A Study of State and Local Efforts Toward Land Development in the Alluvial Valley of the Lower Mississippi River, Including Flood Control, Land Drainage, Land Clearing, Land Forming.* Little Rock: Pioneer Press, for Delta Fund and U.S. Department of Agriculture, Economic Research Service, 1961.

Hays, Robert G. *State Science in Illinois: The Scientific Surveys, 1850–1978.* Carbondale: Southern Illinois University Press, 1980.

Hays, Samuel P. *Conservation and the Gospel of Efficiency, the Progressive Conservation Movement, 1890–1920.* Cambridge: Harvard University Press, 1959.

Hibbard, Benjamin Horace. *A History of the Public Land Policies.* 1924. Reprint, Madison: University of Wisconsin Press, 1965.

*History of Greene and Jersey Counties, Illinois.* Springfield: Continental Historical Co., 1885.

Hoffman, W. H. *City Directory of Quincy, Illinois, 1930–1931.* Quincy: Hoffman City Directories, 1930.

Howard, Robert P. *Illinois: A History of the Prairie State.* Grand Rapids, Mich.: William B. Eerdmans Publishing Co., 1972.

Hull, William J., and Robert W. Hull, eds. *The Origin and Development of the Waterways Policy of the United States.* Washington, D.C.: National Waterways Conference, 1967.

Kelley, Robert L. *Gold vs. Grain.* Glendale, Calif.: Arthur H. Clark Co., 1959.

———. *Battling the Inland Sea.* Berkeley: University of California Press, 1989.

Leopold, Aldo. *Report on a Game Survey of the North Central States.* For the Sporting Arms and Ammunition Manufacturers Institute. Madison: Democrat Printing Co., 1931.

Lynn, Ruth Wallace. *Prelude to Progress, the History of Mason County, Illinois, 1818–1968.* Havana: Mason County Board of Supervisors, 1968.

Marion Steam Shovel Company. *Circular No. 21—Dredges.* N.p., ca. 1906.

———. *The Name and the Machine,* Catalog 188. Cleveland: Caxton Co., ca. 1918.

Martin, Charles A. E., ed. *Historical Encyclopedia of Illinois and History of Cass County.* Vol. 2. Chicago: Munsell Publishing Co., 1915.

Massie, M. D., Capt. *Past and Present of Pike County, Illinois.* Chicago: S. J. Clarke Publishing Co., 1906.
Moore, Jamie W., and Dorothy P. Moore. *The Army Corps of Engineers and the Evolution of Federal Flood Plain Management Policy.* Natural Hazards Research and Applications Information Center, Special Publication 20. Boulder: University of Colorado, 1989.
Murphy, Daniel W. *Drainage Engineering.* New York: McGraw-Hill Book Co., 1920.
*National Cyclopaedia of American Biography.* New York: James T. White & Co., 1931.
Nolen, John H. *Missouri's Swamp and Overflowed Lands and Their Reclamation.* Report to the 47th Missouri General Assembly. Jefferson City, Mo.: Hugh Stephens Printing, 1913.
O'Connell, James C. *Chicago's Quest for Pure Water.* Public Works Historical Society, Essay No. 1. Washington, D.C.: Public Works Historical Society, 1976.
Oglesby, R. T., C. A. Carlson, and J. A. McCann, eds. *River Ecology and Man.* New York: Academic Press, 1972.
Palmer, Benjamin Whipple. *Swamp Land Drainage with Special Reference to Minnesota.* Bulletin of the University of Minnesota, Studies in Social Sciences 5. Minneapolis: University of Minnesota, 1915.
Parmalee, Paul W., and Forrest D. Loomis *Decoys and Decoy Carvers of Illinois.* DeKalb: Northern Illinois University Press, 1969.
Pickels, George W. *Drainage and Flood Control Engineering.* New York: McGraw-Hill Book Co., 1925.
Prince, Hugh. *Wetlands of the American Midwest: A Historical Geography of Changing Attitudes.* Chicago: University of Chicago Press, 1997.
R. L. Polk & Co. *Memphis City Directory.* Memphis, 1914–32.
Samson, Charles M. *Beardstown, Illinois, City Directory.* Bloomington, Ill.: Pantagraph Printing and Stationery, 1915.
Scarpino, Philip V. *Great River: An Environmental History of the Upper Mississippi, 1890–1950.* Columbia: University of Missouri Press, 1985.
Starling, William. *The Floods of the Mississippi River.* New York: Engineering News Publishing, 1897.
Teele, Ray P. *The Economics of Land Reclamation in the United States.* Chicago: A. W. Shaw Co., 1927.
Thompson, John, and Edward A. Dutra. *The Tule Breakers: The Story of the California Dredge.* Stockton, Calif.: Stockton Corral of Westerners, University of the Pacific, 1983.
Vertrees, Herbert H. *Pearls and Pearling.* Columbus, Ohio: A. R. Harding, 1913.
Waller, Robert A. *Rainey of Illinois: A Political Biography, 1903–1934.* Urbana: University of Illinois Press, 1977.
Weatherford, Gary D. and F. Lee Brown, eds. *New Courses for the Colorado River: Major Issues for the Next Century.* Albuquerque: University of New Mexico Press, 1986.
*Who Was Who.* Vol. 1. Chicago: A. N. Marquis Co., 1943.

## Journals, Proceedings, and Pamphlets

Association of County Surveyors and Civil Engineers of Indiana. *Proceedings, Sixth Annual Meeting, . . . 1886.* Indianapolis: Morning Star Publishing Co., 1886.
Association of Drainage and Levee Districts of Illinois. *First Annual Report of Transactions.* N.p., 1911.

———. *Proceedings of Second Annual Meeting.* Beardstown: Enterprise Press, 1912.
———. *Proceedings of the Third and Fourth Meetings.* Winchester: Press of the Winchester Times, 1913.
"The Attack on the Federal Bird Laws." *Audubon Bulletin,* winter 1916–17, 9–13.
"A. V. Wills & Sons." *Contractors Review,* December 2, 1916, 114.
Baker, I. O. "Some of the Engineering Features of Illinois Drainage." *Association of Engineering Societies* 5:11 (September 1886), 426–29.
Barnes, Paul D. "Illinois Convention." *Izaak Walton League Monthly* 5:9 (April 1927), 54–55.
Bartlett, S. P. "The Value of the Carp as a Food Product of Illinois Waters." *Transactions of the American Fisheries Society* 29 (1900), 80–87.
———. "Discussion on Carp." *Transactions of the American Fisheries Society* 30 (1901), 114–32.
———. "The Decrease of the Coarse Fish and Some of Its Causes." *Transactions of the American Fisheries Society* 41 (1912), 195–206.
———. "Fish Waste, Past and Present." *Transactions of the American Fisheries Society* 47:1 (December 1917), 22–27.
Bell, A. H. "Drainage Districts and the Construction of Drainage Canals." *Engineering News and American Contract Journal* 15 (February 20, 1886), 113.
Block, Marvin W. "Henry T. Rainey of Illinois." *Journal of the Illinois State Historical Society* 65:2 (summer 1972), 142–57.
Bogue, Margaret B. "The Swamp Land Act and Wet Land Utilization in Illinois, 1850–1890." *Agricultural History* 25:4 (October 1951), 169–80.
Boyd, George R. "The Refinancing of Drainage Districts." *Agricultural Engineering* 13:10 (October 1932), 258–59.
Bradford, Ralph F. "A Few Remarks on Conservation in the Illinois Valley." *Transactions of the Illinois State Academy of Science* 24:2 (December 1931), 587–91.
"Brief News of the Chapters . . ." *Outdoor America* 2:9 (April 1924), 44.
"Busy at Havana." *Illinois Fisherman and Hunter* 2:6 (January 1914), 10.
"Carp Culture in Illinois." *Prairie Farmer,* April 24, 1886, 260.
Cave, Edward. "Down at Grand Island by Proxy." *Outers' Book-Recreation,* March 1918, 205–12.
"The Chicago Main Drainage Channel." *Engineering News* 33:20 (May 16, 1895), 314–16.
"Chicago's Drainage Canal." *Steam Shovel and Dredge* 10:10 (October 1906), 354.
"Chicago Urges Bill Authorizing Its Uses of Lake Water." *Electrical World* 83:8 (February 23, 1924), 398.
"Club House Burns." *Illinois Fish and Game Conservation Society and Warden's Journal* 3:8 (August 1914), 6.
Cohen, Nat H. "Concerning Fish Laws in Illinois." *Transactions of the American Fisheries Society* 30 (1901), 133–36.
"Commercial Fishermen Organize." *Illinois Fish Conservation Society News-Letter* 2:7 (June 1, 1913), n.p.
"Commission Wages Successful War Against Gars." *Illinois Fish Conservation Society News-Letter* 2:1 (January 1913), 1–4.
"Conditions That Confront Us." *Illinois Fish Conservation Society News-Letter* 1:3 (March 15, 1912), n.p.
"A Cross-Cut Excavating Machine for Drainage Ditches." *Engineering News* 54:10 (September 7, 1905), 250.
"The Deep Waterway." *Steam Shovel and Dredge* 13:11 (November 1909), 908–10.

Dilg, Will H. "A Message to Rod and Gun Clubs." *Izaak Walton League Monthly* 1:2 (September 1922), 54–55.
"The Dredge Ditch in Drainage Work." *Drainage Journal* 23:10 (October 1901), 263–67.
Drummond, Mary. "The Formative Years of the Illinois Audubon Society." *Audubon Bulletin,* fall 1920, 4–6.
Eifrig, C. W. E. "Taking Inventory." *Audubon Annual Bulletin* 22 (1932), 25–27.
"Electric-Hydraulic Dredge Work on Illinois Levees." *Engineering News-Record* 92:22 (May 29, 1924), 944–45.
"Electricity Replaces Steam in Drainage Pumping." *Electrical World* 66:2 (July 10, 1915), 87.
"Electricity Versus Steam in Drainage Pumping." *Electrical World* 64:6 (August 8, 1914), 275–77.
Elliott, Charles G. "Drainage of Swamps." *Drainage and Farm Journal* 6:6 (June 1885), 19–25.
Ellis, Willard D. "Problems of Financing Land Reclamation." *Agricultural Engineering* 12:5 (May 1931), 167–68.
"The Evil of Spring Shooting." *American Field,* April 14, 1894, 344.
"Exterminating the Gar." *Illinois Fish Conservation Society News-Letter* 3:3 (March 1914), 4–5.
"Facts Concerning the State Sanitary Water Board." *Illinois Conservation* 1:3 (fall 1936), 9–10.
"Famous Resort Gone." *Illinois Fisherman* 1:8 (March 1913), n.p.
"Federal Co-operation for Deep Waterways." *Steam Shovel and Dredge* 13:4 (April 1909), 370–74.
"Federal Regulations and the Fish and Game Conservation Society of Illinois." *Illinois Fish and Game Conservationist* 3:3 (March 1914), 7–8.
"First Annual Convention of the Izaak Walton League of America." *Izaak Walton League Monthly* 1:9 (May 1923), 478.
"Fisheries of the Illinois River." *Illinois Fish and Game Conservationist* 3:2 (February 1914), 7–8.
"Fishermen Accused of Arson." *Illinois Fish Conservation Society News-Letter* 2:8 (September 1913), 14.
Forbes, Stephen A. "The Native Animal Resources of the State." *Transactions of the Illinois State Academy of Science* 5 (1912), 37–48.
———. "Sewage Pollution of the Illinois River." *Outdoor America* 3:5 (December 1924), 35–36.
"For Sale." *Illinois Fisherman* 1:8 (March 1913), n.p.
"The Game-Hog or Pot-Hunter." *American Field,* December 1, 1894, 508.
"Governor Dunne's Special Message on Fish and Game Department." *Illinois Fish Conservation Society News-Letter* 2:5 (May 1913), 7–9.
"Grand Island Lodge." *American Field,* March 8, 1902, 214.
"Grand Island Lodge." *American Field,* June 28, 1902, 599–600.
Hannah, Harold W. "History and Scope of Illinois Drainage Law." *University of Illinois Law Forum* 2 (summer 1960), 189–97.
Harman, Jacob A. "Some Problems in Flood Control." *Transactions of the Illinois State Academy of Science* 24:2 (December 1931), 564–74.
———. "Drainage of Areas Subject to Floods." *Illinois Engineer* 8:5 (May 1932), 3–4.
Harrison, Robert W., and Walter M. Kollmorgan. "Past and Present Drainage Reclamations in the Coastal Marshlands of the Mississippi River Delta." *Journal of Land and Public Utility Economics* 23:3 (August 1947), 297–320.
Henderson, B. "Land Drainage Needed in United States." *Reclamation and Farm Engineering* 8:5 (December 1925), 309–11.

Herget, James E. "Taming the Environment: The Drainage District in Illinois." *Journal of the Illinois State Historical Society* 71:2 (May 1978), 107–18.
Hewes, Leslie. "The Northern Wet Prairie of United States: Nature, Sources of Information, and Extent." *Annals of the Association of American Geographers* 41:4 (December 1951), 307–23.
"How Swamp Lands Are Reclaimed in Illinois." *Farmers' Review,* April 25, 1888, 260.
Illinois Association of Sanitary Districts. *Eighth Annual Report, 1931–1932.* Urbana, 1932.
———. *Ninth Annual Report, 1932–1933.* N.p., May 1933.
"Illinois Fights for Health and Waterways." *National Reclamation Magazine* 3:5 (May 1924), 92–93.
*Illinois Fisherman* 1:5 (December 1912), 13.
*Illinois Fisherman* 1:10 (May 1913), 7–8.
*Illinois Fisherman and Hunter* 2:7 (February 1914), 8.
"Illinois Fish Law." *Illinois Fisherman* 1:3 (October 1912), 12.
"Illinois Lake Purchased by Club." *Illinois Fish and Game Conservation Society and Warden's Journal* 3:9 (September 1914), 12.
"The Illinois Sportsmen's League." *Audubon Bulletin,* fall 1921, 4–6.
"Illinois Warden's Action Relieves Health Conditions." *Illinois Fish and Game Conservation Society and Warden's Journal* 3:8 (August 1914), 4.
"Important Convention." *American Field,* April 26, 1902, 1.
"Improvements of the Chicago River." *Steam Shovel and Dredge* 12:12 (December 1908), 913–14.
"Increase Flow Through Drainage Canal." *Steam Shovel and Dredge* 17:2 (February 1913) 116–18.
Iowa State Drainage Association. *Proceedings of the First Annual Meeting.* N.p., 1908.
Jelliff, Fred R. "The State-Wide Menace of Stream Pollution in Illinois." *Outdoor America* 2:10 (May 1924), 17–19, 78.
"Jesse Lowe, Big Works Contractor, Is Dead." *Contractor* 25:10 (May 10, 1918), 212.
Kerr, J. P. "Chicago's Cesspool." *Outdoor America* 3:5 (December 1924), 36–37.
Kreiling, C. H. "Pumping Plants in Levee Districts." *Illinois Engineer* 5:5 (May 1929), 1–2.
Lakes-to-the-Gulf Deep Waterway Association. "Fourteen Feet Through the Valley." By Harry B. Hawes. St. Louis, ca. 1907.
"League Opens Campaign to Save the Illinois River." *Outdoor America* 3:4 (November 1924), 11–12.
"League Wins Bass Fight in Illinois." *Izaak Walton League Monthly* 1:11 (July 1923), 603.
Lee, Judson F. "Transportation: A Factor in the Development of Northern Illinois Previous to 1860." *Journal of the Illinois State Historical Society* 10:1 (April 1917), 17–85.
"Legislation." *Audubon Bulletin,* winter 1917–1918, 8–9.
Leonard, Frank B. "An Introduction to the Drainage Laws of Illinois." *Illinois Law Bulletin* 1:5 (April 1918), 227–47.
Locke, S. B. "Developments in the Clean Streams Campaign." *Transactions of the American Fisheries Society* 62 (1932), 355–62.
McCrory, S. H. "Historical Notes on Land Drainage in the United States." *Proceedings of the American Society of Civil Engineers* 53:7 (September 1927), 1631–32.
McGee, W. J. "Our Great Rivers." *World's Work* 13:4 (February 1907), 8576–78.
"McWilliams Dredging Co." (advertisement). *National Reclamation Magazine* 1:1 (December 1921), 25.

Meehan, W. E. "The Relations Between State Fish Commission and Commercial Fishermen." *Illinois Fish Conservation Society News-Letter* 1:5 (April 15, 1912), n.p.
Merrill, B. G. "Report of Committee on Membership and Publicity." *Illinois Fish Conservation News-Letter* 2:8 (September 1913), 6–7.
———. "Where to Go Fishing." *Illinois Fish and Game Conservationist* 3:6 (June 1914), 11–12.
"Misleading Legislation." *Illinois Fish Conservation Society News-Letter* 1:3 (March 15, 1912), n.p.
Munch, Gus. "Illinois River Duck Chatter." *Outers' Book-Recreation,* March 1919, 147–49, 187–88.
"The Mussell-Shell Industry of Illinois." *Illinois Fish Conservation Society News-Letter* 1:3 (March 15, 1912), n.p.
Nauss, Ralph W. "Malaria in Illinois." *Illinois Health News* 8:6 (June 1922), 164–65.
"A New Style of Scraper Excavator." *Engineering News* 53:9 (March 2, 1905), 216–17.
"New York Legislature Against Lake Michigan Diversion." *Electrical World* 83:12 (March 22, 1924), 589.
"The Old Illinois Game Law Stands." *American Field,* June 22, 1895, 578.
Pearse, Langdon. "The Sewage Treatment Program of the Sanitary District of Chicago." *Western Society of Engineers Journal* 31:7 (July 1926), 261–67.
Perce, H. Wheeler. "Fair Play." *Illinois Fish Conservation Society News-Letter* 2:2 (February 1913), 1–4.
Phillips, Edward O. "Illinois and the Deep Waterway." *Steam Shovel and Dredge* 13:7 (July 1909), 546–51.
"A Popular Resort." *Illinois Fisherman* 1:3 (October 1912), 4.
Potter, W. G. "Improvement and Utilization of the Rivers of Illinois." *Western Society of Engineers Journal* 31:5 (May 1926), 203–14.
"The Proposed Illinois Law." *American Field,* February 16, 1895, 147–49.
Pulley, F. G. "A Levee Job of 2,260,000 Cubic Yards." *Contractors Review,* December 16, 1916, 131–33.
Randolph, Isham. "Results on Some Great Projects." *National Drainage Journal* 1:9 (February 1920), 1–4.
"R. H. and G. A. McWilliams Awarded Drainage Contract in Arkansas." *Contractors Review,* December 30, 1916, 158.
"Sanitary District Blames Delay on Chicago Tax Tangle." *Engineering News-Record* 109:20 (November 17, 1932), 602–3.
"Save Spring Lake." *Illinois Fish Conservation Society News-Letter* 1:3 (March 15, 1912), n.p.
Schram, Emil. "Refinancing of Drainage, Levee, and Irrigation Districts." *Agricultural Engineering* 16:4 (April 1935), 151–54.
"Seining Objectionable Fish from the Illinois River." *American Field,* February 1, 1902, 98.
"Seize 4 Deer and 500 Ducks." *Illinois Fish and Game Conservationist* 2:10 (November 1913), 13.
"Sen. Beall Introduced Society's Bills in the Senate." *Illinois Fish Conservation Society News-Letter* 2:4 (April 1913), 1–2.
Shaw, A. W. "The Inlet Swamp Drainage District." *Engineering News* 53:4 (January 26, 1905), 89–90.
Sherman, L. K. "Drainage and Levee District Project." *Reclamation and Farm Engineering* 8:3 (September 1925), 212.
"Shipping Live Fish." *Illinois Fisherman* 1:9 (April 1913), n.p.
"Specifications for Levees." *Engineering News* 4:36 (September 1, 1887), 233–34.

"Sportsmen, Congressmen, Audubonites and Others." *Audubon Bulletin,* winter 1916–1917, 26–28.

"State News Comments on Work of Sportsmen's Organizations." *Illinois Fish Conservation Society News-Letter* 2:10 (November 1913), 11–12.

Steffeck, Donald W., Fred L. Paveglio Jr., Frank C. Bellrose, and Richard E. Sparks. "Effects of Decreasing Water Depths on the Sedimentation Rate of Illinois River Bottomland Lakes." *Water Resources Bulletin* 16:3 (June 1980), 553–55.

Stewart, Charles L. "Land Utilization in the Illinois River Basin." *Transactions of the Illinois State Academy of Science* 24:2 (1910–13), 557–63.

"Stop Fall Duck Shooting." *American Field,* March 1, 1902, 191.

"Supreme Court Hears Argument in Chicago Canal Case." *Electrical World* 84:25 (December 20, 1924), 1325.

Sutton, John G. "The Cost of Drainage Pumping." *Agricultural Engineering* 13:5 (May 1932), 123–24.

"Swamp Land Reclamation." *American Contractor,* May 16, 1914, 103.

"Synopsis of Fish Law." *Illinois Fish Conservation News-Letter* 2:7 (August 1913), n.p.

Talbot, A. N. "Sewage Disposal." *Proceedings of the Fourth Annual Meeting, Illinois Society of Engineers and Surveyors,* Bloomington, Ill., January 23–25, 1889, 53–59.

Thompson, David H. "The Fishing Industry of Illinois River." *Transactions of the Illinois State Academy of Science* 24:2 (December 1931), 592–95.

Thompson, John. "The Bay City Land Dredge and Dredge Works: Perspectives on the Machines of Land Drainage." *Michigan Historical Review* 12:2 (fall 1986), 21–43.

———. "Commemorating the Large Steam Dipper Dredges of the Panama Canal." *Nautical Research Journal* 43:2 (June 1998), 92–103.

"Thompson Lake at Stake." *Illinois Fish Conservation Society News-Letter* 1:11 (October 1912), n.p.

"To Reclaim Swamps." *American Contractor,* November 2, 1907, 74.

"Trammel Nets Doing the Same Old Illegal Work." *Illinois Fish and Game Conservationist* 2:10 (November 1913), 12.

Turner, Lewis M. "Plant Succession on Levees in the Illinois River Valley." *Transactions of the Illinois State Academy of Science* 24:2 (December 1931), 94–102.

———. "Grassland in the Floodplain of Illinois River." *American Midland Naturalist* 15 (1934), 770–86.

"Unified Public Utilities in Central Illinois." *Electrical World* 61:22 (May 31, 1913), 1146–56.

"Urges Consolidation." *Illinois Fisherman* 1:10 (May 1913), n.p.

Waller, Robert A. "The Illinois Waterway from Conception to Completion, 1908–1933." *Journal of the Illinois State Historical Society* 65:2 (summer 1972), 125–41.

Wheeler, R. A. "Report on the Status of Flood Control in Illinois." *Illinois Engineer* 12:4 (October 17, 1907), 56–65.

"Where to Go." *Field and Stream,* May 1913, 78–82.

"Where to Go." *Field and Stream,* August 1913, 412–15.

Wik, Reynold M. "Steam Power on the American Farm, 1830–1880." *Agricultural History* 25:4 (October 1951), 181–86.

"Work of the Illinois Fish Commission." *Illinois Fish Conservation Society News-Letter* 1:5 (April 15, 1912), n.p.

"Would Have Game Preserves Revert to the People." *Illinois Fish Conservation Society News-Letter* 3:5 (May 1914), 10.
"Would Take a Backward Step." *Illinois Fish Conservation Society News-Letter* 3:2 (February 1914), 5.
Zartman, P. E. "Duck Hunting Reminiscences of Thirty Years Ago." *Outdoor America* 2:3 (October 1923), 68–70, 126.

## Dissertations and Theses

Deleuw, Charles E. "Problems in Design and Construction of Large Land Drainage Districts." Civil Engineering thesis, University of Illinois, 1912.
Dunn, Joel E. "Reclamation of Land by Levying." Civil Engineering thesis, University of Illinois, 1906.
Haungs, Howard C. "Engineering Methods for the Reclamation of Overflowed Lands in Illinois." Civil Engineering thesis, University of Illinois, 1916.
Howell, Cleves H. "Reclamation of Agricultural Land by Diking." Civil Engineering thesis, University of Illinois, 1905.
McDowell, Gary Lane. "Local Agencies and Land Development by Drainage: The Case of 'Swampeast' Missouri." Ph.D. diss., Columbia University, 1965.
Philip, William B. "Chicago and the Down State: A Study of Their Conflicts, 1870–1934." Ph.D. diss., University of Chicago, 1940.
Thompson, John. "The Settlement Geography of the Sacramento–San Joaquin Delta, California." Ph.D. diss., Stanford University, 1957.
Winsor, Roger Andrew. "Artificial Drainage of East Central Illinois, 1820–1920." Ph.D. diss., University of Illinois at Urbana-Champaign, 1975.

# Index

*Page numbers in italics refer to figures and tables.*

John Thompson is a professor emeritus of geography at the University of Illinois, Urbana-Champaign, where he served as both a professor (1967–92) and chair (1966–75) of the department and as the director of the Center for Latin American and Caribbean Studies (1964–67). Born in Peru to citizens of the United States, he received his academic education at Stanford University (B.A. and Ph.D.) and the University of California, Berkeley (M.A.). His research interests in the historical geography of wetlands drainage and drainage technology began in 1956 and became a focal interest after 1972. His research has been in California, Illinois, and other parts of the Midwest.